Architecture Education in the
World: Reviewing and rethink
forward

Edited by:

Madhumita Roy
Jayashree Deshpande
Nandini Kulkarni

Architecture Education in the Changing World: Reviewing and rethinking the way forward

Proceedings of the COA International Conference on the future of Architecture Education (ICFAE 2023) February 15-17 2024, Vellore, India

Edited by:

Madhumita Roy

Jayashree Deshpande

Nandini Kulkarni

Routledge
Taylor & Francis Group

First published 2025
by Routledge
4 Park Square, Milton Park, Abingdon, Oxon OX14 4RN

and by Routledge
605 Third Avenue, New York, NY 10158

Routledge is an imprint of the Taylor & Francis Group, an informa business

British Library Cataloguing-in-Publication Data
A catalogue record for this book is available from the British Library

ISBN: 9781032944944 (pbk)
ISBN: 9781003571018 (ebk)

DOI: 10.1201/9781003571018

Typeset in Sabon LT Std
by HBK Digital

Contents

About the Editors

Madhumita Roy

Madhumita Roy is a professor & former Head in the Department of Architecture, Jadavpur University. She was the founder Director of the School of Planning and Architecture Amity University, Kolkata. She has served in various capacities in different Government Departments like Kolkata Municipal Corporation, West Bengal Housing Board(WBHB), West Bengal Housing Infrastructure Development Corporation (WBHIDCO) & New Town Kolkata Development Authority (NKDA) before joining Jadavpur University. She was associated with the planning of the New Town project since beginning. She has acted in the capacity of Govt Architect in NKDA for approving the building plans online for the first time in the country as a part of the Research and Developmental work from JU. She is the recipient of Young Teacher Award from AICTE and many other prizes and awards throughout. She has successfully completed a number of State and Central Government funded R & D projects. She was also involved with lot of institutional consultancy projects namely New Township Kolkata, **Writers' Building**, Cyclone shelters funded by World Bank, Way-side Motels-'*Pather Saathi*', *Soujanyo Bhavan* (State Reception Complex), etc. She has published in number of national & international level journals, conferences & seminars. She has presented papers in many international conferences and visited many countries for her research related works. She has acted as expert member in various committees formed by WBUT, UGC , NAAC etc. She has acted as one of the members in a High- power committee formed by Defence Minister, Govt of India. Presently she is the member of State Environment Impact Assessment Authority-WB, Ministry of Environment Forest & Climate Change. Her research interests include: 1) the use of GIS techniques and IT interface in developing online sanction systems connected with Architecture, planning related areas as a part of ease of doing business; 2) developing building rules and regulations for new and existing cities incorporating energy efficient aspects; 3) determining energy demands of various urban forms; 4) evaluating the impact of urbanization on biodiversity; and 5) Design and development of sustainable architecture, e.g. Bamboo construction, BIPV structures, etc.

Since my joining in 2001, I have been actively involved in Research work in Building, Energy and Environment. Research proposals were sent to AICTE and UGC. With the fund received, I developed a lab titled 'Energy Efficient Built Environment Lab' in the department, carried out work in the lab and completed my doctoral thesis in 2004. The most original contribution of the Ph.D thesis is determining the reference scale of integration Solar active features into buildings. Based on the outcome of the thesis, initiatives were taken up through WBREDA /MNRE and TERI through R & D projects in developing the present GRIHA rating system and some features were also introduced in the Building Rules of New Town Kolkata Development Authority. I started guiding Ph.D scholars. I took initiative in reviving the State Fellowship programme in the department. In 2006, I received the prestigious funding under the scheme 'Career Award for Young Teachers' in Architecture category from AICTE for the first time in the country. In 2006, I got an opportunity for developing the 'On Line Software for sanctioning Building Plans '(OPS) for New Town Kolkata project. The system was successfully developed & implemented in NKDA. The system is still running in NKDA with the help of the Research Fellows who worked under my guidance in the lab. In 2009, I was assigned with another R & D project of installing a structural Bamboo Building by NMBA, DST with solar roofing aiming to make a perfect Green building example. I have had continuous research journey for the past nineteen years. Presently two important research projects are being assigned -1) Impact of urbanization on Birds' Habitat from Department of Environment-Govt of WB. 2) Adaptive Built Environment solutions in smart cities through harvesting of renewable resources and data synergy funded by RUSA.

Jayashree Deshpande

Jayashree Deshpande is an architect, sociologist and educator based in Pune, India. At the University of Pune, a topper, at the undergraduate level in architecture and a Masters' degree holder in Sociology, Jayashree is the recipient of several awards and scholarships. She has been involved in teaching for more than three decades and has guided numerous students at the undergraduate and postgraduate levels in Architecture.

Jayashree has presented at national and international conferences and workshops and has published numerous peer reviewed papers on Architectural Education, Urban Sociology, Housing, Building Sciences and Sustainable Technology. She has reviewed research papers for Scopus Indexed journals. Jayashree has participated in a UNU research project for integrating sustainability in the curriculum of built environment courses. She has authored a compilation on "Housing and Housing Finance options in India" and documents 'Energy efficient and Environment friendly building practices in India'.

At the University of Pune, Jayashree was the Chairperson of Board of Studies in Architecture, Member, Faculty of Engineering and Member, Academic Council. She has been associated with several universities in India in the advisory category, for the subjects of architecture, education and design. Ar. Jayashree was the Chairperson and member of the Executive Committee of Indian Institute of Architects, Pune Centre, and is a member of national and international professional bodies. She has been felicitated by the Indian Institute of Architects for her extraordinary contribution to the field of Architectural Education, being instrumental in setting up Master's and Doctoral Courses in Architecture at the University of Pune.

Presently, she is the Director, Academic Unit of Council of Architecture, New Delhi, located at Pune. She is involved in conducting Training Programmes and Continuing Education Programmes for architects and teachers in Architecture, publication of books, Awards programmes for thesis at Under Graduate and Post Graduate levels and Heritage Documentation, Research in Architecture etc. She has conducted several training programs on 'Introduction to Energy Conservation Building Code' for the Bureau of Energy Efficiency, Ministry of Power, and 'Universal Accessibility' for the Dept. of Empowerment of Persons with Disabilities, Ministry of Social Justice and Empowerment.

Nandini Kulkarni

Dr. Nandini Kulkarni, a distinguished academician and researcher, serves as the Director at Symbiosis School of Planning, Architecture, and Design. With a Doctorate focused on "Investigating the Impact of Quality of Life on Peri-urban Areas of Metro Cities," Dr. Kulkarni's research expertise bridges the fields of urban planning, architecture, and the socio-economic development of peri-urban regions. Her international recognition is built on a robust portfolio of academic and research contributions. Dr. Kulkarni holds a Master's degree in Urban Planning from VNIT, a Bachelor's degree in Architecture, and a Bachelor of Technology in Interior Design, blending multidisciplinary approaches to address contemporary urban challenges. She has an extensive record of Indexed publications in leading journals and has authored *The Urban Edges*, a book exploring humane approaches to urban fringe development in Indian cities.

Dr. Kulkarni's research is complemented by her numerous patents and design registrations, demonstrating her innovation in architectural, planning and desgn solutions. Her work frequently intersects with applied research, focusing on improving the quality of life in urban-rural fringe areas and developing sustainable models for urban growth. With over 25 years of professional experience in both practice and academia, Dr. Kulkarni has mentored many young professionals and led various research projects that have had a tangible impact on urban policy and design. She has received multiple prestigious awards, solidifying her status as a leader in the field of urban and peri-urban development.

Her involvement in international conferences and collaborations further reflects her expertise and commitment to advancing knowledge in the global architectural and urban planning community.

List of Figures

List of Tables

List of Panelists and Reviewers

"*Samunnati Vistarah*" - Council of Architecture's (COA) International Conference on "Architecture Education in the Changing World - Reviewing and Rethinking the Way Forward".

Dates: 15th, 16th, & 17th February 2024
Venue: VIT Campus, School of Architecture, Vellore, Tamil Nadu, India.

No.	Names of Panellists for moderation of paper presentation sessions
1	KR Sitalakshmi, Chennai
2	Nandini Kulkarni, Nagpur
3	Devang Parekh, Rajkot
4	Ananthakrishna, Bangalore
5	Ramadevi Nandineni, Manipal
6	Rajiv Mishra, Mumbai
7	Monsingh Devadas, Chennai
8	Prasanna Desai, Pune
9	Ranjini Bhattadthripad, Cochin
10	Senthil Kumar, Chennai
11	T. Loganathan, Chennai
12	Jayalakshmi, Chennai
13	Manish Chakraborty, Kolkata
14	Abhijit Shirodkar, Hyderabad
15	Jothilakshmy, Chennai
16	Pratheek Sudhakaran, Cochin
17	Aparna Surve, Mumabi
18	Mainak Ghosh, Kolkata
19	Radhika Nagpal, Sonepat
20	Ketan Kimmatkar, Nagpur
21	Vandana Sehgal, Lucknow
22	Ranee Vedamuthu, Chennai

Sr. No	Names of the Paper and Abstract Reviewers
1	Shilpa Sharma, Mumbai
2	Vaishali Latkar, Pune
3	Harveen Bhandari, Chandigarh
4	Kanika Bansal, Chandigarh
5	Mamatha Raj, Bangalore
6	Mohammed Ali Sharieff, Naples, Italy
7	R. Shanthi Priya, Kattankulathur
8	Sourovee Dutta, Kolkata
9	Suchandra Burman, Kolkata
10	Vijayalaxmi Iyer, Vijayawada
11	Binumol Tom, Kottayam
12	Anita Yammiyavar, Bangalore
13	Rama Subrahmanian, Bangalore
14	Pratheek Sudhakaran, Cochin
15	Mainak Ghosh, Kolkata
16	P. Shabitha, Trichy
17	Sanjib Nag, Kolkata
18	Meghal Arya, Ahmedabad
19	Priya Choudhary, Nagpur
20	Uttam K Roy, Roorkee
21	Arpan Dasgupta, Kolkata
22	Gaurav Raheja, Roorkee
23	Debashish Das, Kolkata
24	Prajakta Baste, Nashik
25	Sameer Deshkar, Nagpur
26	Anand Wadwekar, Bhopal

Sr. No	Names of the Paper and Abstract Reviewers
1	Shilpa Sharma, Mumbai
2	Vaishali Latkar, Pune
3	Harveen Bhandari, Chandigarh
4	Kanika Bansal, Chandigarh
5	Mamatha Raj, Bangalore
6	Mohammed Ali Sharieff, Naples, Italy
7	R. Shanthi Priya, Kattankulathur
8	Sourovee Dutta, Kolkata
9	Suchandra Burman, Kolkata
10	Vijayalaxmi Iyer, Vijayawada
11	Binumol Tom, Kottayam
12	Anita Yammiyavar, Bangalore
13	Rama Subrahmanian, Bangalore
14	Pratheek Sudhakaran, Cochin
15	Mainak Ghosh, Kolkata
16	P. Shabitha, Trichy
17	Sanjib Nag, Kolkata
18	Meghal Arya, Ahmedabad
19	Priya Choudhary, Nagpur
20	Uttam K Roy, Roorkee
21	Arpan Dasgupta, Kolkata
22	Gaurav Raheja, Roorkee
23	Debashish Das, Kolkata
24	Prajakta Baste, Nashik
25	Sameer Deshkar, Nagpur
26	Anand Wadwekar, Bhopal

Foreword

It is with great pleasure that I introduce you to the COA international conference, *"Architecture Education in the Changing World"* In collaboration with School of Architecture VIT, Vellore, Tamil Nadu. This is an opportunity for all of us to come together and address some of the most pressing challenges and opportunities in architectural education today.

In an era defined by rapid technological advancement, shifting societal priorities, and evolving professional landscapes, architecture education must continuously adapt to remain relevant and impactful. Technological advancements have significantly transformed teaching practices in architecture, introducing new knowledge and perspectives. Over the past decade, digitized and interactive learning methods have reshaped education, and UNESCO's 2030 Agenda for Sustainable Development offers valuable guidelines for the 'Futures of Education.' The recent pandemic exposed the need to rethink architecture education in response to growing complexity and uncertainty. As the field navigates a "transitional emergency model," this conference was aimed to explore the evolving trends in architecture education to align with the demands of both the profession and academia. This year's conference was strengthening to explore these dynamics through a series of thought-provoking sessions that align with our key sub-themes:

- **Relevance of Architecture Curriculum to Practice and Research:** Ensuring that what we teach in our institutions remains pertinent to real-world practice and cutting-edge research.
- **Role of Regulatory Bodies:** Understanding the vital role that governing bodies play in shaping architecture education and practice standards.
- **Emerging Technologies in Architecture:** Examining the ways in which innovations such as AI, digital fabrication, and sustainability technologies are revolutionizing both the design process and the built environment.
- **Contemporary Concerns in Architecture:** Addressing critical global issues such as climate change, urbanization, and social equity, and how these concerns are influencing both education and practice.
- **Innovative Initiatives:** Showcasing pioneering approaches in pedagogy, practice, and community engagement that push the boundaries of what is possible in architecture.
- **The National Education Policy (NEP) and its Implications on Architectural Education:** Discussing how India's NEP influences the structure, content, and outcomes of architectural education.

The collective insights and expertise shared at this conference has not only helped shape the future of architecture education but also fostered a deeper understanding of how our discipline can contribute meaningfully to society. I am confident that the discussions, collaborations, and exchanges over the few days must have provided new perspectives and inspire innovation.

I would like to extend my sincere thanks to the organizing committee for their dedication in bringing this event to life and to our sponsors and partners for their invaluable support. Moreover, I am deeply grateful to the distinguished speakers, panellists, and participants for contributing their knowledge and expertise to this important dialogue. As we embark on this journey together, I encourage all of you to engage fully, question deeply, and collaborate openly. The future of architecture education lies in our hands, and this conference is a step toward creating a shared vision for the profession in a changing world.

Prof Abhay Purohit
President CoA

Preface

The contemporary world is undergoing swift and profound transformations - social, environmental, technological, and cultural. These shifts are reshaping every aspect of human life, and the field of architecture is no exception. As we stand at the crossroads of tradition and innovation, the need to critically evaluate and adapt our approach to architecture education has never been more urgent.

The Council of Architecture (COA), the apex body regulating the education and profession of Architecture in India, has envisioned a perspective plan for Architecture Education in India which emphasizes research and development in academic institutes. Diversity in academic programs, collaborations and capacity building to address future needs is an important part of this vision. This International Conference *"Architecture Education in the Changing World: Reviewing and Rethinking the Way Forward,"* by COA aimed to provide a platform for bringing out research and innovative ideas to shape education for the future of the built environment in the context of emerging technologies and environments and the National Education Policy. The conference, served as an important platform for reflection, dialogue, and action. It aimed to explore intersections, bringing together educators, practitioners, researchers, and students from around the world to engage in a dialogue that transcends traditional boundaries and encourages interdisciplinary collaboration, to explore how architectural education can and should evolve in response to the global challenges.

The themes of this conference invited deliberation on established pedagogies, new technologies, and the pressing need for sustainability in design. They encouraged perspectives on the role of the architect in society and the forms of knowledge and skills that future professionals will require. As the discipline of architecture continues to respond to the complexities of contemporary society—whether through technological advancements, environmental imperatives, or social transformations—so too must our approach to imagining the future and educating the next generation of architects to shape a better, more resilient world.

The papers and presentations included in these proceedings reflect a diversity of viewpoints, each contributing to the overarching goal of rethinking how we teach architecture in a rapidly changing world. From discussions on sustainability, digital tools, and design thinking, to reflections on history, theory, and global practice, this volume serves as a testament to the dynamism and expanse of contemporary architectural education.

We are grateful to the eminent speakers, advisors, reviewers, panellists, participants and the organizing committee for their valuable contributions and tireless efforts in making this conference a success.

We hope that these proceedings will serve not only as a record of the conference but act as a stimulus for continuing dialogue, innovation, and progress in the field of architecture education.

Jayashree Deshpande
Conference Co-ordinator, Editor

Acknowledgment

The Council of Architecture (COA) would like to express gratitude to the esteemed speakers Architect Abdul Muluk bin Abdul Manan, Architect Jayantha Perera, Architect Mallika Bose, Architect Sanjeev Vidyarthi, Architect Chu Yang Keng, Dr. Shakila Shamsu, Architect Mohammed Zakiul Islam, Dr Sajal Chowdhury and Architect Milind Kollegal for their valuable insights and perspectives during the conference.

We acknowledge all the panellists and reviewers for providing valuable knowledge to all authors and attendees. We are grateful for the contribution of time and effort for making our event a success.

We thank all the authors and presenters for their engaging and informative presentations on the conference themes. They have provided new perspectives and inspired deeper thinking in these areas.

Thanks to our advisers, the Executive Committee and members of COA for insightful inputs and members of the Conference Committees for meticulous planning and dedication to create a platform for meaningful discussions, networking, and knowledge sharing. We truly appreciate the efforts in making this event a remarkable success.

Heartfelt thanks to all delegates and heads of schools of architecture for active participation and engagement throughout the conference. The contributions have made this event truly memorable.

Special thanks to School of Architecture VIT, Vellore, Tamil Nadu for collaboration and support to conduct the conference.

Dr. Nandini Kulkarni
Member, Core Committee, Editor

Title of the Conference

"*Samunnati Vistarah*" - Council of Architecture's (COA) International Conference on "Architecture Education in the Changing World - Reviewing and Rethinking the Way Forward".

Dates: 15th, 16th, & 17th February 2024
Venue: VIT Campus, School of Architecture, Vellore, Tamil Nadu, India.

Chief Patron: Ar. Prof. Abhay Purohit, President, COA
Patron: Ar. Gajanand Ram, Vice President COA.

Advisory Committee:
Ar. Lalichan Zacharias, Member, Executive Committee, COA.
Ar. Punit Sethi, Member, Executive Committee, COA.
Ar. Nand Lal Chandel, Member, Executive Committee, COA.
Ar. P. Vaitianadin, Member, Executive Committee, COA.
Ar. R Ramesh Kumar, Member, Executive Committee, COA.

Conference Organizer: Shri. R.K. Oberoi, Registrar COA
Conference Convenor: Ar. Madhumita Roy, Professor, Dept. of Architecture, Jadavpur University, Kolkata
Conference Co-ordinator: Ar. Jayashree Deshpande, Director, Council of Architecture, Training and Research Centre, Pune

Core Committee:
Dr. Nandini Kulkarni, Nagpur
Dr. Roopal Deshpande, Nagpur
Ar. Shubhada Chapekar, Nashik
Ar. Poornima Deshpande, Nagpur

Organizing Committee
Dr. Madhumathi, Vellore
Dr. Sharmila Jagadishan, Vellore
Ar. Dinesh, Raghavan, Vellore
Ar. Nilesh Pore, Pune

Publication Committee
Ar. Madhumita Roy, Kolkata
Ar. Jayashree Deshpande, Pune
Dr. Nandini Kulkarni, Nagpur
Dr. Roopal Deshpande, Nagpur
Ar. Poornima Deshpande, Nagpur
Ar. Shubhada Chapekar, Nashik

Conference Themes

1 Relevance of Architecture Curriculum to Practice and Research

The curriculum, being the academic core of the field, needs to consciously evolve. The curricula should reflect the changing nature of architecture practice as well as be research-driven. Architectural practice is undergoing major transformations and Architects are now working in a more collaborative and research driven interdisciplinary environment. The curriculum development needs to be addressed through insights into practice and academics through intensive trans-disciplinary research.

The Architecture curricula across countries and cultures have woven many aspects of society, reflecting its changing nature and including sustainability as a thrust area. These in turn have been important guidelines in transacting the curriculum. This session of the conference focuses on new and innovative ways of thinking about architecture curriculum that are better suited to the challenges of the future. This session invites papers on various approaches to curriculum development in architecture and innovative ways of pedagogical response suited to changing practice trends and importance of research to the architecture program.

This theme addresses the following areas-

a) Research in developing a futuristic curriculum for Architecture
b) Impact of Curricular Trends in Architectural Practice.
c) Influence of Culture on Architecture Curricula.
d) Cross-Border trends in developing a common global curriculum for Architecture Education.
e) Pedagogical response and preparedness of institutions post-pandemic.
f) Resilient Education Systems
g) Needs and aspirations of the changing student community.
h) Emerging areas of learning relevant for Architects to lead changes post-pandemic
i) Response and preparedness of faculty either through self-initiative or institution
j) The applications in practice by way of policies, design interventions etc
k) The practice of architecture as an extension of learnings through education and research

2 Role of Regulatory Bodies

Regulatory bodies are crucial to architecture education, as they set the standards for education and monitor the adherence to the same. They also have the authority to issue accreditation policies and norms for the institutes to follow. In India, the Council of Architecture is the apex body, which, over the years, has upheld its standards to compete globally. Many nations may have similar bodies for the profession and education in Architecture. The role that these regulatory bodies can play, especially in view of India's G20 Presidency, is an important aspect to be considered. Architecture can transcend political barriers and work towards building sustainable infrastructure for the future, through the collaborative efforts of these bodies.

This theme explores the existing practices of governing bodies, their response to the advent of new practices and knowledge areas and systems in the following areas:

a) The Role and Impact of the Regulatory Bodies for Architecture in their countries
b) The Role and Impact of the Regulatory Bodies for Architecture in their region
c) Eligibility criteria and the entrance exams conducted for admissions to Architecture programs.
d) Accreditation systems, policies, and practices to achieve global competency
e) Prescribed norms and standards for institutions
f) Regulatory bodies- a collaborative effort towards a sustainable future across countries

3 Emerging Technologies in Architecture

Technology at all levels, including IT, Construction and software is rapidly evolving, with new techniques and tools being developed every passing moment. This theme was conceived to provide a platform for discussing the high pace of technological transformations in architecture by the professionals and architecture institutions.

The session looked beyond addressing changing technologies in construction of buildings and deals with the evolution in the design, modelling, processing of Building information modelling (BIM) and the advanced drafting software. The conference also anticipates discussions on the influence of state-of-the-art technology, Artificial intelligence (AI), Augmented Reality (AR) and Virtual Reality (VR) in the architecture curriculum, pedagogy and practice and the use of futuristic technology like drones, 3-D printing, twinning etc in construction.

This theme appreciated perspectives on:

a) Multi- Disciplinary Pedagogical approaches using emerging technologies.
b) Impact of emerging technologies on the practice of Architecture
c) Innovative teaching methodologies concerning IT tools such as Artificial intelligence (AI), Augmented reality (AR) and Virtual reality (VR).
d) Dissemination of learning materials through various interactive techniques like gamification etc.
e) Modelling curriculum in the wake of innovative building materials and new processes brought in by the technology revolution.

4 Contemporary Concerns in Architecture

This theme addresses Sustainable Development Goals and Net Zero Targets for a sustainable future. Issues such as energy efficiency in buildings, sustainable approaches, and universal design including equitable design towards inclusiveness can be explored in this idea. In addition, other relevant SDG's were explored, for e.g. social justice, poverty and hunger, gender equality, clean water and sanitation etc.

This theme provided a platform to speak about the role of Architects and Architecture beyond construction. The papers under this theme addressed the following:

a) Approaches towards achieving Net Zero targets
b) Achieving comfort through built environment vis-à-vis climate change
c) Participatory approaches to sustainability, environment and climate change
d) Interpreting SDG no 11 -Sustainable cities and Communities
e) Inclusiveness, access to opportunities, community building, social justice and participatory governance
f) Architecture curriculum towards preparedness of professionals for the emerging challenges (can address any of the SDG's)
g) World food equity through Architecture

5 Innovative initiatives

Architecture education has paved the way for architects to wear multiple hats and effortlessly conduct multiple tasks. This characteristic has enabled many architects to branch out into other disciplines. Architecture education concentrates upon innovation and making a concept viable and presentable; as a result, architects have become crucial members of start-ups, as team leaders, collaborators, and facilitators, and have entered associated and disparate fields.

This conference session received papers, suggestions, and presentations about the skills architects develop during education that enables them to explore and succeed in a variety of disciplines. Topics explore multi-disciplinary research, innovative approaches and ideas linking various disciplines to the built environment. Researchers have addressed, new knowledge areas deemed essential for the practice of architecture.

6 NEP and its implications on Architectural education

Reformations in the National Education Policy (NEP) will result in transformations in the teaching-learning process at all stages of education. The policy has the potential to make landmark changes to the way architecture is taught in the country. The students entering higher education will be equipped with knowledge, and understanding of areas of architectural concern like sustainable development, Indian or traditional knowledge systems, regional cultures, languages, etc. This will have a profound impact on the way architectural knowledge will be deciphered. With the focus being shifted more towards practical studies and hand-on learning rather than just on-paper theories, the policy is also expected to substantially strengthen the linkage between architecture schools and the construction industry. This theme invited discourse on the implications of NEP on the education of architects and, therefore, the profession, education institutes, and teachers. It explored the impact of NEP on the profession, restructuring of resources in teaching, and the changes in teachers' mindset and teaching approaches.

Details of Keynote Speakers

Architect Abdul Muluk bin Abdul Manan
Member of LAM MAPSM, The Centre of Sustainable Design and Built Environment (COS-BE) and PAM Education Committee, Malaysia

Topic:

Architect Jayantha Perera
Honorary Fellow -SLIA
President Chamber of construction Industry Sri Lanka
Chairman Green building council of Sri Lanka

Architect Mallika Bose
Associate Dean for Research, Creative Activity, and Graduate Studies, College of Arts & Architecture
Interim Director, Stuckeman School of Architecture and Landscape Architecture
Professor, Dept. of Landscape Architecture, The Pennsylvania State University

Architect Sanjeev Vidyarthi
Professor and Department Head
Co-Director, Master of City Design Program Urban Planning and Policy
University of Illinois, Chicago

Architect Chu Yang Keng
Founder of IX Architects, Singapore
PPE Examiner, Board of Architects, Singapore

Dr. Shakila Shamsu
Former OSD (New Education Policy),
Department of Higher Education, Ministry of Education (MoE)

Architect Mohammed Zakiul Islam
Head and Professor, Department of Architecture, Bangladesh University of Engineering & Technology, Dhaka

Dr. Sajal Chowdhury
Founding Director (Bangladesh Chapter): ZEMCH EXD Network
Founding Initiator: "Researchitect" Platform
Chief Advisor: "Atut Bandhan" Bengali Magazine on Built Environment

Architect Milind Kollegal
Chief Consultant, Mape Connoisseurs, Hyderabad.
Member of Council of Architecture, New Delhi.

1 Experiments in architecture pedagogical approaches and student responses

Kamlesh Gade[a]

Assistant Professor, School of Architecture, AIKTC, New Panvel, Maharashtra, India

Abstract

The architectural pedagogical process, especially in design studios, has always oscillated between two well-known approaches: one of supplying information in order to expedite the process of designing and the other of challenging students to come up with their own solutions through deduction. Each of the processes may have its own costs and benefits, although it has not been amply demonstrated what those might be, or how and when to choose the method in the studio in order to elicit a certain response from the students. In this paper, we outline the process and result of an experiment conducted on a batch of students where contrasting pedagogical approaches were tested on isolated, but homologous sets of students of Semester 2, dealing with the same design problem. Assessment was carried out on individual developmental parameters by external jurors who were blind to the process and the results were compared with each set, in order to ascertain impacts and results of the teaching method on students. The aim was to ascertain if student performance can be mapped quantitatively based on a pedagogical approach. The results show that the different approaches in design studios can have measurable impacts on different parameters of performance and the same can be further corrected to achieve favorable student responses.

Keywords: Architecture, design studio, pedagogy, quantitative analysis

Introduction

Design pedagogy has always been a point of contention for a long time. Design studios operate on personal interpretations of students. It is commonly understood that there are multiple right answers to the same question, and students could arrive at the answers via many routes. These routes often depend on the methods followed by the studio faculty. The studio becomes the main medium of architectural design education, and the conversation between student and the tutor becomes the means of this education (Ciravoğlu, 2014). Methods of teaching often fall within schools of thought and are practiced by teachers in architecture schools based on their personal philosophies or experience. However, very little is ever demonstrated in the form of empirical evidence to the efficacy of these methods. It is in the interest of design schools to closely observe the effects of specific teaching methods on student responses and how they may vary from student to student based on their personal temperaments.

One such popular contention is between the methods concerning the supply of information to students. The obvious and direct method here is the dissemination of the information or data through a lecture or through direct discussion. This is the classical framework that sees the teacher in the role of a provider who through direct dialogue imparts knowledge of concepts and systems

[a]kam.gade@aiktc.ac.in

DOI: 10.1201/9781003571018-1

to the students. There is, however, a contrasting opinion among many faculty of design that believes that the information should not be disseminated but rather deduced by making the students arrive at it through a series of exercises or through a process. Pedagogical research in architecture also outlines the role of a teacher in architectural education, specifically, not as a provider but as someone who creates knowledge through discourse. Moreover, this must be done without the student feeling that the guide is designing the solution for them (Goldschmidt et al., 2010). Many disagree with this form of teaching as it may be seen as re-inventing the wheel or a waste of precious studio time that may as well have gone into the design process, whereas many believe the former way of teaching robs the students of a mind that can think critically and arrive at certain conclusions through reason.

Faculty invariably align with one or the other of these methods without adequately demonstrating in a quantitative manner why one may be chosen over the other as a superior method or as one having a proven effect or benefit. This is usually done as a result of personal beliefs, philosophies or simply because they may have been taught in a certain way.

In this paper, we take a look at the process and result of an experiment conducted within an architectural design studio, where two homologous sets of students from the same batch are exposed to these two contrasting teaching methods, their performances are measured both qualitatively through faculty observations as well as quantitatively through scoring by external jurors.

Methodology

The experiment was conducted in the second semester design studio in a class of 50 students. The class was so chosen as the impact of teaching methodology on younger students can be considered higher as the students at that stage are more dependent on the faculty input and less autonomous in their design approach. Students in higher semesters may also carry a cumulative experience of previous design semesters and design pedagogies imparted to them by previous teachers.

The students in the class were separated into two groups through an online random order generator. Each group was then analyzed for an equal distribution of students from all scoring categories. (viz. Struggling, leading) in equivalent numbers. This removed the chances of end results being affected by students of any kind in any particular group (Figure 1.1).

Each group throughout the semester was subjected to contrasting methods of teaching; however, both groups had the same end goal in terms of analysis and design. The classes were conducted by different sets of professors with clearly outlined pedagogical approaches. Care was taken to keep students from mixing and exchanging information to avoid data contamination. This was made easier, because the mode of teaching was completely online due to COVID-19 restrictions.

At the end of the term, the students were assessed across various parameters through external assessors and mapped to analyze how the pedagogical approaches affected the performance. Additionally, the existence of a control group was assumed to exist, consisting of students that showed least attendance in both groups as these students would not have received either approach.

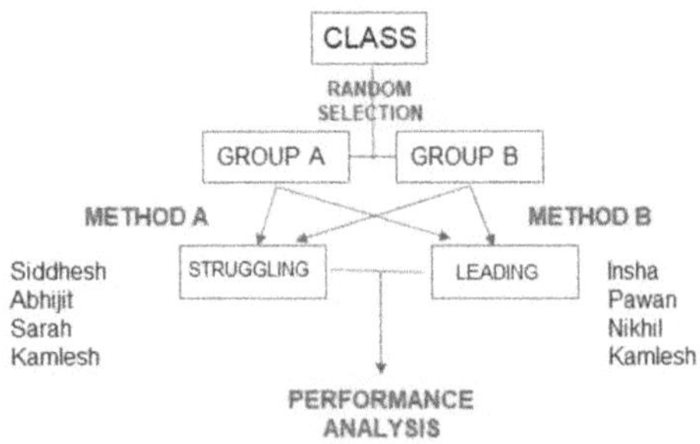

Figure 1.1 Methodology
Source: Author

The Studio Problem

The exercise for the first semester was where the students adopted a process driven approach at handling design, where the primary focus was on creating an aesthetic experience on site through the introduction of basic design elements of planes and surfaces. The intent for this semester was to introduce a parallel approach that relied on the process of problem identification and solution. We understand that the intent of architecture is two-pronged, that is first to provide an aesthetic experience to the user through the creation of experience of space, enclosure, movement, and form; and secondly to solve problems of habitation, shelter, protection and build a space for thriving.

The semester will deal with the skills associated with mapping, site-based issue identification and solution, user group analysis and design logic. All students will be presented with a real site, with a 0.5 – 1 km radius. This will be followed by a preliminary study of access networks, features, urban character, utilities and user groups.

This will be followed by identification of an intervention site of not more than 25 sq. m., fitting into the categories of public urban utilities, kiosks, small scale social infrastructure or public space. This insert will be designed through the lens of the user profile identified in the previous exercise, in accordance with their demographic needs, social aspirations and physical requirements.

The design will exploit research in the fields of (but not limited to) ergonomics and anthropometry, climate responsiveness, universal access, social sciences including gender, age, wealth disparity, et al. For the concluding assessment, all students must produce sheets explaining pre-design study and research, relevant drawings explaining the design, and any analysis justifying the efficacy of the design.

Intuitive Approach Versus Instructional Approach

It is well documented how in studio discourses the language used by the studio guides is key to the students' response to the subject matter and their own assessment of their personal performance in the process (McDonald and Michela, 2022). It then becomes important to understand how certain approaches can alter student behavior, response and performance at an empirical level. Hence, we looked at two contrasting approaches used in the studio environment in a live class environment.

All students across both groups attempted a common design problem where they were to demarcate, study, analyze and finally suggest a design solution to an identified issue in the study area. A common approach in design studios that is a standard method of teaching is to supply the students with zero direction in terms of end products or expected outcomes, challenging the students to find their own answers. This approach is a self-motivated approach and is more intuitive and qualitative as the parameters of study are not defined. This approach was labelled 'Approach A' or 'intuitive approach'.

The second approach or 'instructional approach' was a typical pedagogical approach often used in educational systems where specific instructions are received from the 'teacher' or studio guide and students follow those instructions exactly. This approach leaves little room for exploration on the students' behalf. This was the model followed in Group B.

An example of the diversity of these two approaches can be illustrated. Consider, a student from both groups was visiting site X, in Mumbai and asked to study it. A student from Group A would be told to visit the area, document and photograph 'everything', and speak to 'all' the people on the street, and identify an issue that requires a design intervention. However, a student from Group B, would be told that, they had to document the number of shops in the area, the average density of the place, the list of various businesses operating the area, etc and given the various types of 'issues' that they can study for example, traffic issues, civic issues, social issues, economic issues, infrastructural issues etc. One can clearly see that a group B student is equipped with the list of data points to be collected, whereas a group a student has a more open-ended process of deduction.

The Studio Process

The studio was conducted by a total of seven faculties, divided into each of the approaches equally with one common faculty (author) in both as a moderator. The classes were conducted independently in isolated zoom meetings while following protocols outlined by the approaches. Each week, faculty met to discuss the progress of the studio in each of the groups and decide the information to be disseminated in the next session. As mentioned earlier, Group B was supplied with a set of parameters for study and the same was followed up with individual guides whereas Group A was challenged to create their own parameters for the study and present it in the studio. Special care was taken to ensure that the schedule for both groups remained synchronous in order to rule out any progress resulting from time allowances. The review stages were also kept at the same time.

At the end of the semester, an external review was organized where four jurors assessed the final output of the studio with regards to five parameters viz., mapping, presentation, site analysis, concept and design development. Each parameter marked out of 10 and total to 50.

Mapping
Documentation of site, in terms of features, extents, nature of land use, context, demographics and resulting networks.

Presentation
Included the execution of drawings, level and quality of the same, graphics employed and over-all visual appeal of the output.

Site Analysis
Involved studying the documentary evidence of the site, and creating resultant maps of issues, and site conditions that can invariably point towards design.

Concept
This stage marked the student's ability to think creatively and conceptually towards-design resolution of site issues.

Design Development
Final design evolution and the ability to creatively solve the issues through design in an efficient manner.

The parameters were considered more technical at the beginning and more creative at the end in an increasing manner, parameters of concept and design being the most creative and design driven parameters. At the end of the jury the results were tabulated and analyzed for the presence of patterns relating to the approaches that the students were subjected to.

Assessment and Results

At a glance the results of the external jury showed close to similar performance, where students from Group B may even appear to have done better by a margin. However, on a more broken-down analysis, one can see that students from Group B (Instructional) did substantially better when it came to documentation of data than Group A (Intuitive), however showed significant decrease in creative thinking across less technical categories like concept and design (Figure 1.2).

It is worth noting that the average performance of the leading students to the struggling students in Group A was similar, in comparison the struggling students

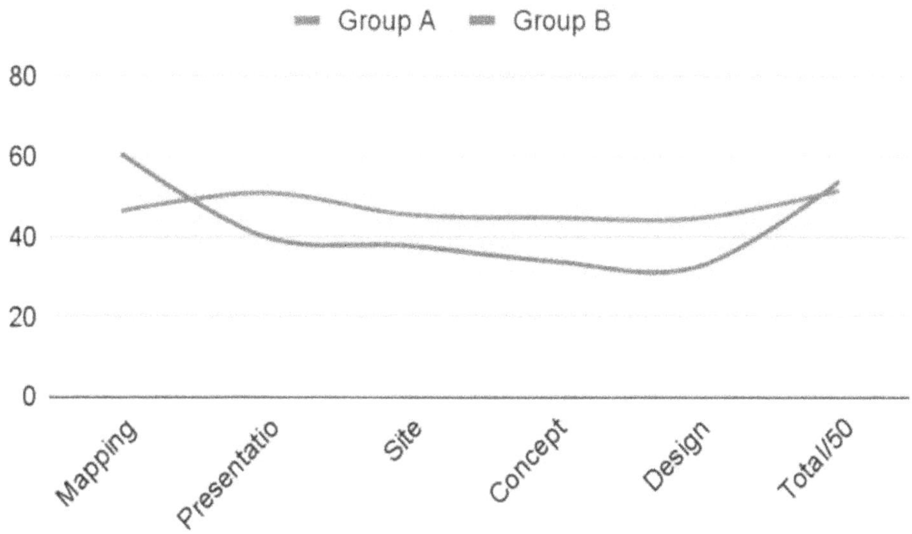

Figure 1.2 Final average marking across categories
Source: Author

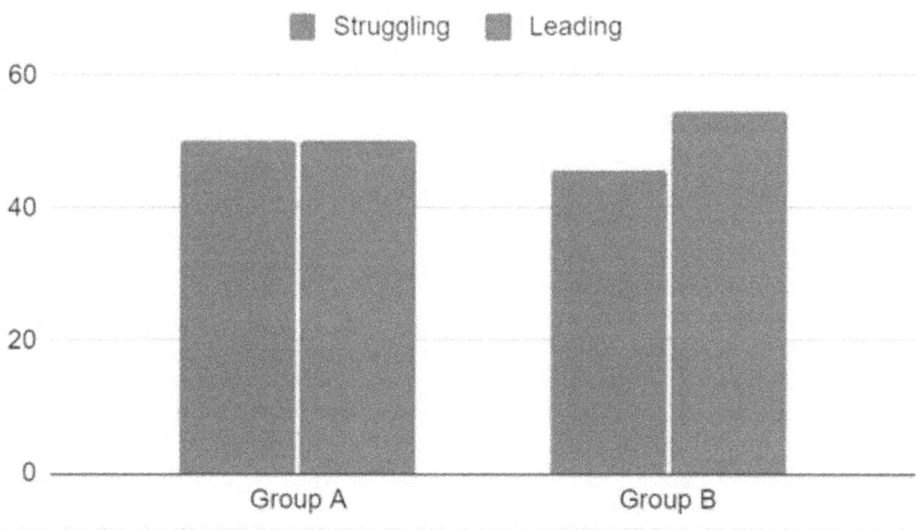

Figure 1.3 Performance of struggling students versus leading students in both groups
Source: Author

of the Group B, continued to lag behind the leading students in a noticeable way (Figure 1.3). This is especially interesting as for the final external jury, fewer struggling students were allowed from group B than from group A. It seems that Group A approach renders the struggling students in a more conducive environment for an increase in performance quality than the latter (Figure 1.4).

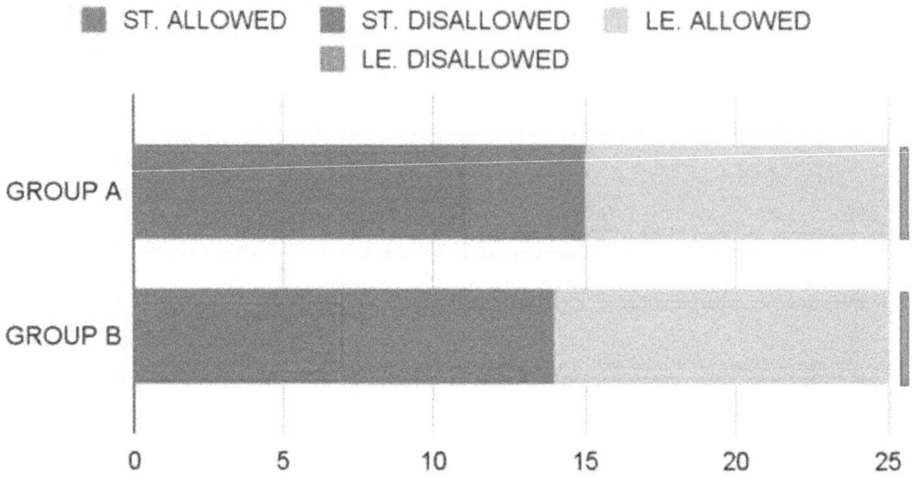

Figure 1.4 Students disallowed at external jury for lack of substantial work
Source: Author

Conclusions

The approach A or intuitive method of teaching design is more challenging academic environment for students. Consider reframing the sentence for clarity. It renders itself useful to motivate the students to find creative solutions and incite more diverse responses. It also allows for all students to be challenged in a more similar fashion, allowing for more equity in terms of student performance across struggling and leading categories. However, for a more technical, documentational etc. performance, an instructional approach or Approach B appears to bear better results, allowing students to focus more on data collection and consolidation.

References

Ciravoğlu, A. (2014). Notes on architectural education: an experimental approach to design studio. *Procedia-Social and Behavioral Sciences*, 152,7–12.

Goldschmidt, G., Hochman, H., and Dafni, I. (2010). The design studio "crit": teacher–student communication. *Artificial Intelligence for Engineering Design, Analysis and Manufacturing*, 24(3), 285–302.

McDonald, J., and Michela, E. (2022).This uncertain space of teaching: how design studio instructors talk about design critiques along with themselves when giving critiques. *Journal of the Scholarship of Teaching and Learning*, 22(1), 48-66

2 Relevance of architecture curriculum to practice and research: Adaptive education – call of the future

AR. Mukta Gokhale–Kulkarni[a]

Associate Professor, Smt. Kashibai Navale College of Architecture, Padmakunj Society, Bibwewadi, Pune, India

Abstract

In response to evolving global challenges in architecture, it's crucial to adapt education to meet students' changing needs. Understanding contemporary students is essential for innovative teaching approaches. Higher education institutions (HEIs) must prioritize this understanding to improve teaching and learning outcomes. This study delves into the concept of effective teaching through the lens of college students, envisioning it as a cornerstone of adaptive education for the future. By exploring students' perceptions of effectiveness and the essential qualities they associate with effective teachers, this research aims to provide insights crucial for shaping teaching practices in an ever-evolving educational landscape. The phenomenological approach was used to examine themes which emerged from the interviews of final year architecture students and post graduate students of Smt. Kashibai Navale college of architecture. The results identified three overarching themes: the student-Teacher relationship, Engagement, and Understanding current trends as the three most important qualities of effective teachers valued by students. The findings highlighted the paramount importance of effective teaching for quality education, emphasizing the need for competent, compassionate, and dedicated educators. Additionally, the study highlighted the evolving perceptions of effective teaching among students, suggesting the necessity for longitudinal research to grasp these changes over time. While this research primarily focused on students' perspectives, it also recognized the need for further exploration into teachers' viewpoints to achieve a holistic understanding. This study contributes to the ongoing discourse on adaptive education, advocating for teaching methods that align with the shifting demands of the future.

Keywords: Adaptive education, effective teaching, phenomenological approach, student perceptions, teaching pedagogy

Introduction

This whole education system revolves around the receptor, the end user, the citizen of tomorrow – The student. As the surrounding environment for students undergoes transformation, the requirements and aspirations of students are also evolving. As these changes unfold, education must adapt, highlighting the pivotal role of teachers as key contributors to the learning process. If educators fall behind in fulfilling their role, a gap emerges, and conversely, if students struggle to keep up, the gap persists. Achieving the ideal equilibrium demands sustained longitudinal research to gauge the dynamic needs of students. This, in turn, enables the continuous upgrading or adaptation of

[a]muktasgokhale@gmail.com

DOI: 10.1201/9781003571018-2

knowledge impartation methods to align with the changing educational landscape. It is of utmost importance to define effective teaching based on the students' viewpoint, as their insights play a critical role in shaping an educational approach that resonates with their unique learning needs and aspirations (Alhija, 2017; Parpala et al., 2011). Recent research on the personal qualities and behaviors of outstanding teachers has unveiled a captivating portrait of effective teaching, offering valuable insights into the intricate facets that contribute to a successful and impactful educational experience (Alhija, 2017; Keeley et al., 2006; Miron and Mevorach, 2014). This study aimed to fully understand effective teaching by directly considering students' perspectives, goals, and experiences. The goal was to gain a complete and integrated understanding of successful teaching dynamics.

Aim

The aim is to understand how college students view effective teaching, including their definition of effectiveness and the important qualities they believe effective teachers should possess.

The study was designed to answer the following research questions:

Research Questions

Q1. What is effective teaching as per college students?

Q2. What are the important characteristics of effective teachers?

Knowledge is built through social interactions, particularly when individuals with more knowledge (teachers) share it. Effective learning occurs when teachers encourage collaboration among students (Vygotsky, 1980). The phenomenological approach was used to examine themes which emerge from the interviews as this method is rarely utilized in existing literature to analyze effective teaching practices. This approach seeks to uncover the underlying structures and meanings of lived experiences, offering insights into how individuals perceive and make sense of the world around them.

The results indicated that specific attributes of effective teachers transcend educational knowledge and are ingrained in the personal characteristics of the teachers themselves.

This study was carried out in two parts –

A. Literature study

B. Student interviews

A. Literature study - Effective teaching, as emerged in the literature study, boils down to two key principles:

1. The technical side involving knowledge and teaching methods,
2. The personal side, which includes passion, the student-teacher relationship, and personal qualities.
3. The conceptual framework of this study was formed by literature in two areas, effective teaching attributes and effective teacher characteristics. Content knowledge, teaching methods and connections between theory and practice were the most important qualities of effective teaching.
   - **Content knowledge** - Content knowledge involves a thorough understanding of the subject matter and a comprehensive knowledge of related disciplines. Effective teachers demonstrate a profound expertise in the subject matter, and students firmly believe that possessing substantial content knowledge is a crucial attribute of effective teaching (Miron and Mevorach, 2014).
   - **Teaching method** - Effective teaching requires not only a strong foundation in content knowledge but also the skill to convey that knowledge in an engaging and comprehensible manner. Teaching and learning are two indispens-

able facets of the educational process. Miron and Mevorach (2014) studied how graduate students in education perceive good teaching. They found that professors' teaching methods significantly influenced students' development of critical and scientific thinking skills especially first year graduate students.

- **Connections between theory and practice** - The link between theory and practice involves the knowledge students gain in a class that is not only relevant to the real world but also applicable to their future endeavors (Anderson et al., 2012).

Effective teacher characteristics - According to common beliefs and literature, the key qualities attributed to an effective teacher are approachability and respect.

- **Approachability** – Approachable teachers are those who display understanding towards students' challenges, provide constructive feedback, and regard each student as a valuable asset. When teachers are approachable, students feel less intimidated about asking or answering questions and are more comfortable admitting when they don't understand a lesson or face difficulties (Miron and Mevorach, 2014).
- **Respectful** – An effective teacher is someone who values diverse styles and opinions among students, treating each individual as an equal. (Miron and Mevorach, 2014). Ethical responsibility demands that teachers take the act of showing respect to students seriously, recognizing it as a crucial and non- negotiable aspect of their role (Anderson et al., 2012).

B. Student interviews - In this phenomenological study, the focal point was the viewpoints of 15 college students, aiming to grasp the essence of teaching effectiveness. The emphasis was on unraveling the individual meanings each participant associated with their specific situations and learning outcomes. Smt. Kashibai Navale College of Architecture was chosen as the primary site to collect data. Using random sampling, participants of 5th year and Post graduate Architecture students were selected for an audio recorded interview. Interview questions were straight forward and included,

Q1. What is effective teaching as per you?
Q2. What are your experiences of an effective teacher?
Q3. What are the characteristics of an effective teacher?

Limitations: The sample size is relatively small and considers only architectural students of final year and post graduate students of architecture branch.

Data Collection

All 15 students took part in a 10 to 15-minute audio-recorded interview. Each transcript underwent a detailed review, and noteworthy excerpts directly related to the students' experience of effective teaching were singled out. This method enables the formulation of meaning from the participants' own expressions. Common themes emerged across all participants' transcripts, reflecting formulated meanings.

Findings and Discussion

The study findings revealed that students held similar perspectives on teaching effectiveness, with personal characteristics of effective teachers resonating strongly among them. All students emphasized the significance of personal characteristics, teaching strategies, adaptive teaching, and

Table 2.1 List of names of students

Sr. No	Name	Gender	Year of study
1	Bhakti	F	Final Year (B. Arch)
2	Akanksha	F	Final Year (B. Arch)
3	Jay	M	Final Year (B. Arch)
4	Dyanda	F	Final Year (B. Arch)
5	Gaurav	M	Final Year (M. Arch)
6	Sayali	F	First Year (M. Arch)
7	Nisha	F	First Year (M. Arch)
8	Abhilasha	F	Final Year (B. Arch)
9	Pranay	M	Final Year (B. Arch)
10	Atharva	M	Final Year (B. Arch)
11	Sayyami	F	Final Year (B. Arch)
12	Yashanjali	F	Final Year (M. Arch)
13	Manav	M	Final Year (M. Arch)
14	Mayuri	F	Final Year (M. Arch)
15	Sakshi	F	First Year (M. Arch)

Source: Author

the crucial aspect of being understood as an integral element in effective teaching. The following themes were identified from the interviews.

Theme 1 – Student Teacher Relationship

Everyone should be treated equally in class. Everyone should feel included. – Dyanda

- **Know the Students:** All 15 interviewees highlighted the significance of professors genuinely getting to know their students. Building a connection with students commences with an understanding of their backgrounds, recognizing that certain teaching methods may not be effective for particular groups. When instructors dedicate time and effort to understand students' learning needs and their lives outside the classroom, adjustments to teaching methods can significantly enhance effectiveness. Without a true understanding of students and a genuine commitment to aiding their learning, teachers cannot effectively impart knowledge.

Sayali gave a wonderful example of how her professor created a personal bond with students:

"He used to bring architect's work books in studios. Whenever we used to explain our design and elements used in it, he used to open the books, show us how that architect has used that similar element in his design and how is it impacting the overall design. It made us learn more and made us go to libraries to do the same. It instilled confidence in us that we are doing something correctly."

- **Approachability:** Establishing a positive student-teacher relationship involves creating an approachable atmosphere where students feel safe to ask questions and share ideas without fear of judgment. It's crucial for educators to offer constructive feedback consistently, be accessible to students, and encourage further exploration of the subject. Additionally, understanding students' learning styles, cultivating a supportive learning environment, and treating students as equals are key elements in building a strong foundation for this relationship, acknowledging and appreciating their unique contributions to the class. As Bhakti aptly stated, 'When a teacher understands our challenges, it's reassuring and motivates us to put in more effort.'

Theme 2 – Engagement

"Teachers should be at par with the current trends. We can help them if they encounter any problems." – Jay.

- **Participation, group work, discussions, hands-on workshops, site visits:** Effective teaching involves employing diverse methods that engage students and encourage active participation. Interviews, group discussions, collaborative projects, and encouraging participation were recognized as highly effective approaches to learning. Gaurav very happily spoke about one assignment, "we were asked to design a board game where different landscape related themes were given. It was a group work and we had to design a board game based on that theme for every age group to play. That assignment stayed with us because it was so fun learning."

  Innovative assignments initiate good participation and rapport with students enabling effective student engagement and learning.

Pooja said, "After every theory lecture we were given specific instructions to explore more information, and then in the next lecture we used to discuss the same. These discussions we remembered more as learnings."

Enhancing student engagement requires teachers to have a strong command of the subject matter and to design assignments that are both meaningful and captivating, developing collaboration and learning. Emphasizing hands-on activities over traditional lecturing and PowerPoint presentations can further stimulate student involvement and comprehension. Incorporating field visits, case studies, and video content into technical subjects can greatly enhance the learning experience, making it more effective and thorough.

Theme 3 – Understanding the Current Trend

"A student wants a teacher to stay updated and impart education about the latest trends in market. Some teachers readily adapt to the situation whereas some teachers don't." – Dyanda

- **Practical and theoretical connection:** Interviewed students unanimously agree that effectively connecting practical knowledge with classroom theory is crucial for good teaching. This approach helps students understand and apply concepts in real-life situations, making learning more meaningful and enhancing skill development.

Teachers who utilize real-life examples to illustrate their points have the ability to encourage students to break away from passive learning habits.

- **Futuristic explorations:** Scope for futuristic explorations need to be made an integral part of syllabus.

"While browsing Instagram, I came across a thesis titled 'Heaven'. It was a futuristic exploration topic, done as a thesis. If we don't imagine and create, though hypothetical, how will we ever be futuristic?" – Jay.

Thought-provoking statements like these prompt educators to reflect on their role in the classroom. It's essential for syllabi to include opportunities for futuristic explorations in design.

Some more key findings also highlight the need for a teacher to be fair while grading, not be biased to some students, in order to be efficient teachers.

Conclusion and Future Research

As times evolve, there is a pressing need to redefine adaptive education strategies. While understanding students is an essential initial step, it represents just the beginning of this broader initiative.

Utilizing individual perspectives and employing phenomenological analysis, college students were placed in a central role as experts in determining what constitutes effective teaching. The three overarching themes from the study included, Student – Teacher relationship, Engagement and Understanding the current trend which addressed the two main questions,

Q1. What is effective teaching as per college students?

Q2. What are the important characteristics of effective teachers?

Effective teaching is paramount for quality education. The presence of competent, compassionate, and dedicated teachers is indispensable for delivering excellence in learning.

Similarly, students perceptions of effective teaching also matters. As students continue to grow and learn, their perceptions of learning and teaching also evolve with time. Longitudinal research needs to be done to understand this change over the time.

While this research focuses on understanding students' perspectives, there is a recognized need for additional research to delve into teachers' viewpoints. This holistic approach would lead to a comprehensive understanding of the situation, offering more concrete insights into adaptive education strategies.

This study has unearthed thought-provoking insights into what constitutes effective teaching from a student's perspective. Certain aspects require further exploration to grasp their significance, potentially leading to significant pedagogical shifts.

Statements such as "There is a curriculum, but there also is no curriculum. When we practice outside or work in offices, we feel we haven't learnt much in college years" prompt deep reflection on necessary changes and their starting point.

The interplay among students, teachers, and real-world experiences forms a distinct triangle. If harmonized, this dynamic relationship could effectively address the gaps identified in the study.

References

Alhija, F. N. A. (2017). Teaching in higher education: Good teaching through students' lens. *Studies in Educational Evaluation,* 54, 4-12.

Anderson, M. R., Ingram, J. M., Buford, B. J., Rosli, R., Bledsoe, M. L., & Onwuegbuzie, A. J. (2012). Doctoral students' perceptions of characteristics of effective college teachers: A mixed analysis. *International Journal of Doctoral Studies*, 7, 279-310.

Keeley, J., Smith, D., & Buskist, W. (2006). The Teacher Behaviors Checklist: Factor analysis of its utility for evaluating teaching. *Teaching of Psychology*, 33(2), 84-91.

Miron, M., & Mevorach, M. (2014). The" good professor" as perceived by experienced teachers who are graduate students. *Journal of Education and Training Studies*, 2(3), 82-87.

Parpala, A., Lindblom-Ylanne, S., & Rytkonen, H. (2011). Students' conceptions of good teaching in three different disciplines. *Assessment & Evaluation in Higher Education*, 36(5), 549-563.

Vygotsky, L. S. (1980). *Mind in society: The development of higher psychological processes*. Cambridge, MA: Harvard University Press.

3 Application based contemporary design education

N. Devi Prasad[a]

Practising Architect, Professor and Design Chair, Vellore Institute of Technology, Tamilnadu, India

Abstract

The process of teaching architecture comprises awareness and fulfilment of the sensorial qualities of a physical environment and the responsible application of technology to optimize the process of creation and its upkeep. While there are several methods to teach the various subjects related to architectural education, the role of the student in engaging with the process is substantially underplayed. The unique nature of architectural education encompassing subjects that are both closely connected to subjective and objective thought processes unfold opportunities of multidisciplinary engagement in problem solving environments that are manifest in regions beyond those of the pureplay built environment and cast the thought process as a springboard for participation in a much larger framework of human activity.

Application Based Contemporary Design Education (ABCDE) process is a concept for enhanced learning outputs incorporating a collaborative and constantly fluid progressive process among teachers, students and practitioners to encourage ownership of process and enhanced intellectual development. The purpose of this paper is to elucidate the thought processes which may eventually lead to spatial products in the environment which engage students more meaningfully in the learning process and are symbiotic with the needs of the human senses. The tools which are discussed and have been implemented in live classroom situations include immersive learning, reverse engineered discovery, ownership, mind mapping and role plays including gamification.

Keywords: Applied learning, architectural education, gamification, immersive learning, mind mapping, ownership, reverse engineered discovery learning, teaching architecture

Introduction

Application based contemporary design education (ABCDE) is a concept for enhanced learning processes incorporation a collaborative and constantly fluid progressive process among teachers, students, and practitioners to encourage ownership of process and enhanced intellectual development.

While the core objective of architectural education is to produce a talent pool equipped with skillsets to design spatial entities embodying functional needs and fulfilment of the experiential qualities of a physical environment, the process of achievement of such an objective could be streamlined through directed thought process flow. The purpose of this paper is to elucidate the thought processes which may eventually lead to directions yielding to spatial products in the built environment which engage students more meaningfully in the learning process and are symbiotic with the needs of the human senses. The method of teaching outlined in this treatise pertains to subjects allied to architectural education and have been applied over a period of eight years from 2015 through 2023 and have seen student cohorts benefitting

[a]andeviprasad.ams@gmail.com, designchair.vsparc@vit.ac.in

DOI: 10.1201/9781003571018-3

substantially from the point of view of knowledge assimilation, analytical thinking and application confidence. The tools which are discussed in some detail include mind mapping, gamification, immersive inferential learning and reverse engineering.

The advent of artificial intelligence into the architectural domain has enabled application-based tools to serve as a source for both performance based evaluations and form finding and could with indiscriminate use subvert the principles behind the spatial, social and cultural elements that define architectural creativity. The processes described in this text are directed towards creating a stronger foundation to evaluate the applications of artificial intelligence in architecture while appreciating its enormous potential to optimize processes.

Education in the Current Scenario

The process of teaching has over the past two decades morphed from a reactive position to a creative one. The rise in entrepreneurial thought and the cross disciplinary nature of work across virtually all sectors has created an environment wherein knowledge is a tool for creative application and not an absolute by itself. In all disciplines, there is no formulated curriculum that can predict the precise nature of work demands but only a thought process flow which can adapt itself to varying requirements. The conventional adherence to stereotyped knowledge dissemination born out of an examination mindset derived from textbook knowledge needs to be replaced by an application-based system evolved out of a teaching method which employs knowledge as a tool to be applied to real world processes at every stage of information sharing. It is thus imperative that while every subject in an architectural curriculum is conducted, the linkage to real world application is established explicitly in the classroom and field beyond. According to Robert Doyle, '*It is an environment that allows students to take some real control over their educational experience and encourages them to make important choices about what and how they will learn*' (Doyle Terry, 2008).

In the case of architectural education, a subject which encompasses both linear and lateral thought, there are subjects born of mathematics and logic and those born of arts and humanities. Some of the conventional subjects of objective nature include architectural graphics, structural design, construction technology, utility services and project management. Design studios, art forms appreciation, history, and culture form some of the subjects designed to improve lateral thought and creative interpretation. The unique nature of the combination of such subjects offers several application opportunities even beyond the realm of the built environment.

Mind Mapping as the Missing Link in the Form-Function Discovery Process

There is a continuous and constant challenge to translate ideation into real form during the design process. In simplistic terms, the process involves prioritization of specific objective and subjective design parameters to create a basis for the discovery of form as pertinent to function and context. The missing link between the raw data of building programs and the form-space experiential output of architectural design can be directionally suggested by mind mapped analyses. The application of mind mapping, popularized by Tony Buzan, is a manner of graphically organizing multiple parameters which are relevant to a centralized theme or goal through a process of prioritization with instant visual clarity. The concept of mind mapping is useful in organizing the thought process through a personalized form of imagery and words which connect to the theme. It is a useful memory recall process and the design process is one which can benefit from such an application tool.

The process of architectural design is complex and includes several factors in its realization including but not limited to user needs and satisfaction, financial diligence, contextual relevance, social symbiosis and environmental responsibility. The organizing of the multiple parameters which contribute to the evolution of the eventual architectural organism need to be prioritized and assembled in a rational but typically intuitive fashion to create distinctly individual entities which manifest the particular design identity. A constitution of diagrammatic visual quality is therefore needed to eventually arrive at the intended process and product. Often such processes occur within the framework of the mind in a subconscious fashion and while not all designers may need a visual representation of the thought processes which flow through their minds, most structured design thought processes would benefit from a visualization of process flow such as those provided by mind maps.

Different building environments need variant priorities in their realization and their emphasis or lack of it during the thought process would greatly determine the directions of the eventual architectural product. A mind map to clarify the design parameters for a design project involving complex building typologies is illustrated below *(see Figure 3.1)*

Geometry and Form Finding-The Concept of Volumetric Creativity

The importance of geometry and the discovery of form through spatial coordinates born of experiential design needs to be given greater importance. While most engineering principles such as structural engineering, service and utility planning are greatly facilitated by some form of geometric spatial understanding, the process of understanding space through geometry is most often researched only during the year of foundation design in an undergraduate program in architecture. Coordinate geometry, parametric analysis and 3D realization of design ideation is born of the mechanics of objectivism of mathematics. Surface planning needs to be replaced by volumetric creativity. The development of thought process flow to intuitively generate volumetric identity can be enhanced through greater applied learning of geometry and architectural graphics. Geometry needs to be taught as an applied subject with live reference to existing building plans and volumes. Students may be encouraged to interpret the geometry of the intricate flooring patterns in Mughal architecture while using scaling methods to understand the real from the drawing world. Scaling and building element proportions can be taught by photo interpretation of real buildings-the

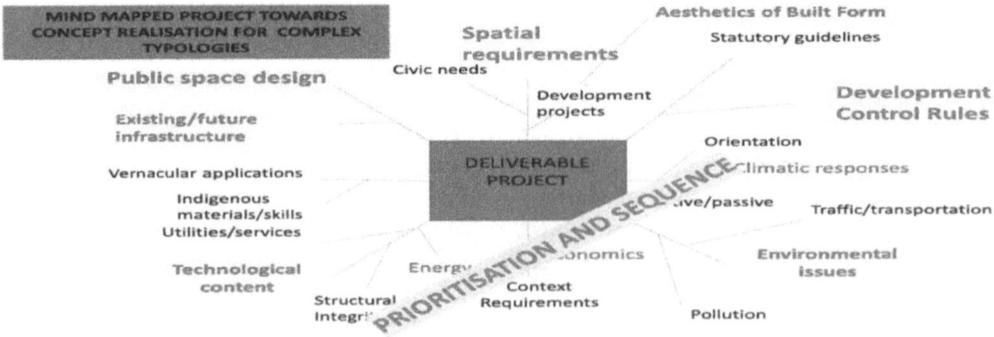

Figure 3.1 Mind mapped project
Source: Author

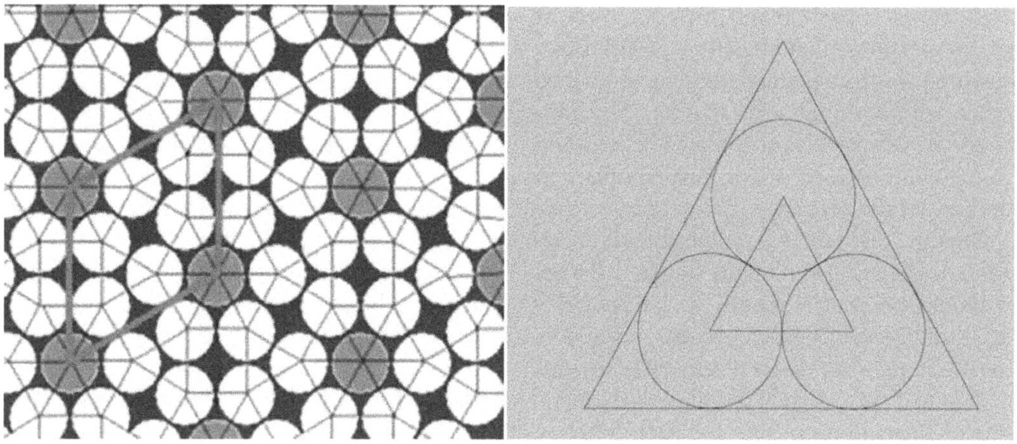

Figure 3.2 Exercises in interpretation of real time mosaic patterns
Source: Author

process similar to live sketching of objects or details wherein there has to be a cognitive relationship between the real image and the drawn figure and a sense of oneness between what one sees, and one draws (*see Figure 3.2*)

Gamification and Role Play in Architectural Learning

As a process for student awareness towards the realities of the practitioner's field, the subject of professional practice is conventionally taught in all learning institutions. While the transactional nature of the profession is well shared with the student fraternity, the vagrancies of the human factor are not adequately explored, and which are perhaps more significant in the realization of the commissioning of projects. There are several institutions which impart some training in the form of 'soft skills', these however tend to be substantially textbook formatted and not based upon real world experiences.

The process of architectural form making and spatial design and the ability to communicate the rationale behind such proposals is a serious factor in determining the directions in which projects are assigned to one designer or the other. Communication skills are an integral part of the scheme of obtaining projects from prospective clients. Psychology and body language and pre-meeting awareness skills are an integral part of such entrepreneurial efforts. There is little doubt that even the most earnest and even rationally sound ideas by designers presented to clients who are opposed to certain ideas due to their personal choices fall by the wayside. A similar fate can befall well intentioned efforts at architectural form making provided that sufficient efforts are not made to understand the perceptions of the client and their priorities. The importance of communication in the overall process cannot be overestimated and the method of communication delivery needs adjustment to suit the situation usually facilitated by a process of putting on the client's thinking cap and perceiving the world from an alternative standpoint. Empathetic communication leads to greater dialogue and usually better results on both the sides of the client and the designer.

As an academic process towards enhancing the skills needed to communicate better and understand priorities of all team players including the user, the financial resource provider, the contractors and the statutory and administrative agencies involved, a cold rehearsal of the future situational circumstances likely to prevail could be enacted in the classroom. These could take the form of students as role players in a forum of the architect as a participant in public interest group meetings, client meetings, vendor discussions, project management reviews and other transactional interactions. The language of communication could be adapted to the comfort level of the participants and the role of the players.

During one of the experiments conducted during a studio on housing design with second year students of the Bachelor of Architecture Undergraduate program, the students were divided into three groups of 13 students each with a design objective of providing housing for each other in each group. The process included an initial questionnaire on their identities, their age and marital status, their extended families and their special needs and professional interests impinging on their domestic lives. With a gamut of requirements ranging from single parents, joint families, celebrities and businesses from home, the students were exposed to various domestic situations involving their engagement with each other and with the overall community. The result was very successful and students learnt to collaborate as a group- it taught them to reach forward with better communication and a skillset to work together. Team bonding resulted and the entire cohort were exposed to a wide range of domestic requirements that they had not experienced before. A further dimension to the process was an inclusion of part assessment of each other's works to gauge if the designs were in keeping with the requirements of the client. The architectural form, conventional functional qualities and visual symbiosis as a community development were assessed by the teaching faculty.

Immersive Learning and Ownership as Learning Tools-Creative versus Reactive

In the process of teaching, the traditional methodology of teaching has been reactive teachers profess, and students react. In a world wherein the rate of technological change was limited, the rate of knowledge acquisition and impartment by teachers had kept pace. However, in the current rate of change, which is being experienced in the construction industry, designers are hard pressed to keep pace. Even professional designers are sometimes challenged by technological advancements and in the case of the teacher who is not constantly immersed in the field of construction methodologies, the challenge of obsolescence is furthered, and this creates a constriction in the thinking skills of the students. Teachers need to therefore constantly upgrade their skills in the real-world scenario.

The ubiquitous presence of the internet and access to digital tools has led to substantial agility of reaction. Just as several experiments have been made in creating variant forms of design studio engagements, the same should also be explored in technical subjects. Students research the internet while engaging in finding solutions to specific tasks outlined by teachers. While studying the subject of professional practice, students could interact with practicing architects and project managers. Real time assessment of publicly available information such as builder promoted projects enables students to understand the complexities of larger projects. In these cases, students could be asked to browse the internet for available information on developer projects and analyze the projects in terms of planning parameters, financial data, cost estimates and project management scenarios. The sense of *ownership* by students molded on real time projects chosen of their own free will yield engagement with their assignments in a far superior intellect-driven way than any reactive-to-teacher choice exercises. The process also ensures that the intricate queries of the students and their interpretation of various learning modules is addressed by the teachers, who, incidentally, need to upgrade their own skillsets to meet student needs. There is thus a greater push towards intra faculty and industry professional engagement. The process of ***ownership*** in the choice of learning and the subsequent ***immersion*** is a crucial factor in the deep enhancement of the research and learning process. A corollary of a more widespread basis for learning necessitates that teaching faculty also needed to constantly upgrade their knowledge and the process of learning. The teaching and learning process is therefore more evolved and dynamically aligned with ***contemporary*** design and technology evolutions.

In a studio on Interior Design, the same process was employed in a cohort of 40 students to design the interior of a residence of a famous personality using the perceived mindsets of their clients and aesthetic inclinations. Forty different designs emanated with suitable explanations for the suggested visual and spatial vocabularies employed. ***Ownership*** of the projects and personal research into the preferred choices of their clients produced highly creative results.

Reverse Engineering as a Mind Stimulation Tool

Another important experiment conducted in the building technology lab on steel construction was to encourage students to choose relevant steel structures and reverse engineer and produce technically correct drawings. Students thus photographed and documented railway platform roofs, steel bridges and overhead walkways and several other structures and carried forward the reverse engineered discovery learning process and thereby understood the structure more thoroughly. In a similar exercise in a construction technology studio, students were asked to photograph real time interiors of

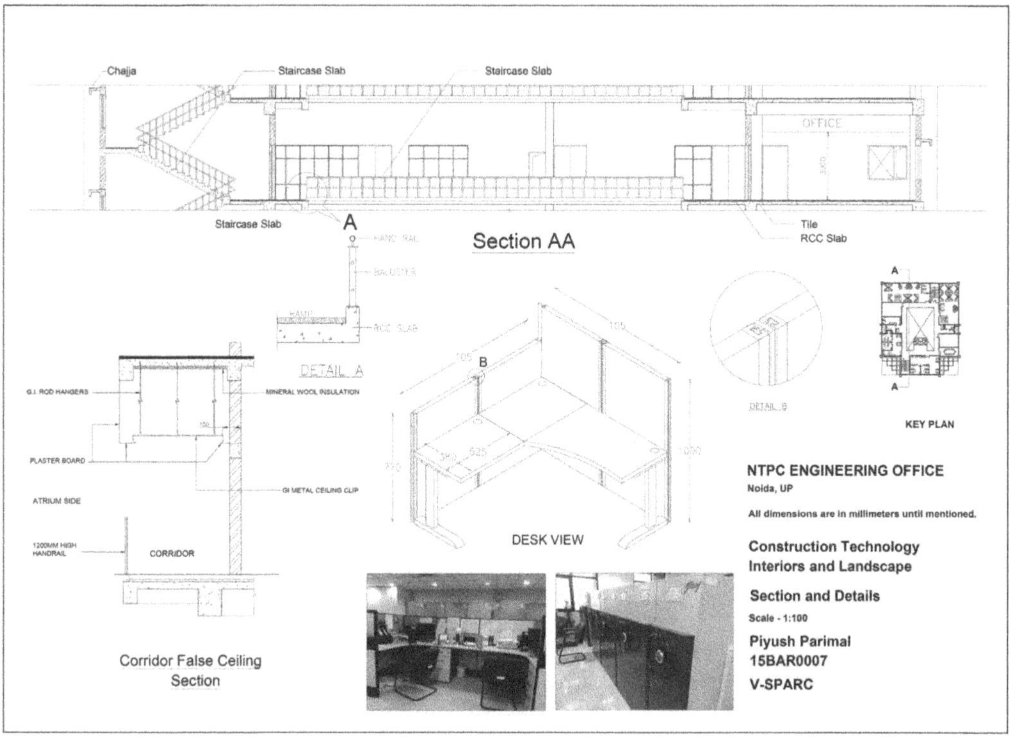

Figure 3.3 Learning through reverse engineered interior details
Source: Piyush Parimal, 2017, Undergraduate student project, Vellore Institute of Technology

existing restaurants, textile shops, boutique watch shops, office spaces, etc. and asked to research and understand the construction details and materials which had been employed as well as understand brand names and product manufacturers (*see Figure 3.3*).

The challenge of these immersive exercises lies in the knowledge available with teachers to deal with student queries on contemporary products and services and hence, industry professionals need to be constantly engaged with institutions. This has been facilitated of late with the 'Professor of Practice' position as well as visiting faculty who are constantly in support of most architectural institutions.

Faculty Internship as a Meaningful Knowledge Updating Process

In an effort to keep faculty constantly updated on the latest developments in technology, construction processes and design thinking and thereby to more meaningfully contribute to the processes of reverse engineering and broad-based resource learning, faculty members need to intern themselves in the offices of practicing professionals to keep abreast of developments in the architectural field. Such internships can be mandated for a period of six to eight weeks for every five years of teaching and research.

Conclusion

The challenge of teaching architecture lies in constantly keeping abreast of technological and design innovations and creating new ways to attain educational objectives. The need to sustain student interest in learning is through applied learning demonstrated by an immersive inferential process based upon ownership and choice and the free usage of reverse engineered discovery learning. Structuring of thought processes using techniques like mind mapping and engaging in spatial understanding techniques derived from a sound understanding of 3D geometry would bridge the gap between ideation and product. Artificial intelligence could enable the process of optimizing the final outcomes within the oversight of the core elements of architectural design.

References

Doyle Terry, (2008-p. xv), Helping Students Learn in a Learner-Centered Environment, Stylus Pub.

Piyush Parimal, 2017, Undergraduate student project, Vellore Institute of Technology, India.

4 Bridging the gap of architecture profession and academia through work, earn and learn model

AR. Suchetha Mathew Thazhathil[a]
Ph D Scholar, National Institute of Technology, Calicut, India

Abstract

Architecture education has evolved over centuries from a form of art to an integrated discipline that combines art, science, technology, culture, and even psychology. However, it has not been able to keep pace with the rapidly evolving technology that the world has been witnessing over the last few decades with the advent of information technology. The construction industry is rapidly expanding in India, and the architecture curriculum must align with the needs of the contemporary era. Typically, teaching methods can be teacher-oriented or student-oriented, with architecture education predominantly favoring a student-cantered approach which should foster strong connections with the industry. The current architecture curriculum places a strong emphasis on design studios whereas the practice encompasses design, planning, and management. The practice emphasized curriculum ensures that students stay aligned with the latest industry trends and developments. The new NEP 2020 allowing choice of subject and increased flexibility to learners is very much required for the multidisciplinary approach of architecture profession. To adapt practice-based learning/ teaching, we need to redefine the way the architecture institutions function. Institutions should work as a consortium of academia, interdisciplinary professionals and technology. The proposed model works very close to the practitioners and the students and will give opportunities for students to choose their field of interest.

Keywords: Architecture education, NEP, practice emphasized curriculum, profession-academia interaction, student centric learning

Introduction

Traditionally Architects have been associated with the design of buildings, spaces and urban planning. However, in today's world, the scope has substantially increased and covers Interior Design, Zoning/Code compliances, Redesign Services, Planning, sustainable design, Conservation, Cost estimation, Construction Management, Urban Planning, Damage Assessment, Landscape, Sustainability Certification, Structural Engineering, Specialty Consulting, Energy Modelling, MEP Engineering, Practice Based Research and Building Commissioning (AIA, 2020). In addition, there is advanced technology available for assisting architects in addressing these complex elements scientifically and more comprehensively which is an integral part of the architect's works. Traditional drafting boards have given way to computers and complex software. Therefore, the role and activities of architects have significantly transformed.

The transformation in the scope of Architecture practice calls for significant changes in Architectural education as well. With the large portfolio/coverage Architecture needs to cater to, it is no longer possible to cover the entire list of topics in classrooms.

[a]asuchetha_p230207ar@nitc.ac.in

DOI: 10.1201/9781003571018-4

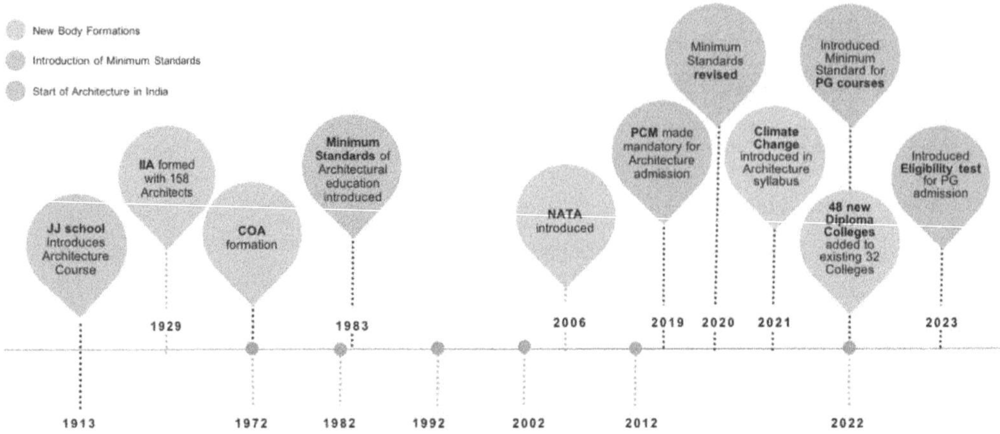

Figure 4.1 Major milestones of architecture education in India
Source: COA annual reports for 2016–17 to 2021–22, coa.gov.in

Climate change has become an important consideration in every activity undertaken today, from pollution control to carbon footprint and energy conservation. Almost every activity listed above, that today's Architects are required to perform, will have due consideration of climate change and sustainability. This will require the academic work to analyze these aspects in greater detail with input from other faculties.

Technology has aided in dissecting the complex issues in a very scientific and systematic manner. These technologies come with very friendly to complex user interface and outputs. Therefore, this is another aspect that the modern-day Architecture education needs to cater for. These technology applications start from simple 3D modelling to complex simulation programs addressing larger complex issues.

Aim

Improve the architectural professional condition by resilient education system.

Objectives

- To improve the business condition of Architectural profession
- To increase the employability of Architects
- To penetrate Architects' services to a large population
- To improve interest and admission to Architectural course

Literature Review

The formal 5-year architecture education started in India in 1913 at the Sir J. J. School of Architecture, Mumbai. The main purpose was to understand the traditional Indian construction methods. Another major milestone was the formation of the Council of Architecture (COA) in 1972, to regulate the architecture education and practice in India. Major events in Architectural education in India is shown in Figure 4.1.

Architecture education has gone through changes and is currently facing a major challenge due to the severe drop in number of students applying for admissions to architecture institutions.

The main reasons causing this drastic change are several but largely driven by the alterations to eligibility criteria, duration of the course and high tuition fees. The summary of number of institutions offering architectural education and annual registrations, total and gender wise for the last six years where this crisis has been more evident is shown in Figure 4.2 above.

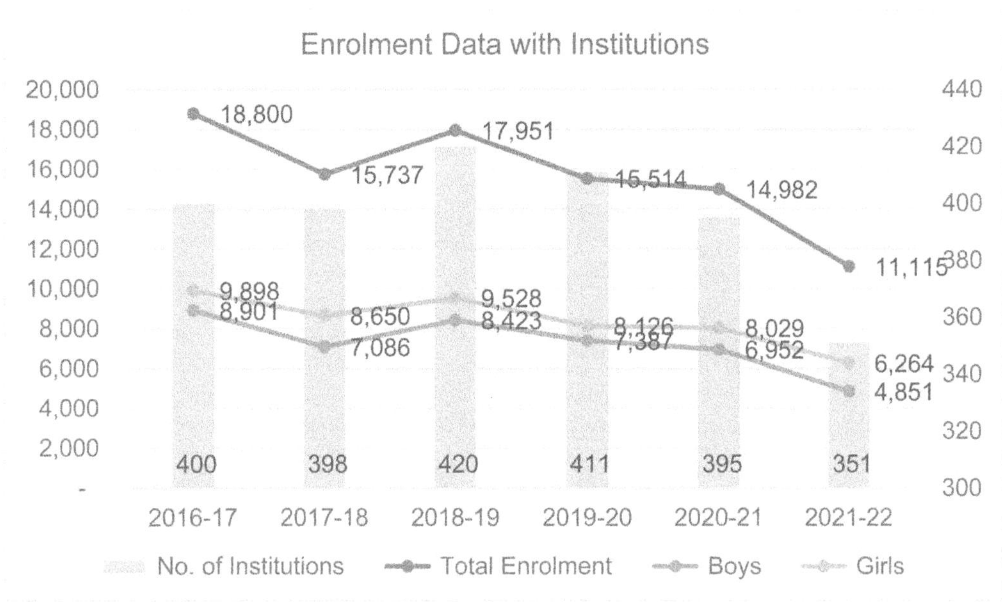

Figure 4.2 Enrolment data of architecture institutions in India
Source: Council of Architecture, annual report 2021–22, coa.gov.in

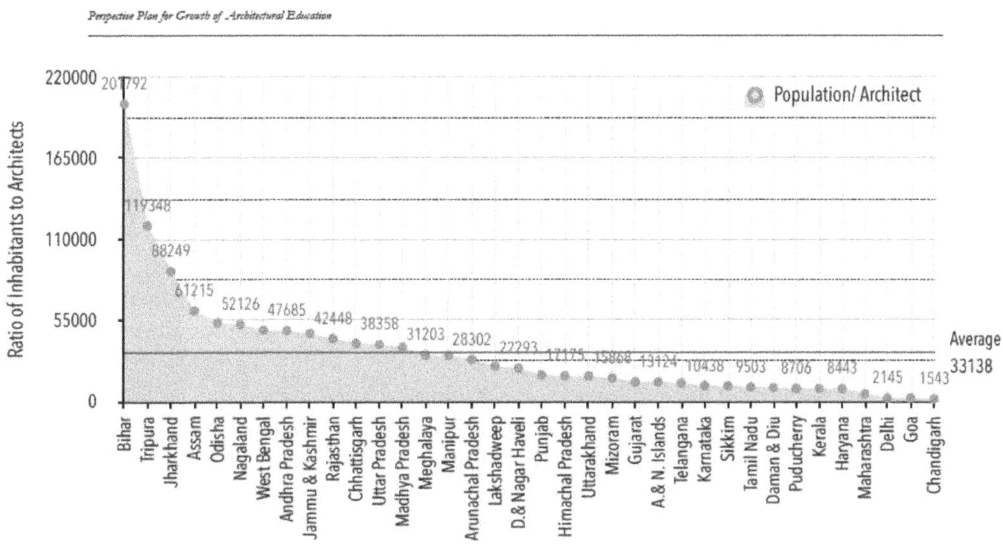

Figure 4.3 Population penetration state wise
Source: COA report perspective plan for growth of architectural education 2020

There is a significant variance in the penetration of Architects from state to state (COA, 2020). While Chandigarh has penetration of about one Architect for every 1543 people, Bihar is on the other extreme with one Architect catering for 201, 792 people, see graph in Figure 4.3 below. This clearly indicates that Architectural services are hardly used in Bihar while Chandigarh may have a more considered approach for all development activities.

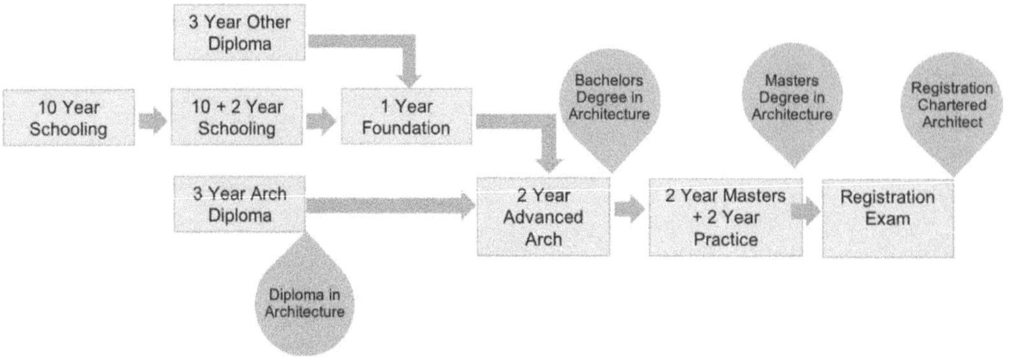

Figure 4.4 COA recommendation of NEP

Source: Architecture education: Way ahead in pursuit of education reforms–interim report 2023

The same variation is also seen in the number of Architectural institutions available in each of the states across India with Bihar having only one institution for the population of 119 million whereas Chandigarh with 1 million population has one Architecture College. Maharashtra has the highest number of institutions at 103 for a population of 120 million. The number of NATA centers also reflect a similar pattern with Bihar having only five centers for a population of 119 million and Maharashtra has centers in 23 places for a 120 million population.

The New Education Policy (NEP) announced by Government of India is definitely aligned with the complex scope that today's Architecture needs to cater by bringing in more flexibility in terms of courses and subjects of choice. However, the implementation needs careful consideration with close interaction of academia, practicing Architects, Industry and allied disciplines. Flowchart of the NEP as proposed is included in Figure 4.4 for ease of understanding. The NEP report demands the brightest professional to be retained in academia, which is not the case today, especially in Architectural field. Contrary to this, the COA report clearly identifies a shortfall in qualified faculty for Architectural course. Previous education policies of 1986, 1992 & 2009 were focused on accessibility and equality. However, the (NEP 2020) focuses on flexibility and identifying unique capabilities with no hard separation between different disciplines. The promotion of a multi-disciplinary approach is meant to lay an emphasis on conceptual learning that is respectful of diversity and local context. Activity-based learning, introduction of master instructors with profound local knowledge are also other outlines of the NEP, making it far more complex and result centric. The NEP also calls to mitigate opportunity cost and fees for higher education with flexibility to enter, exit and re-enter.

COA has started registration of architects from year 1975 but the registration and renewal is not taken seriously by architects, see Figure 4.5 below while Figure 4.6 shows the various segments that qualified architects are engaged in.

Growth of Industry

The Indian GDP has been growing consistently over the last three decades from a meagre USD 270 billion in 1991 to USD 3.385 trillion in 2023. The growth rate also has been good for the last three decades.

A significant portion of the GDP comes from Construction activities with the construction industries share at 8.19% of GDP in 2022-23. It is worthwhile to note that the y-o-y growth registered by construction industry is higher than the overall GDP growth, the latest being at 13% for 2023.

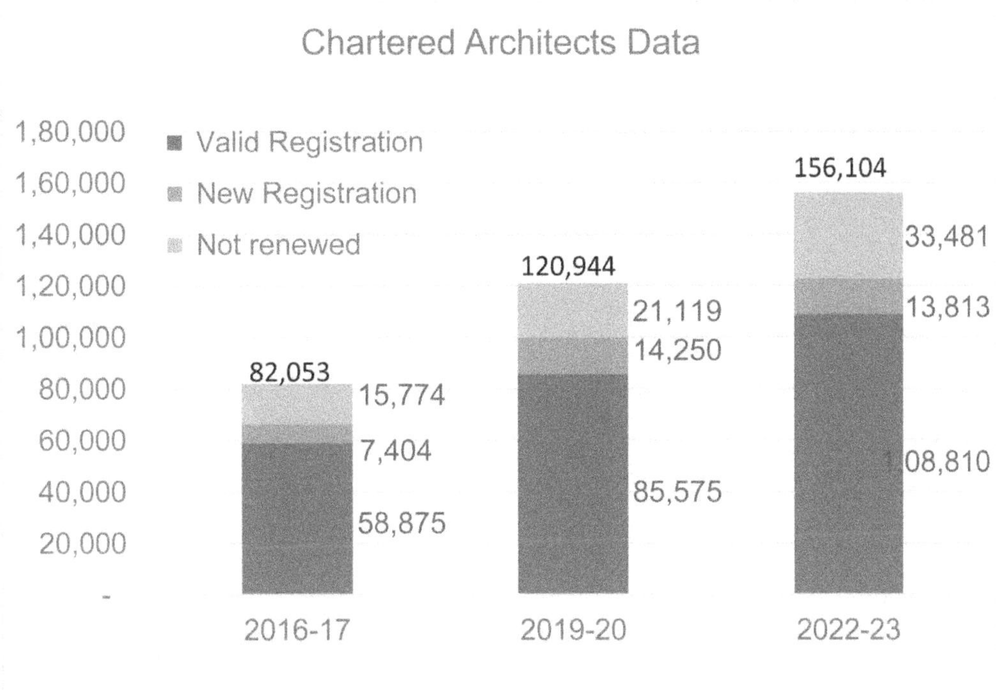

Figure 4.5 Registration of architects with the COA
Source: Annual report COA 2016–17 to annual report 2022–23, coa.gov.in

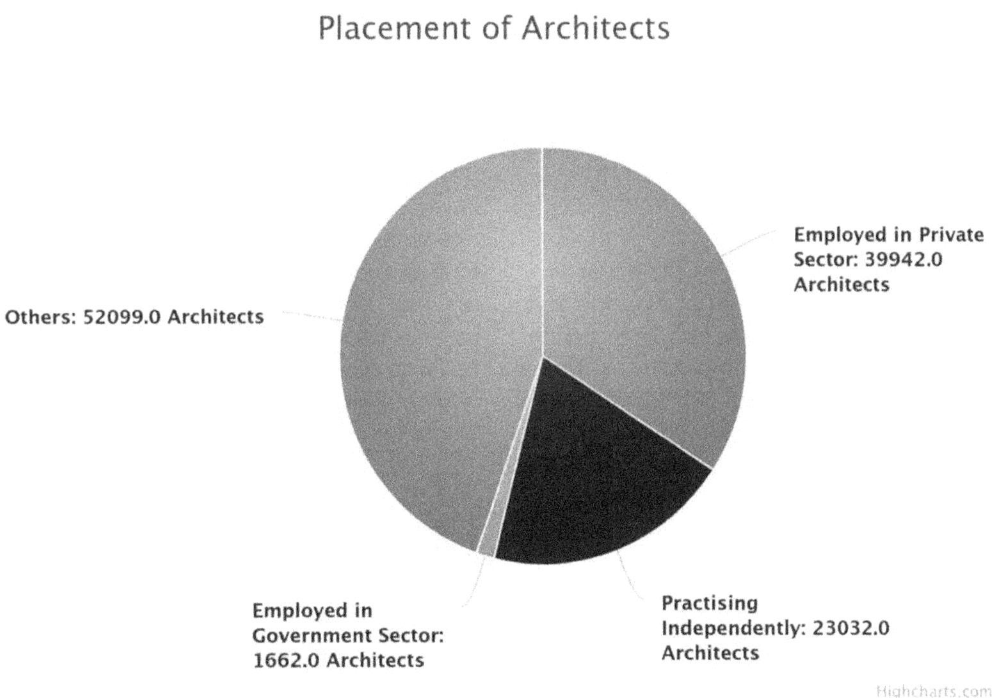

Figure 4.6 Architect's practice trends
Source: COA website

Table 4.1 Methodology

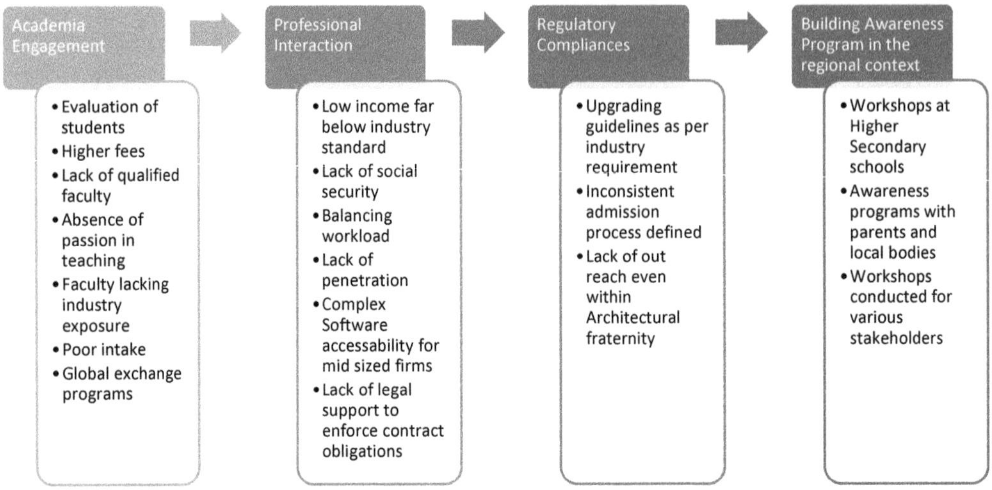

Source: Author

India is expected to be among the fastest growing construction markets, with a high growth rate, with major targets till 2030, along with US, UK, China and Indonesia. Housing demand is expected to lead the sector, as per an independent report. This will lead to considerable increase in the Architectural services.

Research Question

Will integration of practice and academia improve the Architecture profession's business condition?

Methodology

Table 4.1 below summarizes the major factors influencing architectural education and practice, both supportive and detrimental in four categories. It is important that these key factors be harnessed in a manner that a seamless integration happens between academia, practicing architects, regulatory bodies and community at large.

Results and Implications

Syllabus

The first introduction of minimum standards in Architectural education was introduced in 1983. Under these provisions, the five year course was split in two stages with first three years in Stage 1 and last two years in stage 2. There was only marginal difference in course content though, which is evident in Figure 4.7 below. Thereafter, there has been hardly any major revision in the syllabus of B. Arch course in last three decades while the industry has seen significant transformation in terms of demand, awareness and scope as discussed above. The work culture had a significant transformation calling for extensive changes in curriculum requirements. As discussed earlier, AutoCAD and 3-D software have taken over manual drafting and therefore curriculum program needs to move on to these tools more rapidly. In addition, with the wide variety of practice fields that can be pursued in profession, students should be given the opportunity to pursue subjects of their choice and aptitude rather than one size fit for all.

Tuition Fees

Architecture courses are usually perceived as an expensive course compared to other contemporary courses. This is evidenced in the Table 4.2 below with the fee structure recommended by COA for Architecture course. This is very high

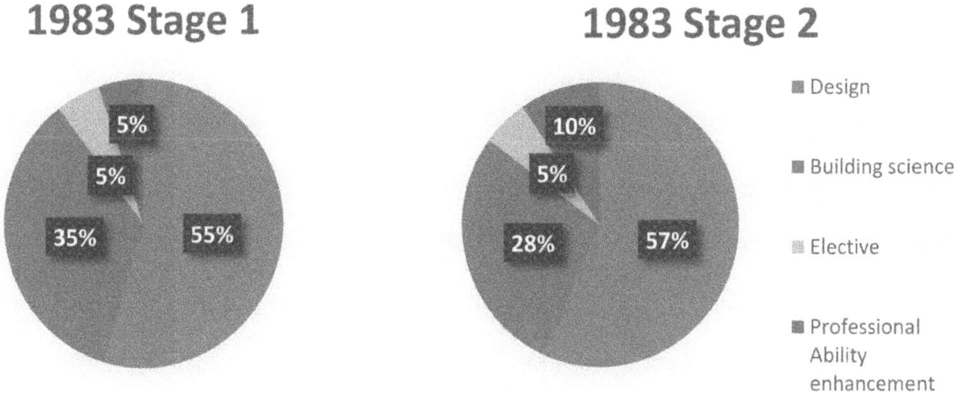

Figure 4.7 Course content of B. Arch in India
Source: COA, minimum standards of architectural education regulations 1983

Table 4.2 COA recommended fees structure

Description	Rural	Urban	Metro
For institutions following 6th Pay Commission	₹ 156,528	₹ 164,933	₹ 180,338
For institutions following 7th Pay Commission	₹ 192,200	₹ 201,977	₹ 218,754

Source: COA recommendation to fee regulatory Authorities of India, ANNEXURE-1, dated 29/12/2023

compared to the fees applicable for other technical courses.

In addition, the salary commanded by young professionals in architecture field is way below industry standard which dampens the interest in taking up architecture course. This leads to lower intake and also lack of talented students taking up the course.

Comparing the cost to fees structure of Architecture course with other technical courses, one of the key differences is that technical courses have multiple courses in the same institution and thereby the administrative and overhead costs are better distributed as compared to Architecture course. This is a genuine problem and with direct impact on the fee structure, there are possible mitigation measures which could be considered. A few possible solutions are discussed below.

Incorporating NEP

The institution will include three different cells, academic cell, research cell and an Institute's professional cell. There will be multiple entry and exit opportunities for the students. The students with an architecture diploma, can join in the second year without needing to complete the foundation level while students without an architecture diploma will have to start with the first year.

Students will be given the opportunity to join the research or project cell during their second and third year of the course to enhance their practical experience.

Those who wish to join for a post-graduation degree, will be guided by practicing architects during their fourth and fifth year. The co-working space allotted in the project cell will bring industrial interaction to students/interns. The co-working space can provide professional infrastructure and professional software on a shared platform which will be more economical for starting a profession. The graphical representation of this proposed scheme is shown in Figure 4.8 below for ease of understanding.

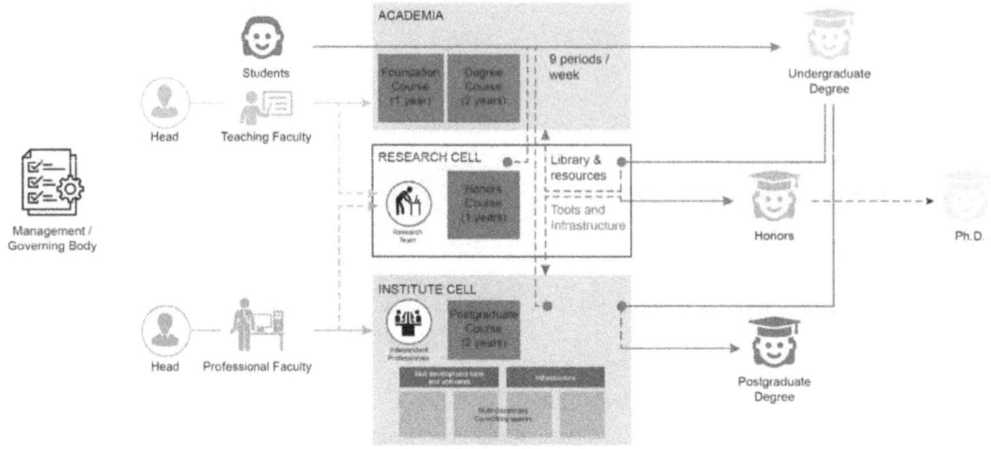

Figure 4.8 Learning and working model
Source: Author

Diversify Architectural Course
Bringing in the diversity in architectural course whereby the institution can offer multiple related courses like construction management, interior design, resource conservation including water, cost estimators, 3D designers are so forth. The institute design cell can take up related projects.

Integration of Real Projects
The educational institution can collaborate with local civil bodies/organizations to bring real-world projects into the learning environment. These projects could span various disciplines and industries relevant to the postgraduate courses offered by the institution. By engaging in commercial transactions with these real projects, the institution has the potential to generate revenue. This revenue can be utilized to subsidize the fees of the courses, making education more accessible and affordable for students.

Complementary Services and Community Penetration
In addition to commercial projects, the institution can offer complementary services to the local community. This could include workshops, seminars, or consultation services related to the expertise of the institution. These services not only contribute to community engagement but also increase awareness and penetration of the institution within the local area. As the institution becomes more ingrained in the local fabric, the contextual relevance of its programs grows. This heightened relevance not only attracts more interest in the courses offered but also cultivates a sense of ownership and pride among community members. This increased awareness can lead to higher interest in the courses offered. The positive impact on the local community can also result in word-of-mouth recommendations, further boosting the institution's reputation. As awareness grows and interest in the institution's programs increases, there is likely to be a rise in demand for admissions. The real-world application of knowledge through engagement with local projects can make the courses more attractive to potential students. Figure 4.9 below shows the above in graphical format for ease of understanding.

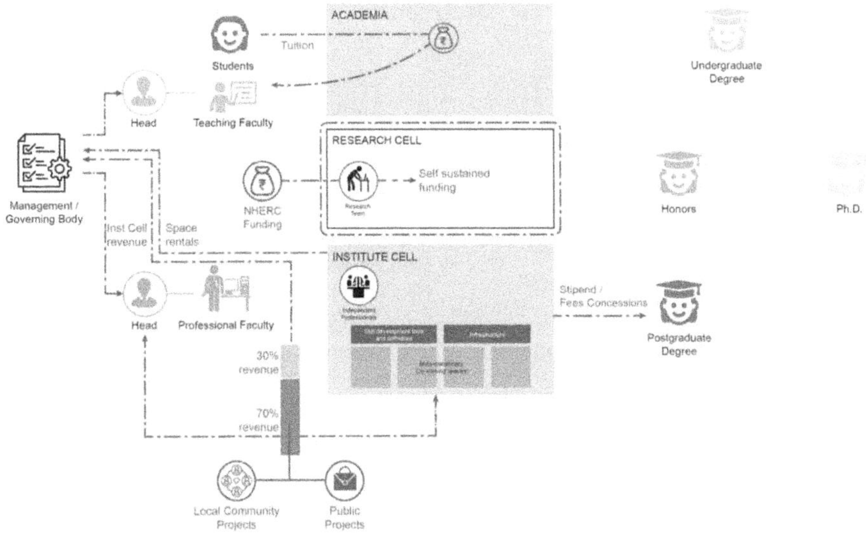

Figure 4.9 Learn, work & earn model - involvement of students in real projects
Source: Author

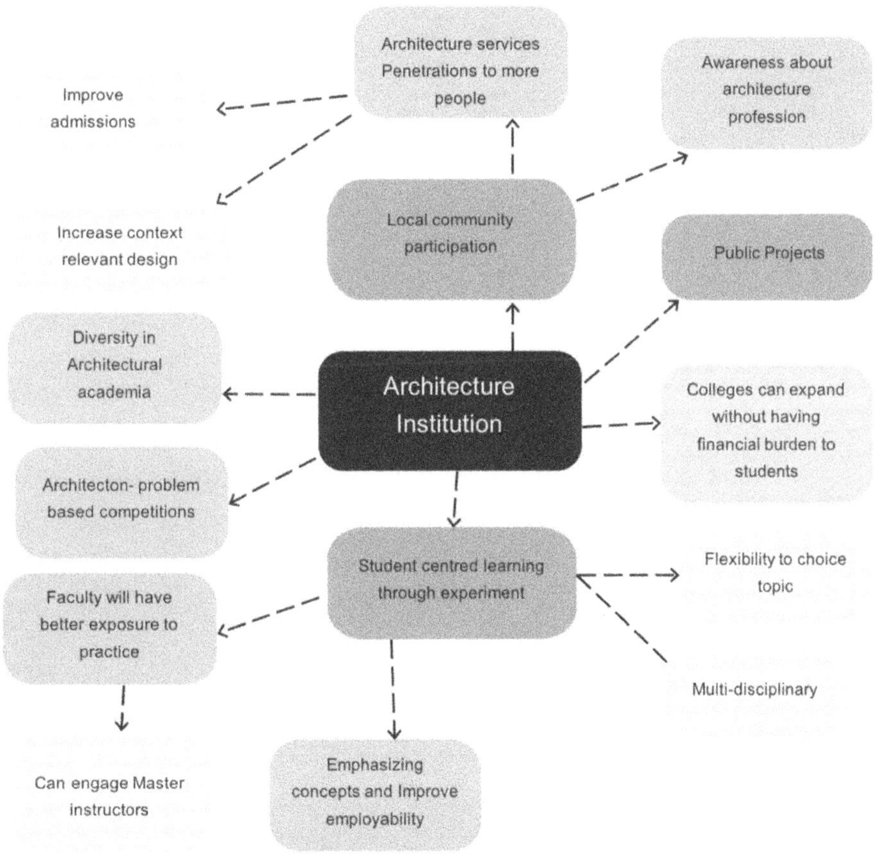

Figure 4.10 Proposed model
Source: Author

Conclusion

The implementation of NEP will be helpful for aligning the course with the industry requirements and can also help in improving the interest for the architecture course itself. Architectural institutions need to be more collaborative with and rooted in the community where they are located to increase awareness and ensure closer interaction. An experienced architecture professional's active participation with institutions in their locality/region and their assistance on real projects undertaken by institution will increase the architectural awareness among the local community.

References

American Institute of Architect, Firm Survey Report 2020.

COA Annual Reports for 2016–17. To 2021–22.

COA (2020). COA Report Perspective Plan for Growth of Architectural Education 2020.

Minimum Standard of Architectural Education regulations 1983 published by COA.

New Education Policy (NEP) (2020). Published by Government of India.

What do Numbers Reveal about Architecture Admissions and the State of Education? Srivatsan. A, Bhatt Chirayu (2023).

5 Bridging the gap between education and practice through research: An approach towards sustainable built environment

Arundhati Wategave[a]

I/C Principal, Appasaheb Birnale College of Architecture, Sangli, Maharashtra, India

Abstract

In nurturing the efficacy of architectural and planning spheres, the proposed research raises as the key force, putting the base aimed at the subsequent generation of Builders, Architects, and Designers devoted to fostering sustainability in people, societies, metropolises, and towns. This study is an essential investigation that employs techniques, instruments, and tactics to foster understanding, knowledge, and insights. This analytical investigation turns the procedure into concrete ideas, notions, and tactics. In planning and design practice, the methodical design of space into a meaningful place creates a strong basis for projects at different dimensions. Research not only provides a macro- and micro-level case-specific foundation, but it also provides precise project guidance. Within the field of architectural education, research is essential to forming the next wave of planners, designers, and architects who hope to make a positive impact on sustainable communities, cities, and towns. A solid foundation is created for the planning and advancement of the built environment through the incorporation of research into the educational framework with the study of the role of research and architecture education integration and investigation of challenges and barriers. This fundamental understanding serves as the cornerstone for developing an influence that is sustainable in terms of the social, political, economic, and environmental spheres. Research needs to be presented as a tool that turns students into real-world producers of sustainably built environments that bridge the knowledge gap between practice and theory. Thus, the importance of research will not just be limited to postgraduate studies but will start at undergraduate level, impacting young planners and architects. The case study highlights the challenges faced and initiatives taken by Appasaheb Birnale College of Architecture, Sangli, Maharashtra, India in integrating research, emphasizing the need for strategic initiatives to create a research-oriented educational environment, and promoting sustainability in the environment, economy, and society.

Keywords: Architectural and planning spheres, education, practice, research, sustainable built environment

Introduction

In the dynamic and constantly changing fields of planning and architecture, the symbiotic relationship between practice and education is crucial in molding the professionals who create our built environments. The need to bridge the gap between theoretical understanding and real-world application is increasing as the global society places a greater emphasis on sustainability. In addition to bridging the gap between theory and practice, this research paper examines the transformative potential of research as a catalyst for the development of sustainable built environments (Umang et al. 2021).

[a]drapwategave@gmail.com

DOI: 10.1201/9781003571018-5

The interface between research and the real-world complexities of space planning and design is frequently marked by a gap that impedes the smooth transfer of theoretical understanding to useful implementation. Acknowledging research as a potent instrument, this paper aims to clarify its effectiveness in closing this disparity, especially by supporting a change that incorporates research into undergraduate education. This article aims to support educators, professionals, and legislators in implementing a comprehensive approach that incorporates research into the very fabric of architectural education, thereby producing a generation of architects prepared to tackle the complex challenges of the built surroundings sustainably. It also aims to emphasize the significance of research in this pursuit.

The importance of a sustainable built environment has become increasingly significant in the face of social dynamics, economic upheavals, and environmental issues. This essay aims to emphasize the value of research in this endeavor and provide guidance to practitioners, educators, and legislators on incorporating research into architectural education. Architecture and planning involve the study of constructing, modifying, and understanding space, from macro to micro levels. They address cultural, ethical, artistic, socioeconomic, and environmental factors that impact our quality of life. Urgent research is needed to address the challenges posed by climate change, urbanization, globalization, and social transformations. Additionally, the fields of architecture and planning are rapidly expanding with the emergence of new media and technologies, making research critical for ongoing knowledge base extension, understanding development, and enhancing architecture education, learning, and practice.

Literature Review

The Role of Research in Architectural Education

Literature review gives the worldwide scenario through examples on how research is the key to enhance practice if it is incorporated throughout education.

A study conducted by Gregor and Legény (2019) at the Faculty of Architecture at Slovak University of Technology in Bratislava (FA-STU) found that the university fosters creativity, problem-solving skills, practical abilities, artistic and scientific outcomes, and critical thinking through projects and activities. Umang et al. (2021) identified research areas, posed questions, collected data, developed ideologies, and prioritized content for architectural research. A study by Roberts (2007) categorized research into design as consulting, design project work, and research instruction, examining the relationship between research and instruction in architecture. The Welsh School of Architecture prioritizes design based on evidence. Tvedebrink and Jelić (2021) proposed a research-based design paradigm for architecture education after investigating how a new master's course affects students' abilities to conduct health-related research. The EAAE Charter on Architectural Research emphasizes ethics, responsibility, inclusivity, and diverse practices for societal impact. Hwang et al. (2020) incorporated advanced AI and research-based learning techniques. Zairul (2020) conducted a systematic review of student-centered learning (SCL) approaches in design studios and identified five clusters: cooperative education, inquiry-based education, project-based education, problem-based education, and educational technology, offering guidelines for educators working in design studios. According to Hershberger (1970), architecture holds significance for men through convention, use, purpose, and value. Nyka et al. (2020) explored the integration of the CDIO framework into Gdańsk University of Technology's architectural engineering curricula for its potential benefits and impacts on architectural education and design research.

Enterprise architecture (EA) provides numerous benefits to organizations and projects by improving compliance evaluations, management propagation, and knowledge exchange (Foorthuis et al. (2016). In a recent study by Ummihusna and Zairul (2022), 19 peer-reviewed publications on immersive learning technology (ILT) in architectural education were examined with an emphasis on year-two studies. The most popular ILT was virtual reality, and it was found to predominantly evoke a state of motivation. The results of the study may encourage the adoption of ILT in higher education. The architecture education program proposed by Alakavuk (2016) included technical and design courses, as well as expression-based courses with building construction courses that cover structural systems, materials, and stability. A case study conducted in Izmir, Turkey in 2013 focused on earthquake architecture and earthquake-resistant buildings, requiring students to consider environmental data, semantic paradigms, and digital technologies (Morales-Beltran 2015).

Objectives and Methodology

Objectives -The research objectives focused on evaluating research integration in undergraduate architectural education, identifying benefits and challenges, and exploring students' readiness for sustainable design practice.

1. To discuss research integration in education and practice.
2. To understand Indian and International perception of education and practice.
3. To identify how research can become a part of education will give success in practice.

Methodology – Two different methods at a broader level are applied to this topic, the Conceptual research method and the descriptive method.

The conceptual method is adopted to reinterpret the idea of bridging the gap in architecture education and practice through research that will lead toward a sustainable built environment.

The descriptive method is used to discuss the existing scenario of education and practice. The purpose of this method is to report what is happening. In this method, comparative method and correlational method have been used. In the correlation method correlation between education,

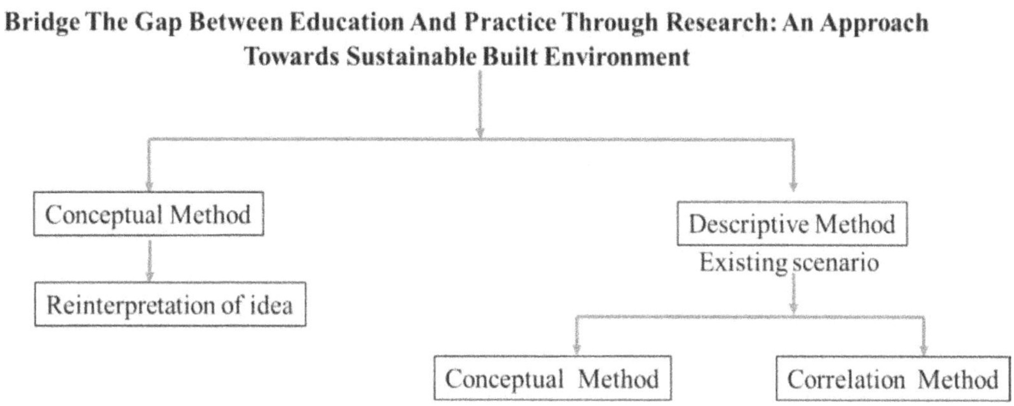

Figure 5.1 Methodology

practice, and research has been discussed and in the comparative method, Indian perception and International perception have been discussed.

Correlation between Research – Education, and Practice

Research – Research is a crucial tool for producing knowledge. It involves critical inquiry and synthesizing the process into thoughts, concepts, aspects, or strategies, which generates insights and understanding (EAAE Charter on Architectural Research 2022). Research can be structured, unstructured, or even random, but the outcome will always be contextual, streamlined, and specific. This helps define the clear path and approach of any work. Such research creates a strong foundation that is case-specific at both macro and micro levels, leading towards the accurate direction of the project or design. However, knowledge gained through research is only the baseline to design and develop any built environment. To create sustainable societies, communities, cities, and towns, research should be part and parcel of architecture education. It is vital to develop the next generation of architects, planners, and designers who strive for sustainability.

Research and education – It is important to introduce research to Architecture students from the beginning, both in terms of stages and scales. This will help them develop their methodologies, strategies, and philosophies, and ensure they are confident in their knowledge, process, and methodology when they enter the real world of the profession. By integrating education and research thoroughly into the practice, students will be well-prepared for their future careers (Importance of Research in Architecture 2024; Bridging the Gap: Profession and Academics in Architecture 2024).

Research should be introduced at the early stage of B.Arch., as well as at the final year stage and post-graduate level. In the early stages, each theory or studio subject should be planned and designed based on various methods of research. Although there is no need to introduce the Research Methodology subject at an early stage of B.Arch. (i.e. during the first and second years), different methods and techniques of research should be introduced in each theory and studio subject. This will enable students to understand the various research concepts at various levels, through the perception of architecture. At the upper level of undergraduate, research methodology kind of theory subject should be introduced, which will help students imagine and understand the various research concepts at various levels through the perception of architecture. Research should be the core of education from bottom to top, with linkages to the actual discipline of architecture. When research becomes an integral part of architecture education, it will have a very positive impact on architectural practice, which will provide a sustainable lifestyle to society.

The shift of research in academics from the graduate level to the undergraduate base level will have a direct impact on architecture and planning practice, which will emerge with physical, social, economic, and environmental sustainability impacts the built environment. Such education becomes the more ethical base for architectural practice.

Research helps us formulate the gap between existing situations for any project from all angles and helps to select the data collection method, data analysis, and interpretation of data with the situation and project. Conclusions, findings, and outcomes of research-based education will develop in-depth field examples and actual experiences of methods and procedures.

Research and practice - Research is a way of thinking that involves analyzing the various aspects of regular professional work, understanding and establishing guiding principles, and developing and testing new hypotheses that can improve practice and the field. To discover answers and explain our findings, research involves asking questions about what needs to be done and systematically examining and analyzing data to make necessary improvements for better professional services.

Research-based practice – Professionals use refined methods and procedures based on studies to provide effective services for clients.

Research process – Research is a necessary and valuable process that involves gathering, evaluating, and interpreting data and information in a controlled, methodical, legitimate, critical, and empirical manner. Research skills are essential for students who want to effectively serve in their profession, and it should be mandatory for all professionals to improve their research skills and professional knowledge. (Importance of Research in Architecture 2024)

Architecture has three stages of study, which include the process, product, and performance. Research is critical as it serves as the foundation for new design concepts and spans the entire cycle from conception to implementation. The first section of the abstract is the site analysis, which provides an overview of the location and project type viability. The second phase, which can be classified in various ways, begins with research. The types of information that are gathered first include how, why, what, when, and so on. Next, it moves on to the typology and functions, including material exploration, climate, pertinent developments, building techniques, etc. Thus, the entire procedure is a form of investigation done to generate data, understanding, and insights based on skills, methods, and tools suitable for the architectural domain.

Before creating anything, understanding, analyzing, synthesizing, and concluding are the basic stages of the journey that an Architect should have for any successful project. It describes the way every person participates in architectural research, either through direct involvement or indirect involvement. Therefore, it is strongly recommended that a "Research program" be a part of the architecture curriculum at every stage directly. This systematic investigation can bridge the gap in existing knowledge and create a better futuristic built environment. There is a discernible difference between the results of projects based on research and those without research. Research provides facts, recommendations, and conclusions to make the design acceptable and practical, while the design offers suggestions for addressing current and future problems. Research is an approach to any project to understand and relate to the past aspects of the same for creating a new future. Research provides the process with direction and a background to the context. It is not only about discovering and documenting, but it also helps design more relevant and appropriate solutions.

Perception towards Education, Research, and Practice

Indian Perception

In India, the architectural profession does not consider research as a distinct aspect of the design process. Instead, designers rely on case studies to fully understand the project and customer needs. Additionally, industry surveys are conducted to gather information on user needs. However, the current process lacks a systematic approach to performing these tasks. If a suitable procedure is established, practitioners can recognize researchers as an essential component of the design process. Architecture, research, and design are interrelated in that architecture can be used to analyze problem components, connections, and functional solutions. Research studies provide solutions for both current and future problems, while also making designs more practical and functional. In India, the architectural profession does not consider research as a distinct aspect of the design process.

Instead, designers rely on case studies to fully understand the project and customer needs. Additionally, industry surveys are conducted to gather information on user needs. However, the current process lacks a systematic approach to performing these tasks. If a suitable procedure is established, practitioners can recognize researchers as an essential component of the design process. Architecture, research, and design are interrelated in that architecture can be used to analyze problem components, connections, and functional solutions. Research studies provide solutions for both current and future problems, while also making designs more practical and functional.

International Perception

The Royal Institute of British Architects (RIBA) employs case studies and interviews to address important questions about the nature of research in practice. The study uses interviews to create case studies and provides insights into how practicing architects currently view research as an aspect of their work. RIBA has made an effort to learn about the different ways that architects define, acquire, conduct, and apply research across a range of practice sizes and forms. The study collected data to identify the areas of research architects are most interested and involved in, and the benefits research expertise offers to architects' business success and quality of work (How Architects Use Research 2014).

Case Study: Fostering Research Integration in the B.Arch. Program at Appasaheb Birnale College of Architecture (ABCA) Sangli, Maharashtra

ABCA is facing challenges when it comes to integrating research into its bachelor's level curriculum. Traditional architecture education places a heavy emphasis on design-centric coursework, which makes it difficult to incorporate research elements into courses. The academic calendar's tight schedule also limits students' ability to dedicate sufficient time to research, impacting the quality of their output and engagement. Additionally, there is a perceived disconnection between academic research and practical application, which creates a psychological barrier. To overcome these challenges, ABCA needs to implement faculty development programs, make pedagogical shifts, allocate resources for research infrastructure, and integrate research into the broader curriculum. By doing so, ABCA in Sangli can ensure a more robust integration of research into its B.Arch. program, which will equip students for sustainable architectural practice.

Key Findings

Architects consider research to be a crucial aspect of their profession, focusing on the specific requirements of each construction project, as well as its functional and technical aspects. Post-occupancy evaluation is becoming increasingly important, employing studies from various fields such as design theory, sociology, policy, sustainability, materials, goods, and methods. Architects recognize the potential of research as an additional practice area and component of a wide range of services. There is a growing recognition of the significance and role of research in knowledge-based professions. It has the potential to contribute to project processes and can be a separate field of practice. ABCA, Sangli effectively integrates research, emphasizing the need for intentional efforts to create a more research-oriented learning environment, despite obstacles such as conventional wisdom, financial limitations, and academic schedules.

Conclusion

The process of designing space using a particular methodology is crucial for planning and designing projects of any scale. This procedure involves using both experimental and intuitive, as well as experienced and tested, knowledge to tackle challenges and issues regarding human activity and the real and virtual environments in which it takes place. By stepping away from the traditional education of architecture, we can relate architecture to planning and designing spaces at both macro and micro levels. Through research, we can develop the next generation of architects who will contribute to sustainable built environments using socially, environmentally, economically, and physically friendly approaches. Research is the key to applying theories into practice.

References

Alakavuk, E. (2016). Integration of building construction courses in the architecture education program. In SHS Web of Conferences (Vol. 26, pp. 01036). EDP Sciences.

Foorthuis, R., van Steenbergen, M., Brinkkemper, S., and Bruls W. A. G. (2016). A theory building study of enterprise architecture practices and benefits. *Information Systems Frontiers*, 18, 541–564. https://doi.org/10.1007/s10796-014-9542-1.

Gregor, P., and Legény, J. (2019). Role of research in architectural education at FA-STU. *World Transactions on Engineering and Technology Education*, 17, 140–145.

Hershberger, R. G. (1970). Architecture and Meaning. Journal of Aesthetic Education, 4(4), 37–55. https://doi.org/10.2307/3331285.

Hwang, G.-J., Xie, H., Wah, B. W., and Gašević, D. (2020). Vision, challenges, roles and research issues of artificial intelligence in education. *Computers and Education: Artificial Intelligence*, 1, 100001.https://doi.org/10.1016/j.caeai.2020.100001.

Morales-Beltran, M. (2015). Integrating architectural and seismic design of buildings in teaching practice. 3rd Turkish Conference on Earthquake Engineering and Seismology. October 14-16, 2015, Izmir/Turkey (https://www.tdmd.org.tr/pdf/pdf2015/TDMSK_243.pdf)

Nyka, L., Cudzik, J., and Szakajło, K. (2020). The CDIO model in architectural education and research by design. *World Transactions on Engineering and Technology Education*, 18, 85–90.

Roberts, A. (2007). The link between research and teaching in architecture. *Journal for Education in the Built Environment*, 2(2), 3–20. DOI: 10.11120/jebe.2007.02020003.

Tvedebrink, T. D. O., and Jelić, A. (2021). From research to practice: is rethinking architectural education the remedy? *HERD: Health Environments Research and Design Journal*, 14(1),71–86. Doi: 10.1177/1937586720953529.

Umang, R., Karmakar, V., and Sharma, S. K. (2021).Importance of research in architecture. *International Journal of Engineering Research and Technology (IJERT)*,10(01).

Ummihusna, A. and Zairul, M. (2022). Investigating immersive learning technology intervention in architecture education: a systematic literature review. *Journal of Applied Research in Higher Education*, 14(1), 264–281. https://doi.org/10.1108/JARHE-08-2020-0279.

Zairul, M. (2020). A thematic review on student-centered learning in studio education. *Journal of Critical Reviews*, 7(2), 504–511.

Website References (Active as of 20-06-2024)

[1] Bridging the Gap: Profession and Academics in Architecture. https://architecture.live/bridging-the-gap-profession-and-academics-in-architecture/

[2] EAAE Charter on Architectural Research. https://www.eaae.be/about/statutes-and-policypapers/eaae-charter-architectural-research/

[3] How Architects Use Research. https://riba-prd-assets.azureedge.net/-/media/GatherContent/How-Architects-Use-Research/Additional-Documents/HowArchitectsUseResearch2014pdf.pdf?rev=79b85021a25b4ead80fc8a426b0efa0a

[4] Importance of Research in Architecture. https://www.re-thinkingthefuture.com/rtf-architectural-reviews/a5879-importance-of-research-in-architecture/

6 Renovation in architectural assessment process to lead out inheritance potential of NATA-qualified architectural students

Kalyani A. Salvi[a]

Professor at CTES College of Architecture, N. G. AcharyaMarg, Chembur, Mumbai

Abstract

Architecture has varied professional alliances. Successful architects imbibed with their creative mindset and skills without qualifying degree or diploma certification. Admission to architectural education always starts by cracking the entrance examination of NATA. It helps to verify the inner competence of the candidate which is confirming the future education in architecture. One can be an Interior designer, licensed professional for building approval, retrofitting expert, restoration expert, set designer, filmmaker, advertiser, site coordinator, good presenter, entrepreneur, product designer, landscape, graphic designer, facility manager, project coordinator, photographer, academician, theatre, cost estimating, scheduling, sales and marketing professional, sculptor, writer. Architectural teaching has the challenge of finding and developing these hidden creative intentions in students using different assessment technologies. Teaching in architecture needs to change its teaching assessment. During education, students' potential needs to be cultivated and nurtured by identifying his/ her skill set. This research paper will examine the necessity of using better assessment technology in education to guide professional career practices.

After successive learning of basic theory and technology, architectural education in India shall direct students further as per current market needs. Successive learning of five semesters of architectural academics, the progress report shall reveal a clear picture of the student's skill set category. This will help architectural students in the next year to develop his/her skills for professional development. Research is to find better technology for skill assessment in architectural education. This will develop reports for the next education level in architecture. Such freshly graduated architects will be absorbed directly into Indian architectural practice.

Keywords: Architectural professional, assessment technology, Indian job market, NATA, skill set

Introduction

Trends in architectural professions are rapidly changing. It is not limited only to designing; it has widened from designing a brief to completion of the project. The role of an architect has been now beyond approvals of municipal drawing submissions. Development in infrastructures, real estate, commercials, and health care demands more architectural professionals at different levels. As per the Council of Architecture record, there has been a noticeable demand for architectural colleges after 2012. Awareness of this profession has been growing, a survey conducted by the Council of Architecture about the Population in States and Union Territories and the Number of Architects and Colleges, shows an increasing demand for the architectural profession. It is a total of five years and is a ten-semester course. Other professional degrees are easily pursued in three or four years. Architecture graduates start their careers at the age of 23. When an architect enters the professional

[a]kalyaniamols.ctescoa@gmail.com

DOI: 10.1201/9781003571018-6

arena, he takes another few years to understand the demand of the professional market and individual professional growth. New architects must learn the required skill set after graduation, to set up a practice in the growing demand of the market. It may be appropriate to consider introducing three-year degree programs in Architecture in smaller cities and rural areas. Council of Architecture (2020) is one of the alternatives. Past few years, the demand for creativity has expanded with growth in advertisement. Research in architectural education as per current architectural practices needs to be inculcated throughout the semester pattern. It is also stated in 'Perspective Plan for Growth of Architectural Education' published by the Council of Architecture to foster research and development in architectural education to fulfil the required skill set in the Indian market. Architectural institutions need to develop suitable curricula in thrust areas and build capacity for the changes that are likely to take place. This present research paper aims to find out need of change in architectural educational system to fulfil professional demands in practice.

Background

Architectural project or allied career is a research journey. Each project starts with the collection of data, coordinating with team members, referring literature, and working on details till the completion of the project. It is a creative progressive work. It is not possible to work on all the allied branches of architecture practices. You can be a master of any one of them to indulge your passion. A person can be mature in the research journey by investigating on required skills. The methodology of writing a research paper is the same as the methodology of working on any architectural project to complete as per the client's brief. Apart from the architectural field other allied careers like landscape, interior, writing, design, sales, and marketing in architecture are rising out of the market demand. These all are creative fields if it starts with artistic minds to develop any of fields.

Methodology

A questionnaire survey was conducted to find out the gap between education and practice. It is based on checklist given in the book Bell, J. to get unbiased honest responses. Analysis of subject required to retain as per the old syllabus. Research methodology to analyze and develop an architectural syllabus from the first semester of the B.Arch. till the sixth semester of the B. Arch. Degree course. Sentence needs reconstruction for clarity. This produces results to identify hidden skill sets in the students. To fill professional gap in the practice, a year which is two semester courses will be added later after the sixth semester strictly to develop inquisitive ability in the skilled areas of the student. There is never enough time to learn humanities, construction techniques, structures, and environmental science to its deepest knowledge, so required knowledge set is added from first to six semesters. However, more time is needed to develop the skilled passion of the students to learn the process of any development. A new semester pattern to develop student's inner skill set is suggested. This new pattern will be compared with conventional educational pattern in architectural education. It is compared with development in the practice of students with modern and conventional techniques. Graph is supported to show growth of fresh graduate architectural students in professional practice. A new pattern is suggested in assessment process to enhance inbuilt skill set to lead students into varies allied practices.

Data Collection – Questionnaire Survey (2015 to 2020 Batch Students)

Q1. Year of passing of student
Q2. Are they into architectural practice?

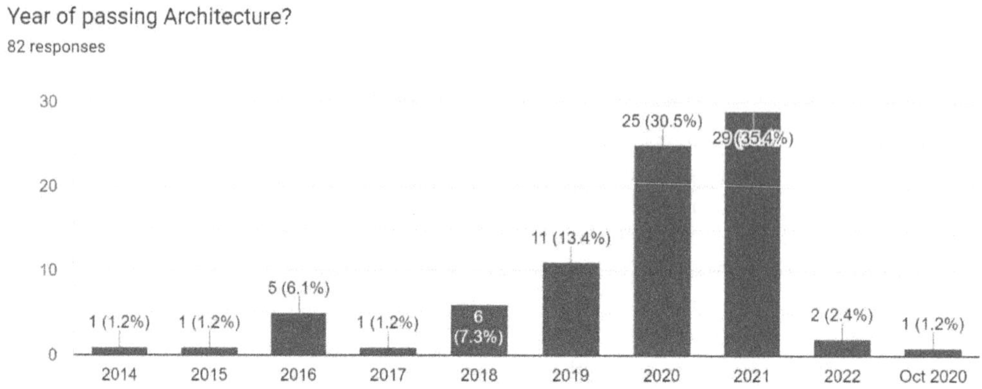

Figure 6.1 Survey of 82 students passed out from the 2014-15 academic year. Graphical percentage showing results of students who are working in professions other than architecture

Q3. Describe their role in the professional job market.
Q4. List subjects taught in architectural education which are helping students in their professional practice.
Q5. Do you think development of creative intellectual is needed in architecture?

A survey has been conducted to collect data for the architectural job market. Students who passed from 2015 to 2020, have five years of data to know the use of architectural education in professional practices. More responses are collected from 2020 and 2021 batches. These batches faced extreme challenges in setting their profession into architectural practices. Sentence requires clarity. Their responses are crystal clear to judge changes for fresh architects due to pandemic recessions in construction sectors. These freshly graduated architects struggled for their job roles as architects. They set their career in allied areas of architectural professions. They learn new skill sets to get absorbed into the professional marketplaces. Digital marketing, Architectural journalism, commercial real estate design, Graphic design and operations, visual and experience design, and real estate investors are a few allied areas. Figure 6.1 shows responses from Architects who are working currently in fields other than architecture. Most of them are from 2018 onwards so trends in the professional markets and the necessity of creative personnel are in demand for the Indian creative market.

Figure 6.2 showing a of passed students working in the allied field of Architecture.

It concludes, most of the 80% of students are working in the field of architecture. Responses received on the working areas are given below.

Students are working as Project Manager, BIM specialist, Interior designer, Planning and execution, Designer for fabrication strategies, computational architect, BIM engineer, Design coordinator, Project coordinator, entrepreneur of small start-up designer firm, Execution head, handling of sites, 3D renderer, sales and marketing, designer for industrial products; jobs which are not directly related to architecture education. These are allied career paths stemming from architecture education. These are not pure architecture. Graduates are working more than designing and drawing, making use of the basics of knowledge gained during their architectural education.

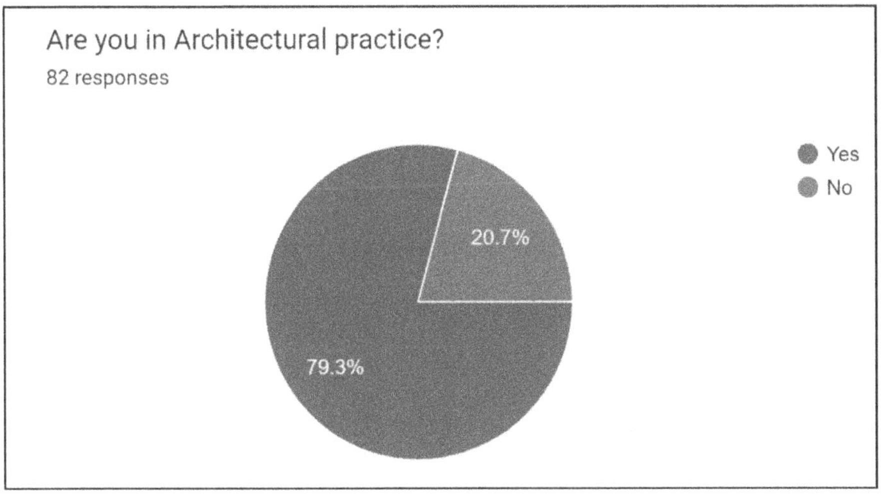

Figure 6.2 Survey shows responses received from the 82 students about practice in architecture

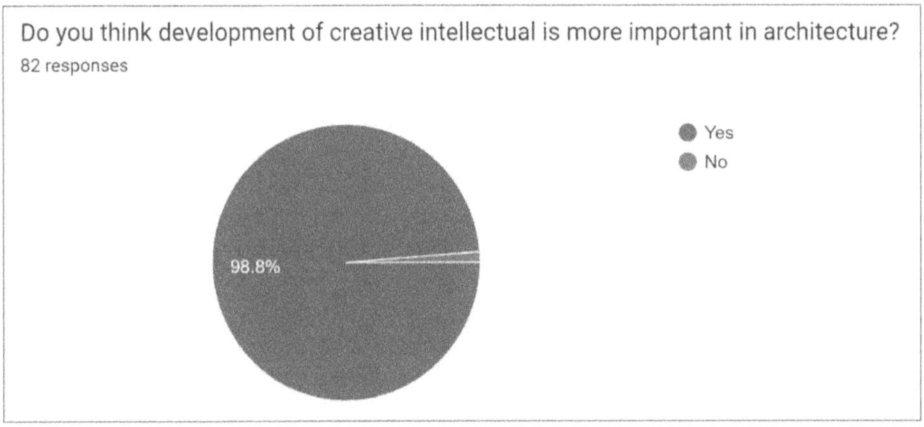

Figure 6.3 Survey shows responses received from the 82 students about believing creative intellectual is essential for working **professionals**

These architects are connected to architecture through creativity. They learned life skills more than what they had studied during 5 years of degree courses to set themselves as an architect in the above practices. They are practicing other than designing, drafting, and planning which reflects on the grading system they scored. While practicing in the field of architecture, students believed it is important to have creative intellectual working in architectural and allied architectural fields as per Figure 6.3 survey feedback.

Creativity is essential involvement in all the fields where these students are working. 98% of students who graduated as architects are using their creative intellect in the professional practices of architecture and allied architectural fields.

The Figure 6.4 release survey output showing none of the subjects is useful in architectural professional practice. Students are practicing architecture, based on the basics of all the subjects they learned while pursuing architecture. They are working in allied areas by developing their

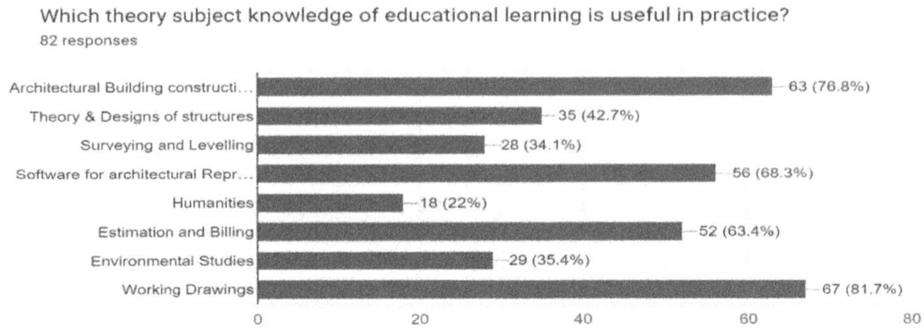

Figure 6.4 Survey of 82 architectural students is using knowledge gained during their five years of education. Most of these subjects are taught during the first to six semesters

passion and managing the needs in the creative field. Based on these responses, it is observed that the architectural education system is helpful but not essential to work in architecture. For their careers to flourish, it is observed that graduates are required to develop creative skills. Job markets in allied and architectural fields currently demand skilled persons who can fit into the creative field of design areas. . However, it is equally important that such interdisciplinary studies be guided by a thorough knowledge and understanding of the cultural conditions and motivations – both historical and contemporary, as well as local and global – which are shaping architecture around the world (Cris Abel, 2004). Further research has been undertaken to develop a system to find out hidden skills while pursuing architecture. Literature review – It includes data from different techniques in education to display results. Results reveal skills the student will develop in the last two semesters of degree education. Different assessment techniques are given in educational technology. In architectural education, commonly assessments are based on a mix-type to evaluate students as per another analytical program. Diagnostic assessments are based on knowledge gained by students. Formative assessments are checking the progress of knowledge gained by the students. Summative assessments are forcing students to gain a set of knowledge and mandate to pass them. Criterion-refereed assessments are based on measuring student's progress against their peers. The non-refereed assessments are like standardized admission tests which are generally conducted to test ability. All these assessments are useful when there is knowledge, analytical, and comprehensive information available to check their effectiveness. Amongst all, impassive assessments are based on the comparative assessment of the students amongst his earlier and set goals. It applies to all creative, musical, and athletic skills. A student has taken an admission or already passed a non-referred assessment to check his/ her ability. This previous assessment is displayed in the results in the form of an analysis of a wide range of assessments based on feedback. Assessment for learning – as opposed to assessment of learning as discussed above – aims to ensure that assessment empowers all learners so that they can benefit from education in the long term, but it is often overshadowed by competitive assessment (Hughes, G., 2014). We keep the first assessment as a background one to assess the quality performance of the student. Like musicians, athletes set their practice by referring to quality icons in their fields to reach there and they keep working on the fields where they are lacking behind. In this process, the coach most important to assess him into different parameters. Progress reports show students the best quality skills to work in any creative field. Learners at different levels of creative fields in the same subjects. The teacher has a thoughtful analysis of the different parameters of each subject to check the level of a learner. Case studies will be based on two architectural subject assessments to compare their current and suggestive Impassive kind of assessment.

Case Study

Building Technology subject and Humanities are considered case studies for the development of assessment patterns for architecture students from Semester 1 to Semester 6. It is considered a total architectural degree course for four years. A total of six semesters out of eight semesters will produce a progressive report analysis for each subject. Seven and eight semesters will give them as a choice to reach their best level to perform in the analyzed creative field to reach their maximum potential.

Building technology subject – A conventional way of marking is based on sheets (knowledge-based), written papers, and reports which are marked based on the knowledge they earned. Table 6.1 shows an Analysis of the method of marking from semester 1 to semester 10 for architectural building construction and material subjects. It is purely a knowledge-based assignment. It is more necessary to learn different detailing techniques in Architecture rather than theory knowledge.

While NATA qualified students shall have qualities to write, present, create, and analyse. These qualities are listed in Table 6.2 while assessing Architectural building materials and construction subjects with creative parameters to assess students.

Ipsative assessment system applied to the entire architectural degree course. Verb is incorrectly conjugated. Rewrite for clarity. Technical and theoretical knowledge subjects are considered as a part of semester 1 to semester 6. After learning theory lectures, students are expected to work on presentation panels, hands- on workshops and seminars. They will be working more on progressive techniques. These ways of working will give teachers to judge them on creative intellect. They will be using all the communication skills, and creative skills to show their intellect.

This Example of Ipsative marking for a subject will show a cumulative report of the student. Semester 1 to 5 report needed to be submitted to the external board to arrange a semester examination system for the set of students based on the report submitted.

The result of the Semester 6 examination will confirm the skill set of the student. It shows the cumulative set of attributes to select an elective for the last two semesters. This will be generated from the university examination to guide students for the elective.

Table 6.1 Conventional marking system

Architectural building construction & materials subject assessment in semesters	Sem 1 – Int.	Sem 2 - Int.	Sem 3 - Int.	Sem 4- Int.	Sem 5 - Int.	Sem 6 - Int. Ext.	Sem 7- Int.	Sem 8 - Int.	Sem 9 - Int.	Sem 10 - Int.
	Marking (Int.- Internal and Ext.- External) – Paper, Reports, Sheets, PPT									
Knowledge	✓	✓	✓	✓	✓	✓	✓	✓	✓	✓

Source: https://archive.mu.ac.in/syllabus.html

Table 6.2 Proposed ipsative marking system by Author based on university syllabus requirement

Architectural building construction and materials subject assessment in semesters	Sem 1 – Int.	Sem 2 - Int.	Sem 3 - Int.	Sem 4 Int.	Sem 5 Int.	Sem 6 Ext.*
Int. – Workshop, Models, Panels, Hands-on, Seminars						
Writing	✓	-	✓	✓	✓	✓
Presentation skills	✓	✓	✓	✓	✓	✓
Creativity	✓	-	✓	✓	✓	✓
Composition	-	✓	-	✓	-	-
Hands-on	-	-	-	-	-	-
Data collection	✓	-	✓	✓	✓	✓
Data interpretation	-	-	-	-	-	-
Modelling	-	-	-	-	-	-
Application	-	✓	-	-	-	-
Detailing	✓	✓	✓	-	-	✓

Ext* - External marking based on the assessed set of skills based on internal markings from semester 1 to semester 5. It is purely to upgrade students' inbuilt set of skills which will help to choose electives in Semesters 7 and 8

Source: https://archive.mu.ac.in/syllabus.html

Graphical Table 6.1 based on goggle survey conducted by Author shows comparative growth in the field of allied architectural field

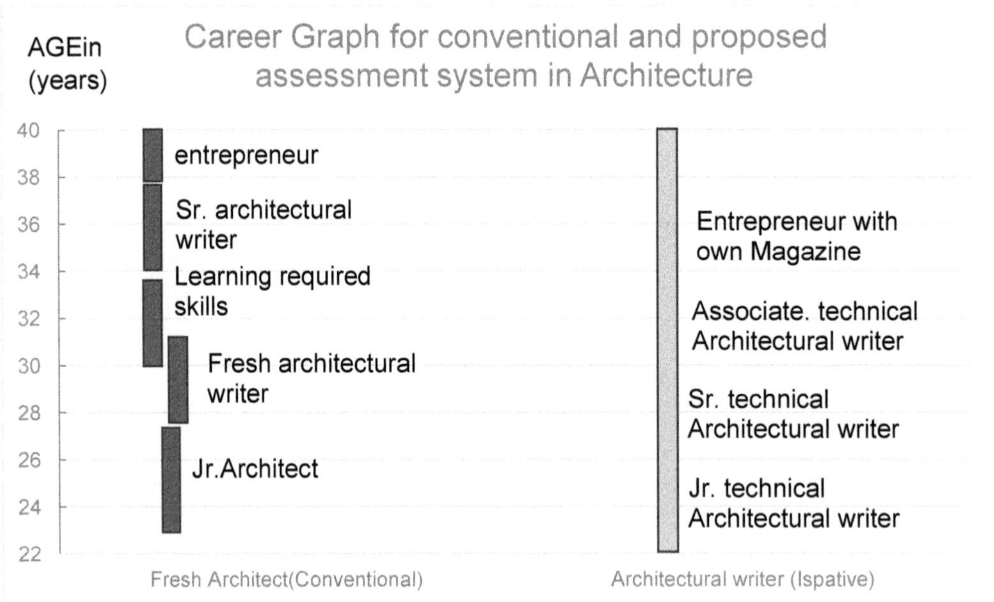

The above result of the student will generate elective options for the student. That will reflect student's strong skill set, journalism in architecture for newspapers, magazines, or journals. Further during last two semester, required knowledge and training will be part of his education. Immediately after passing, student can start own professional journey in journalism for architectural magazine.

Observation

At the early age of their career, students can start with architectural, specification writing, Digital writing, construction specification writing for a company, technical writing, documentation writing, and associate editing, and further many prospects are joining in allied fields of architecture.

Suggestions

It is observed theory subjects are important to start architectural practice in any arena. All architectural students confirmed their creative ability by passing NATA entrance. During education, it is believed to find out and develop students' skill sets with technical knowledge. This will have a lot of prospects to develop in the future in professional practices. Student at an early age in careers gets analyzed reports. This report will display student's prominent creative sets which will be set for electives. During one year of electives, students will learn all required technical and application-based knowledge to learn focused skills. It is going to be helpful for Indian markets to enhance creative practicing streams with the best creative people.

References

Bell, J. third edition, Doing your research project, pg.no.144

Council of Architecture (2020).August 2020, Council of Architecture, Perspective Plan for Growth of Architectural Education., Pg, 26

Cris Abel (2004), Elsevier publications, Architecture, technology and process, Pg.10

Hughes, G. (2014). Ipsative Assessment: Motivation Through Marking Progress Pg 15-30

7 Architecture graduate employability readiness: The gap between academia and practice

Shanu Raina[a]

Associate Professor, Shristi Manipal Institute of Art, Design and Technology, Bangalore, Karnataka, India

Abstract

The employment issues confronting architecture graduates in India are diverse, stemming from a mismatch between the educational curriculum and the changing demands of the business. This study intends to investigate, analyze, and solve this crucial topic by studying graduate workforce readiness while bridging the gap between academic and industry demands in the architectural area. The fundamental goal of this study is to identify the complicated issues faced by both architecture graduates and practitioners. By concentrating on this gap, the research hopes to uncover critical areas that require attention and development to improve architecture graduates' employability chances in a competitive labor market. The study used a quantitative approach, with designed survey questionnaires sent to architecture graduates and employers associated with architectural businesses. Data was collected between November 2023 and December 2023, and a thorough analysis was carried out using the Statistical Package for Social Studies (SPSS). The research relied heavily on descriptive statistics to examine and interpret the obtained data, revealing light on the various viewpoints of graduates and employers on employability skills and industry expectations. Finally, this study emphasizes the critical need for a significant revolution in architectural education to bridge the gap between academic teachings and industrial expectations. The findings highlight the importance of graduates acquiring appropriate skills and abilities that suit the changing requirements of the employment market. Addressing this gap is critical for the effective integration of graduates into the workforce and the architectural discipline's long-term growth and development within the larger economic environment. As a result, implementing initiatives to improve academic-industry alignment is critical to increasing the employability of architecture graduates and contributing to the resilience of the architectural profession in an ever-changing global context.

Keywords: Architecture education, employability skills, perception of employability

Introduction

In India, pursuing an undergraduate degree in architecture is crucial for obtaining a professional license, which allows individuals to practice as architects in the country and it becomes imperative for architectural education in preparing future practitioners to tackle the complex challenges of the built environment. Additionally, students need to focus on career development at every level of education to develop career maturity. The employability of architecture graduates is a complex and multidimensional issue, with a focus on how graduates are prepared for professional work (T. Mari, 2019) and is closely linked to the acquisition of employability skills, as emphasized by Prabowo and Yunus (2023). It is assumed that the universities don't produce graduates with the appropriate employability skills and

[a]raina.architect@gmail.com

DOI: 10.1201/9781003571018-7

this highlights a need to investigate the gap between what is taught in educational programs and the practical skills needed for employment in the architecture field (Oluwatayo and Opoko, 2016). It is essential to ensure that educational curriculums are aligned with industry-specific competencies to improve graduates' job prospects. Architecture graduates face a variety of challenges in their quest for employment, including architecture graduates having a lower rate of employment than their peers in other built environment disciplines, according to a study done by Poon (2016). Other challenges include realizing design visions, meeting client and market demands, adhering to employment criteria, and considering other factors. These challenges highlight the importance of graduate work preparedness, which encompasses a range of skills and characteristics that are essential for success in professional roles. These skills and characteristics include transferable abilities, essential competencies, and general traits that are considered necessary for gainful employment.

It is essential to critically evaluate the mismatch between industry evaluations of work preparation and the standards set by higher education institutions. This evaluation is important to ensure that graduates meet and exceed industry expectations without difficulty. Additionally, in the age of Industry 4.0, preparing graduates for the workforce goes beyond traditional skills. Graduates must also be proficient in technology, possess effective communication skills, collaborate well with others, be adaptable, think critically, and be creative. The need to build an updated framework for architectural education programmes becomes an urgent need (A. Varma, 2021)

This comprehensive data clearly explains the job situation for architecture graduates, digging into real estate education and requesting feedback from a variety of stakeholders, including employers. It emphasises positive findings about graduates' computer literacy, communication skills, and analytical acumen, resulting in a complete and integrated analysis of the factors impacting employability in the area of architecture.

What is Employability?

Employability is a multifaceted concept with various definitions (Hogan et al., 2013). In the field of employability, scholars have proposed various definitions to capture its multifaceted nature. According to (J. Hillage, 1998), employability is a dynamic capability that empowers individuals to navigate the labor market and realize their full potential through continued employment. (Lefresne, 1999) takes a probabilistic approach and defines employability as the likelihood of obtaining a job or transitioning to employment within a specific demographic and temporal context. (Harvey, 2001)focuses on the individual and sees employability as a graduate's ability to secure a satisfying job. Heijde et al. (2006) broadens the concept by emphasizing that employability is a continuous process that involves the acquisition, creation, or fulfilment of work through the application of competencies. (E. Berntson, 2006)introduce the human capital perspective, which suggests a positive relationship between education, competence development, job tenure, and perceived employability. (E. Berntson, 2006) adds a subjective element to the definition by conceptualizing employability as an individual's perception of opportunities for obtaining new, equivalent, or improved employment. This exploration of employability definitions reveals its complex layers and contributes to a better understanding of its implications in the scholarly landscape.

According to Chu and Reynolds (2021), employability skills can be categorized into three themes as follows:

- Fundamental skills such as communication (E. Berntson, 2006), managing information, using numbers, thinking and problem-solving;

- Personal management skills such as demonstrating positive attitudes and behaviours, being responsible and adaptable, learning continuously, working safely; and
- Teamwork Skills such as working with others and participating in projects and tasks.

According to Sakamand Jamaluddin (2022), employability skills are crucial for graduates to stay competitive in the job market and sustain employment, as emphasized by Prabowo and Yunus (2023). The concept of employability is multifaceted, encompassing the ability to acquire and maintain employment, continuously perform work efficiently utilizing one's competencies, and a set of essential skills necessary for individuals to thrive in the job market. However, the definitions and operationalizations of employability skills may vary.

Employability in Architecture

Employability in architecture is a multifaceted concept that encompasses various skills and attributes essential for success in the field. Architecture is a major profession that affects the development of the country (Suhaili et al. 2015), and it is true more so for developing countries like India. As highlighted by (J. Thompson, 2022), academic perspectives on employability skills in architecture shed light on the crucial skills valued in the profession and the potential impact of external factors such as the COVID-19 pandemic. Additionally, (J. Thompson, 2022)provide insights into the perspective of employers on architecture graduates' employability skills during their employment, emphasizing the practical application of skills in the professional environment. Furthermore, Caven and Diop (2012) present a comparative study of intrinsic rewards in the architecture profession, offering valuable insights into the non-monetary aspects of employability and career satisfaction in architecture. Moreover, (F.B.A. Halim, 2019)) explore the relationship between employability skills and self-efficacy of students with learning disabilities in the vocational stream, indicating the importance of personal efficacy in conjunction with employability skills for success in architecture. Careers in architecture are characterized by higher levels of job insecurity and lower levels of remuneration than other equivalent professions (Caven and Diop, 2012). Despite continued efforts by universities, obstacles and tensions remain in academic discourse, policy, and practice around graduate employability (Bridgstock and Jackson 2019).

Aim

Investigate and address the gap between academia and industry demands in architecture employability.

Objective

1. Research Question: What are the primary challenges contributing to the misalignment between the skills acquired during architectural education and the demands of the industry?
   Objective 1: Identify challenges faced by graduates, understand industry expectations, and analyze the disparity between skills and industry requirements.
2. Research Question: How do architecture graduates perceive their "work-readiness" upon graduation, and what factors influence this perception?
   Objective 2: Assess the preparedness of graduates to enter the workforce, and examine factors influencing readiness.
3. Research Question: To what extent is current architectural education aligned with the specific competencies demanded by the industry, and where are the key areas of misalignment identified by graduates and employers?
   Objective 3: Evaluate the alignment of architectural education with industry competencies and identify areas needing curriculum improvement.

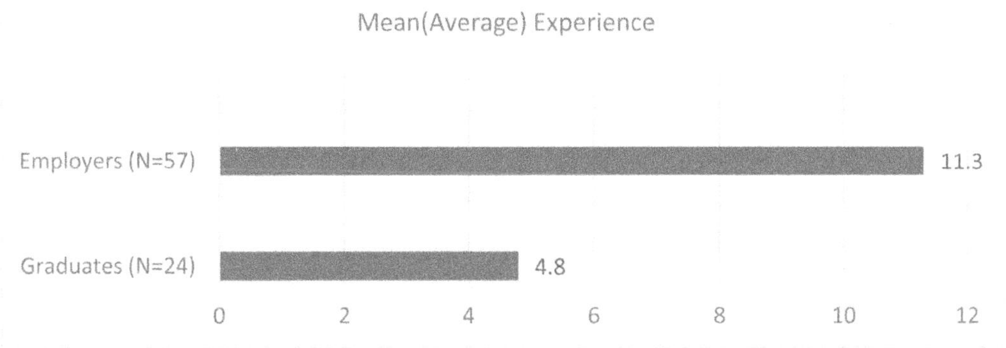

Figure 7.1 Mean (average) experience of respondents
Source: Authors

4. Propose actionable curriculum changes, and suggest measures to improve graduate employability.

Research Methodology

Survey Questions and Methods

The overall objective is to identify areas that require improvement to enhance the employability prospects of architecture graduates in a highly competitive job market. To achieve this goal, the research uses a quantitative methodology. The study uses a tailored survey questionnaire designed for two groups: architecture graduates and employers associated with architectural businesses. The purposeful sampling was adopted for the survey (N = 81), it aligns with the employability skills outlined by Higher Education Institutions (HEIs) in India. An online web questionnaire was developed in October 2023. Data collection took place from November 2023 to December 2023. The Survey consists of three sections (A-C). Section A evaluates the architecture course assessment; section B evaluates the graduate preparedness assessment; and section C explores architectural skills and development assessment. The collected data undergoes a statistical analysis using the Statistical Package for Social Studies (SPSS), which employs descriptive statistical measures like frequency, mean, scores, and standard deviations.

Respondents Profile

The survey targets a diverse sample of architecture graduates from different institutions across India and employers from various architectural firms engaged in graduate recruitment. The mean (average) of experience of architecture graduates was 4.24 years (Figure 7.1) and the mean (average) of total years of experience of employers was 13.8 years Figure 7.1). The study upholds ethical considerations by obtaining informed consent from participants and maintaining confidentiality. However, the study acknowledges certain limitations, including the small sample size (N = 81) of graduates (N = 24) and employers (N = 57), and the natural subjectivity in self-assessments and perceptions. In conclusion, this comprehensive methodology is set to provide empirical insights into the dynamics between academia and industry in the realm of architecture, contributing to a nuanced understanding of graduate employability readiness.

Result

In (Figure 7.2), comparing graduates (N = 24) and employers (N = 57) in the field of architecture education, mean and standard

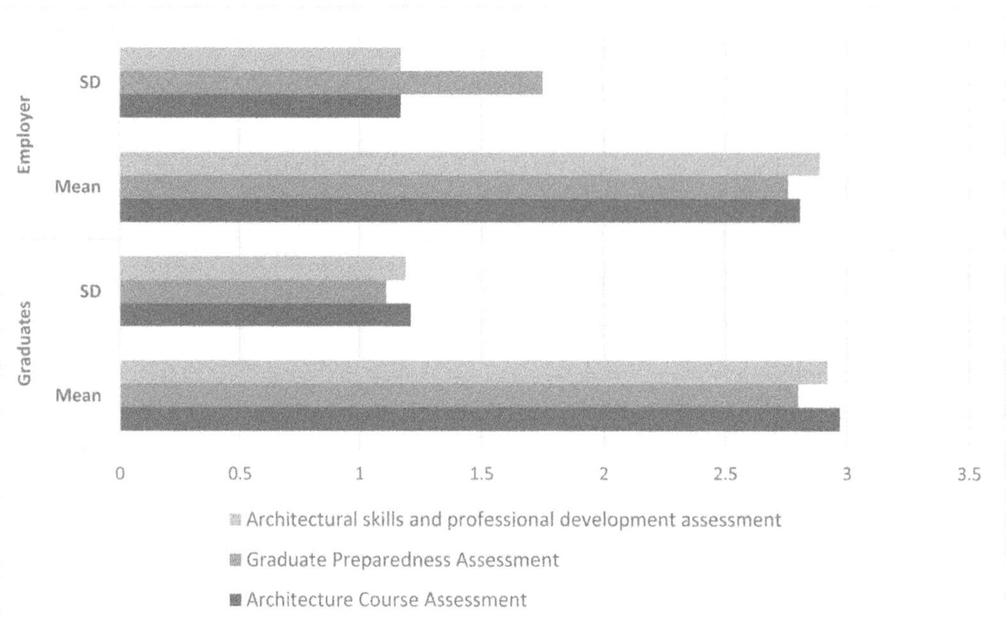

Figure 7.2 Comparative assessment scores - graduates vs. employers
Source: Authors

deviation values are presented for three key assessments. For the architecture course assessment, graduates rated their understanding or performance slightly higher (Mean = 2.97, SD = 1.21) than employers (Mean = 2.81, SD = 1.17). In the case of the graduate preparedness assessment, graduates again provided slightly higher ratings (Mean = 2.8, SD = 1.11) compared to employers (Mean = 2.76, SD = 1.75). Similarly, for the architectural skills and professional development assessment, graduates' mean scores (Mean = 2.92, SD = 1.19) were marginally higher than those of employers (Mean = 2.89, SD = 1.17). These findings suggest a trend where graduates generally perceive their performance and preparedness more positively than employers. The standard deviations indicate varying levels of agreement or disagreement among both groups, highlighting the need for further exploration into specific aspects of assessments and factors influencing the variability in responses.

Table 7.1 displays the significant differences in the way architectural graduates and employers perceive architectural course assessment. Graduates are concerned about the curriculum not aligning with the current industry needs, with a substantial percentage disagreeing with this statement and more focus is given on theoretical knowledge. However, graduates generally believe that the course fosters independent learning, while employers have a more positive response. Results clearly show that there is mixed feedback regarding the encouragement of adaptability, indicating potential differences in how graduates and employers perceive graduates' adaptability. A common point of concern among graduates and employers is their reservations about the integration of real-life projects into the curriculum and the absence of adequate marketing and business courses in the curriculum, highlighting a gap in preparing graduates for the business aspects of the architectural profession. To facilitate

Table 7.1 Survey results for three categories (A, B and C)

			Graduates (N = 24)					Employers (N = 57)				
			SA	A	N	D	SD	SA	A	N	D	SD
A	A1x	Emphasises theoretical knowledge over practical skills (%)	33	42	13	13	0	12.3	40.4	24.6	17.5	5.3
	A2	Alignment with current industry needs (%)	4.4	0	29	33	33	1.8	17.5	10.5	54.4	15.8
	A3	Fosters independent learning (%)	4.2	4.2	42	4.2	8.3	3.5	43.9	28.1	17.5	7.0
	A4	Encourages adaptability (%)	4.2	42	21	29	4.2	1.8	26.3	24.6	31.6	15.8
	A5	Integration of real-life projects (%)	4.2	21	33	33	8.3	5.3	28.1	22.8	33.3	10.5
	A6	Inclusion of marketing/business courses (%)	0	0	2	11	11	1.8	3.5	7.0	49.1	38.6
	A7	Critical Thinking (%)	21	54	25	0	0	8.8	42.1	22.8	19.3	7.0
B	B1	Being 'work ready' upon graduation (%)	0	4	25	38	33	1.8	12.3	19.3	49.1	17.5
	B2	Adequate construction knowledge (%)	0	4.2	45.5	42	8.3	1.8	12.3	15.8	42.1	28.1
	B3	Adequate practical skills (%)	8.3	12	37.7	29	13	1.8	14.0	24.6	47.4	12.3
	B4	Adequate theoretical knowledge (%)	8.3	50	25	12.5	4.2	10.5	49.1	24.6	14.0	1.8
	B5	Proficiency in soft skills (%)	17	29	32.7	13	8.3	8.8	40.4	24.6	17.5	8.8
C	C1	Career readiness and teamwork (%)	13	21	38	17	13	1.8	19.3	26.3	35.1	17.5
	C2	Communication (%)	17	33	29	13	8	1.8	40.4	26.3	24.6	7.0
	C3	Technical skills (%)	8	25	33	17	17	14.0	52.6	7.0	19.3	7.0
	C 4	Relevant laws, project management, and sustainable simulation models (%)	0	13	33	29	25	3.5	14.0	24.6	36.8	21.1

Source: Authors

smoother transitions from academia to professional practice in the field of architecture, critical analysis points underscore the need for a closer examination of the curriculum to address discrepancies and enhance the course's alignment with industry needs.

The analysis of the graduate preparedness table reveals several important points (Table 7.2 and Figure 7.3) and it can be implied from the table that a large portion of graduates, approximately 67%, feels neutral or dissatisfied about their readiness

Table 7.2 Ranking of various factors according to the graduates and employer

			Graduates		Employer	
			Mean	Ranking	Mean	Ranking
A	A1	Emphasises theoretical knowledge over practical skills	3.92	2	3.46	1
	A2	Alignment with current industry needs	2.08	6	2.42	6
	A3	Fosters independent learning	3.29	3	3.28	3
	A4	Encourages adaptability	3.13	4	2.74	5
	A5	Integration of real-life projects	2.79	5	2.93	4
	A6	Inclusion of marketing/business courses	1.63	7	1.88	7
	A7	Critical thinking	3.96	1	3.33	2
B	B1	Being 'work ready' upon graduation	2.00	5	3.33	2
	B2	Adequate construction knowledge	2.46	4	2.23	5
	B3	Adequate practical skills	2.75	3	2.54	4
	B4	Adequate theoretical knowledge	3.46	1	3.61	1
	B5	Proficiency in soft skills	3.33	2	3.32	3
C	C1	Career readiness and teamwork	3.04	3	2.61	3
	C2	Communication	3.38	1	3.14	2
	C3	Technical skills	2.92	2	3.56	1
	C 4	Relevant laws, project management, and sustainable simulation models	2.33	4	2.46	4

Source: Authors

to work upon graduation. Most of both graduates (87.5%) and employers (70.2%) have concerns about the adequacy of construction knowledge. The above table also indicates a potential gap between their perceived readiness and employer expectations and highlights the need for curriculum improvements in this area. While graduates have mixed feelings about their practical skills, employers are more positive, with 59.6% agreeing or strongly agreeing that graduates possess adequate practical skills. There is agreement between graduates and employers regarding perceived theoretical knowledge, with approximately 58.3% of graduates and 59.6% of employers indicating satisfaction. However, a significant portion of graduates (46%) feel neutral or dissatisfied with their proficiency in soft skills, while employers have a more favorable perception (49.1%). These findings suggest a critical need for targeted curriculum enhancements to address gaps in construction knowledge, practical skills, and

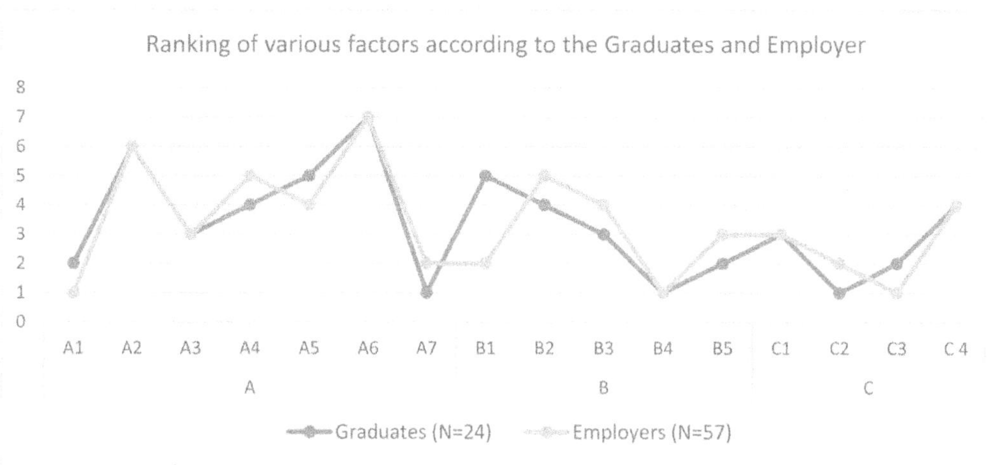

Figure 7.3 Ranking of various factors as per graduates and employers
Source: Authors

soft skills, ultimately ensuring that graduates are well-prepared for the expectations of the architecture industry.

It has been observed that there are some differences regarding the evaluation of architectural skills and professional development. Approximately 38% of graduates are neutral about their career readiness and teamwork skills, indicating potential gaps in their perceived preparedness for collaborative professional environments. On the other hand, employers generally believe that graduates are ready for their careers and possess good teamwork skills. Communication skills seem to be an area of difference, with graduates expressing dissatisfaction or neutrality, while employers generally commend graduates' communication abilities. Both graduates (33% strong agreement) and employers (66.6% strong agreement) positively acknowledge technical skills. However, a significant portion of both graduates (62.5%) and employers (57.9%) express dissatisfaction or neutrality concerning relevant laws, project management, and sustainable simulation models. This highlights the need for curriculum enhancements in these critical areas. These findings demonstrate the importance of aligning educational outcomes with industry expectations to ensure that graduates are well-equipped for the multifaceted demands of the architectural profession.

The ranks and mean values from the architecture pedagogy assessment reflect that graduates value academic knowledge, critical thinking, and flexibility, and employers value practical skills, real-world project integration, and soft skill expertise. The statistics reveal a potential discrepancy between academic focus and industrial expectations. Bridging these perception gaps is critical for improving architecture education, better preparing graduates for the changing needs of the professional world, and ensuring a more consistent approach to educational techniques.

Discussion

The current study has yielded insightful findings about the dynamics between architecture graduates and employers, highlighting notable disparities in their perceptions. It is indicated from the study that graduates show high levels of confidence in academic achievements, which implies a potential misalignment between the skills acquired in educational settings and those sought by employers in the field of architecture. However, employers on the other hand comparatively lower ratings signal a wider range of opinions regarding graduates' readiness and proficiency. Self-perception vs. employer expectations graduates' higher self-assessment scores across various assessments, including Architecture Course Assessment, Graduate Preparedness Assessment, and Architectural Skills and Professional Development Assessment, imply a degree of confidence in their academic journey. This disparity emphasizes the importance of further investigation to understand the specific skills and knowledge valued by employers that might be currently underemphasized in academic settings.

The study which resulted in an evaluation of the architecture course curriculum indicates significant difference between academic priorities and industry needs. The Graduate Preparedness Assessment reveals a considerable proportion of graduates feeling neutral or dissatisfied about their readiness to enter the workforce. A potential gap was noted in the curriculum, particularly in construction knowledge, practical skills, and soft skills graduates may not be acquiring the practical and industry-specific skills that employers consider crucial. In the case of architecture graduates, soft skills requirements differ from other professions and vary based on construction industries or market needs Suhaili et al. (2015). Noteworthy differences in perceived communication skills and concerns about relevant laws, project management, and sustainable simulation models indicate specific areas for improvement in the curriculum. These findings emphasize the necessity for curriculum enhancements to better align it with the industry's needs and improve graduates' employability. Most institutions are modifying their teaching techniques to provide students with skills that will enable them to study throughout their lives Khodeir and Nessim (2020). Oluwatayo and Opoko (2016) suggests using efficient innovative teaching tactics such as learner-centered learning (SCl), blended learning, flipped learning, virtual learning environment (VLE), game-based learning, interactive learning, technology enhanced learning (TEL) and collaborative knowledge sharing.

Conclusion

Despite changes in the field, Architectural education in schools continues to be based on educational approaches that are becoming increasingly distant from professional demands (Masdéu Bernat and Fuses, 2017). The fact that graduates tend to think more highly of their skills and preparedness than their employers could suggest a mismatch between academic training and the industry needs in the field of architecture. Graduates tend to score higher in all assessments, which might indicate confidence in their academic journey, but employers tend to score graduates lower, with a wider range of opinions regarding their readiness and proficiency. It's important to investigate the reasons for this gap in perception so that we can improve the skills and knowledge of graduates to meet the needs of employers. To sum up, the difference in the way graduates perceive their skills and what employers expect from them indicates that architectural education needs to evolve significantly and to do so qualitative studies such as interviews or surveys could help understand the skills and knowledge that employers value most in graduates. If the proposed recommendations are executed, they could help graduates enter the workforce more smoothly. Collaboration between academic institutions and industry stakeholders is essential to align educational programs with the continuously changing needs of the architectural profession. As the architectural field faces the challenges of an ever-changing global context, it is vital to address these issues for the profession's long-term growth and development.

References

A. Oluwatayo, A. O. I. E. O. I., 2016. *How do students perceive their employability readiness: The case of architecture students.* Ota, Nigeria, Covenant University Press, pp. 193-195.

A. Varma, M. J., 2021. COVID-19 responsive teaching of undergraduate architecture programs in India: Learnings for post-pandemic education. *Archnet-IJAR,* 15(1), pp. 189-202.

A.S. Prabowo, A. Y. W. P. R. D. P. C. S. A., 2023. Development of employability skills of vocational school students in Bogor, Indonesia. *Jurnal Pendidikan Indonesia Gemilang,* 9(2), pp. 97-102.

C.V.D. Heijde, B. H., 2006. A competence-based and multidimensional operationalization and measurement of employability. *Human Resource Management,* pp. 449-476.

Creswell, J., 2014. *Research Design.* 4th ed.: SAGE Publications.

E. Berntson, M. S. S. M., 2006. Predicting perceived employability: Human capital or labour market opportunities. *Journal of Economic Psychology,* 27(2), pp. 736-754.

F.B.A. Halim, W. W. M. S. I., 2019. The relationship between employability skills and self efficacy of students with learning disabilities in vocational stream. *Asian Journal of University Education,* 15(3), pp. 163-174.

Harvey, L., 2001. Defining and measuring employability. *Quality in higher education,* 7(2), pp. 97-109.

J. Hillage, E. P., 1998. *Employability: Developing a framework for policy analysis,* London: Department of Education and Employment.

J. Thompson, P. S., 2022. This is what gets people hired!': Academic perspectives on employability skills in architecture and the potential impact of COVID-19. *Journal of Teaching and Learning for Graduate Employability,* 13(1), pp. 127-141.

L.M. Khodier, A. N., 2020. Changing skills for architecture students employability: Analysis of job market versus architecture education in Egypt. *Architectural Engineering,* 11(3), pp. 811-821.

Lefresne, F., 1999. Employability at the heart of the European employment strategy. *Transfer: European Review of Labour and Research,* 5(4), pp. 460-480.

M. Masdeu, J. F., 2017. Reconceptualizing the design studio in architectural education: Distance learning and blended learning as transformation factors. *International Journal of Architectural Research,* pp. 1-6.

P. Suhali, E. A. M. A., 2015. Soft skills construct for architecture graduate in accordance with industry requirements. *International journal of humanaties, arts and social sciences,* 1(3), pp. 119-123.

Poon, J., 2016. Relationships between demograpgics factors and employment prospects of architecture, construction and urban planning graduates. *International Journal of Construction Eductaion and Research,* 13(2), pp. 83-101.

R. Bridgstock, D. J., 2019. Strategic institutional approaches to gradyates employability: navigating meanings, measurements and what really matters. *Journal of Higher Education Policy and Mangement,* 41(2), pp. 468-484.

R.C. Hogan, T. C.-P. R. K., 2013. Employability and career sucess: bridging the gap between theory and reality. *Industrial and Organisational Psychology,* 6(1), pp. 3-16.

S.K.W. Chu, R. R. N. T. M. N. C. L., 2017. *21st century skills development through inquiry- based learning from theory to practice.* 1st ed. Singapore: Springer Nature.

S.Z.M. Sakam, R. J. N. W., 2022. Inculcation of employability skills in technical and vocational; education: the knowledge of employability skills and teaching practices.. *Academic Research in Business and Social Science,* 636(1), pp. 1-9.

Singh, R., 2023. Pedagogy for teaching fire safety through design-based immersion og the national building code 2016 with feedback from students of undergraduate architecture. *The Scientific World Journal,* pp. 1-10.

T. Mari, S. S. S. G. S. K. F. A., 2019. Architecture garduate work readiness: the gap between learning and employability. *IOP Conference Series: Materials Science and Engineering,* Volume 636, pp. 1-9.

V.C. Aven, M. D., 2012. Architecture: a " rewarding" career? an anglo-french comparative study of intrinsic rewards in the architecture profession. *Construction Management and Economics,* pp. 513-523.

8 Delayed council of architecture registrations: Implications and solutions

Parthiban Rajukalidoss[1,a] *and Nina Lazar*[2,b]

[1]School of Architecture, Vellore Institute of Technology, Vellore, India

[2]School of Architecture, Vellore Institute of Technology, Vellore, India

Abstract

The Council of Architecture (COA) is the statutory regulatory body responsible for the education and practice of the Architectural profession in India. Each professional's credentials and systemic validity of registrations are housed in the 'Register of Architects for India' an official database of the COA. Each architect is assigned a registration number that signifies their respective year of professional registration and is authorized to use the title 'Architect'. The initial registration may get delayed due to various personal and technical impediments. Such delay impacts the professional career. The industry recognizes an architect's entire experience, i.e. from their year of graduation of the five-year program, regardless of COA year of registration. However, in academia, these delays result in irretrievable loss of experience. Registered architects do fail to renew their registration many times. But they have the opportunity to rectify the same by paying the fine. Their original registration year remains unaffected, even if there was a lapse, which is comparable to the 'initial delay' in COA registration. Hence, the study uncovers the issues of architectural professionals with delayed registrations, the underlying reasons, and eventual repercussions. Delving into this subject offers a deeper understanding of the challenges faced by the fraternity including both academicians and practitioners, and for improving the professional registration process and remedies for the future.

Keywords: Academia, architecture, COA, industry, professional registration, regulatory body

Introduction

The Council of Architecture (COA) established under the Architects Act 1972 is the supreme authority, endowed with the power to regulate architecture education and profession in India (Architects Act, 1972). Academically trained graduates from COA-approved intuitions are incumbent to establish the statutory bond by formal registration. Currently, the undergraduate level professional program of architecture, i.e. Bachelor of Architecture (B. Arch) 5 years, follows the framework of medicine, law, and charted accountancy. From this point of view, COA is comparable to the All India Medical Council; both institutions are constitutionally endowed to offer unique registration numbers to the respective qualified individuals. Those with valid COA registration numbers are legally entitled to use the prefix architect, practice architecture in India, and also be involved in academic teaching. In this background, the significance of 'professional registration' as an architect is our academic concern.

Aimed at catering to the exponential growth of the nation, the COA is geared up to remodel the framework of architecture education in line with the new National Education Policy, i.e. NEP-2020 (MHRD, 2020). At this junction, experts observe that the 'Council of Architecture's mandate is to improve professional standards, and

[a]parthiban.rajukalidoss@vit.ac.in, [b]nina.lazar@vit.ac.in

DOI: 10.1201/9781003571018-8

its power to regulate education emerges from that context. Without reviewing this connection, merely regulating education is bound to fail' (Srivathsan and Bhatt, 2023). Such statements lead to Catch-22 situations. Hence our attempt to discuss COA, 'delayed professional registration', valid 'years of experience', academia, and industry needs a chronological outline of professional education in India. The following schematic overview presents the evolution of professional education in India under the heads 'native desi-model', 'imperial/colonial-model', and finally the 'post-independence-model'.

Years of apprenticeship with 'master/guru' were the native desi-model of gaining formal professional knowledge in India i.e., guru-kula (Balaram, 2005; Wright, 1952). This old model was 'deemphasized and phased out as licensing and registration requirements set by professional bodies made it difficult for such professionals to become registered' (Mazumdar, 1993). Professional knowledge gained through trial-and-error experiments, long-term interaction, and experiences within a particular geography are novel sources of illiterate tribal communities. Such pristine knowledge capitals may not be attributed to medieval literate communities that draw authority from unverified texts. Therefore, professions such as medicine, law, and architecture in the medieval native model were entangled in the exploitation of the masses' beliefs in superstitions (Wujastyk, 2004).

The earliest normative educational reform may be brought under the term imperial-colonial model trusted by missionaries, orientalists, and colonial officers. The model was advanced by the colonial government, and English was introduced as the medium of instruction (Narayanan, 2015). It is vital to mention that the region-specific knowledge concerning climate, geography, material, and vernacular know-how of bygone 'traditional-guilds-masters' has become collateral damage to these colonial reforms, and some of the losses may be irreversible. Nevertheless, formal education for 'all' (Ansuman, 2013) sprouted in this model; the older system was muddled in discrimination based on social, caste, gender, and economic orders.

The post-independence model is rooted in 'critical reasoning and scientific temper' that is embedded in the constitution of India. Jawaharlal Nehru, the first Prime Minister idealized science. Nehruvian educational model inspired by the West remained unchecked for maintaining quality (Ansuman, 2013). Today, architectural education in India is the product of the post-independence model, under the ambit of the Architects Act, 1972 of the government of India. A host of seventeen schools which are still active, have been offering architecture education in India from before the establishment of the Architects Act and the COA.

The oldest is the Department of Architecture, Sri JJ School of Arts, Mumbai (elsewhere Bombay) established in 1913 and the last school to be established before the formation of COA is the School of Architecture at the University of Visvesvaraya, Karnataka in the year 1967 (COA, 2024b). Modern architecture schools in India were modeled after the British and American academia (Mazumdar, 1993), currently, there are three hundred and eighty schools approved by the COA (COA, 2024b). Prevailing current global architecture education is the product of 'criticism of the late modernist architecture during the 1960s' (Fischer, 2020) that eventually led to the post-modern era of architecture. The current service sector of the 'architecture industry' in India is a product of the post-independence model. Currently, NEP-2020 argues for a clean shift from imperialist-inspired models and argues to devise a nationally rooted model (MHRD, 2020).We are on cross road to another radical change in the foreseeable future, and a short review of the vision and mission of the Architects Act and the COA respectively, will aid in elaborating our subject of concern.

Literature Review: The Structure of the Architects Act and the COA

Architects Act, 1972 establishes the structure and the working realm of the COA in New Delhi. Formal definitions are laid out in Chapter 1, followed by the institutional structure of COA in Chapter 2, 'Registration of Architects' and 'Register of Architects for India' in Chapter 3, and finally aspects such as the penalty for falsely claiming, use of title covered in Chapter 4.

The pragmatic structure and organization of the act is commendable. For example, 'Recognition of qualifications granted by authorities in India and foreign countries' (The Architects Act, 1972) features before introducing the role of the COA in preparation/maintenance of the 'register of architect of India' as well as the councils mandate in regulating architecture education. Historically speaking, the Architects Act recognizes the need to formally induct academically trained professionals into its fold. Simply, the earliest 'registered architects' of India were instated into the newly found COA registry in 1972. The academic degrees of these pioneers did not have the approval of the COA, because the history of modern academic teaching of architecture significantly predates the advent of the Architects Act as well as the COA. This observation helps to understand the problem of our concern namely, 'delayed professional registration'.

The COA is nearing five decades of its inspection. The COA has authored a gradual overall increase in architecture schools/departments over the period as shown in Figure 8.1. The only exception is the decade 2010 to 2019 with 3.5 times increase in the establishment of new architecture schools. Between the years 2010 to 2014, a total of 132 new schools were founded, which is 35% of all architecture schools in India i.e. three hundred and eighty schools (COA, 2024b).

Recognizing this larger canvas our concern is 'delayed COA registration', where we have framed the problems in two major divisions: implications and solutions. By implication, we are discussing architects who have delayed their COA registration and how it impacts them. These 'architect' may be practitioners, academicians, researchers, visiting faculties, and a combination. These professional activities spring from an academic education recognized by the COA. As solutions, we suggest some remedies to the deprived professionals.

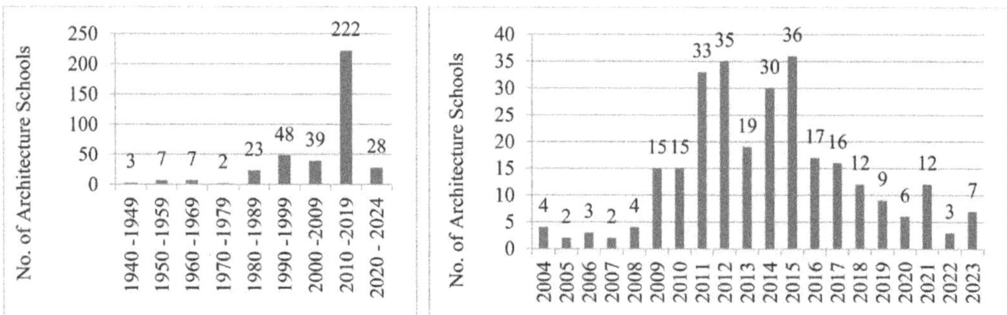

Figure 8.1 Distribution of architecture schools over the years
Source: COA, 2024b

Research Objectives

The research objectives for the current study are: 1) To understand career drawbacks of 'delayed professional registration, 2) To explore remedies for 'delayed registration, 3) To investigate the benefit of counting 'valid years of professional experience post registration' with the COA, and not after graduation from a COA approved academics.

Methodology

Research Framework

The methodology includes three phases, i.e., preparation of questionnaire, data collection through questionnaire survey, and data analysis, as illustrated in Figure 8.2. The methodology is designed to record and analyze the details of the architects and to identify the potential issues faced by them regarding COA registrations and to further provide with few suggestions.

Phase I: Preparation of Questionnaire

The first phase involves the preparation of the questionnaire to record the responses of the architects. The questionnaire is structured in five parts. The first part briefly describes the purpose of the questionnaire survey, i.e. to record and analyze the concerns of the fraternity concerning the delayed registration. The second part records the general details of the respondents, which include Age, Gender, State of current residence, City of current residence, Education (B. Arch), Profession, and Registration details. The third part records the details of the respondents on their Master's degree. The respondents without a Master's degree will be automatically redirected to the fifth part. The fourth section consists of questions related to the

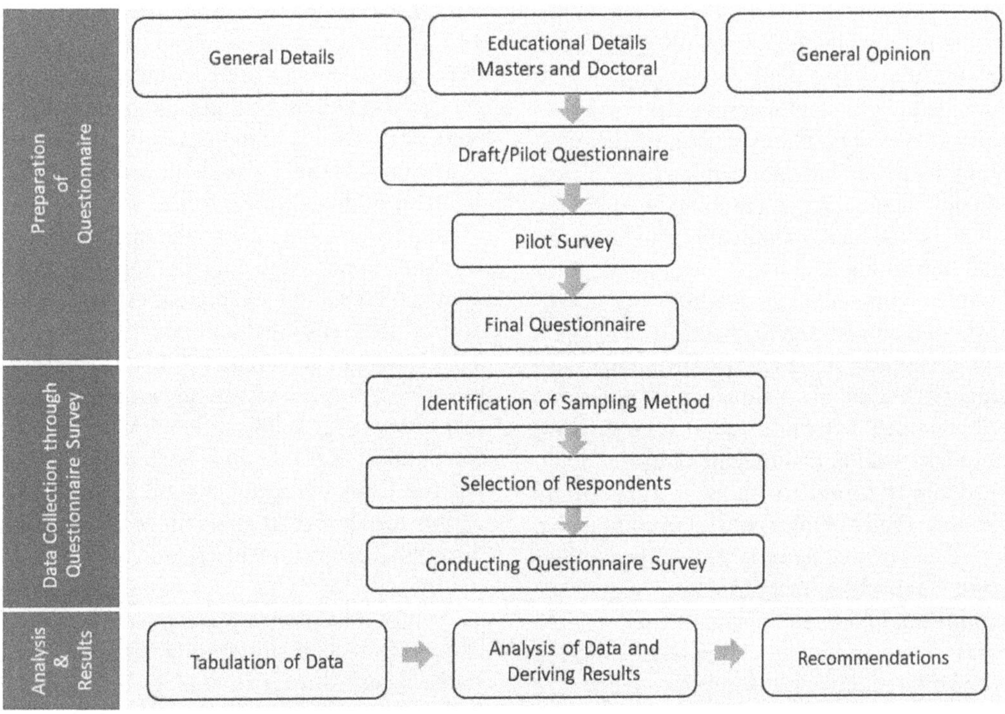

Figure 8.2 Methodology flowchart
Source: Authors

Doctoral Degree. In case the respondent is without a doctoral degree, the questionnaire is redirected to the fifth part. The fifth part records the opinion of the respondents on the relevance of a Master's degree and a Doctoral degree for teaching architecture. The respondent is asked to mark the relevance on a Likert scale. The decision on the number of scale points to be included in the Likert scale is critical. According to Liberg et al. (1997), the optimal number of scale points lies between 5 and 7. Therefore, the questionnaire is designed with a 5-point linguistic scale ranging from low (1) to high (5) relevance. The questionnaire was developed as a self-administered online survey using Google Forms to collect responses from around the country. The questionnaire was circulated among limited architects to identify the critical modifications required and to refine the draft to the final questionnaire.

Phase II: Data Collection through Questionnaire Survey

The data collection involves three tasks: 1) Identification of the sampling method; 2) identifying the pool of respondents; 3) designing and administering the questionnaire survey. The identification of the sampling method was done based on a few considerations. As mentioned earlier the COA is the only statutory that updates and maintains the list of architects. Our target respondents are graduates of COA-approved educational institutions. Hence, it was decided to share the questionnaire with alumni groups of various COA-approved architectural institutions. A combination of judgmental and snowball sampling techniques was found to be most appropriate for this study. Judgmental sampling, also called purposive sampling, is a non-probability sampling technique in which the sample members are chosen based on the researcher's judgment. Snowball sampling is a non-probability sampling technique where the respondents are free to self-recruit future respondents from among their acquaintances. Thus, the sample group is said to grow like a rolling snowball. The respondents were selected through a combination of judgmental and snowball sampling. The online questionnaire link was posted in many architectural alumni groups and forwarded further the initial respondents. The perception of architects on COA registration is then gathered through the survey.

Phase III: Analysis and Results

The data analysis involves several tasks: 1) Understanding the registration status and details; 2) Analyzing the delays in registration and the reasons for the delays and dynamics of those respondents; 3) Analyzing the renewal/non-renewal of registration and reasons for the delay; 4) Valid years of experience as per the COA, to arrive at implications and solutions.

Questionnaire and Data Collection

To the circulated questionnaire a total of 145 responses were received. The general dynamics of the respondents such as age and gender and highest educational qualification were noted and is presented in Figure 8.3.

The pool of the respondents largely (91%; 132 respondents) fall into the age group of 26 to 55 years, who are in the prime of their working years. Out of the total respondents, 62.8% (91 respondents) are male and 37.2% (54 respondents) are female. The responses were received from architects covering a total of eleven states, with major responses from states such as Tamil Nadu, Karnataka, Kerala, and Maharashtra. As per the COA currently, a total of 1,14,161 professionals are registered in thirty-nine different geographical regions of India (COA, 2024a). The respondent dynamics also show that the responses were received from architects residing in thirty-nine cities such as Bengaluru, Chennai, Delhi, Cochin, Trivandrum, Ahmedabad, and Hyderabad to list a few.

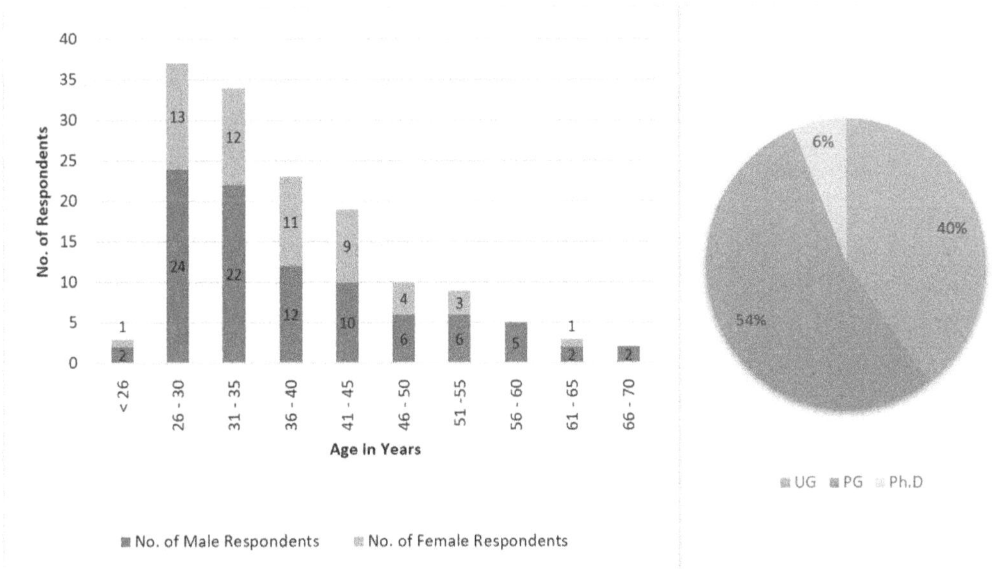

Figure 8.3 Respondent dynamics – age, gender and highest qualification
Source: Authors

Analysis and Results

Use of Professional Title

The subject is framed in such a way that multiple options could be chosen by the respondents to identify their professional titles. Options included: academician/ researcher, designer, planner, practicing architect, visiting faculty, other (space given to be filled by the respondents), and finally the option currently not into any profession. 82 respondents i.e. 56.55% are associated with academic teaching of architecture. Overall 26.8% of respondents exclusively identified themselves as practicing architects and 40% of the respondents chose to identify as practicing architects + another title. Around 33.2% chose not to identify themselves as practicing architects. 25.5% identified exclusively identified themselves as academician/research or visiting faculty and 0.02% individuals identified themselves as academician/research + another title. Among the total 54 are male and 91 are female respondents: 33 female and 64 male respondents used the title practicing architect (exclusive usage); in comparison, 17 female and 21 male respondents used the title academician/researcher exclusively.

Registration Status

The analysis shows that most of the respondents are registered (95.86%; 139 respondents). Few (six respondents) are not registered with COA, including respondents with the highest qualification of UG and PG. One respondent has just applied, two respondents are waiting to receive their degree certificate, two respondents are not in architectural practice, hence not registered and one respondent completed AMIE.

Delays in Registration

Out of 139 who have registered, more than two-thirds (69.78%; 97 respondents) had delayed the registrations and approximately one-third (30.22%; 42 respondents) have not delayed. Sixty male respondents (out of 91) and thirty-seven female respondents (out of 54) delayed the registrations. The details of respondents

Table 8.1 Details of delayed registrations

Years of delay	Respondents having highest qualification as:			Total respondents	Reasons for delay and no. of respondents	
	UG	PG	PhD			
1	13	29	3	45	Not specified	53
2	6	11	1	18	Delay in receiving degree certificate	8
3	3	5	1	9	Lack of awareness	4
4	4	6	1	11	Personal delay	6
5	3	1	0	4	Due to arrears	5
6	1	0	0	1	Didn't see any advantage	1
7	0	1	0	1	Not practicing	2
8	1	0	1	2	Pursuing masters	8
9	1	0	0	1	Avoid sending degree certificate	5
11	0	0	1	1	Staying abroad	2
12	1	1	0	2	Working in different fields	1
13	1	0	0	1	Loss of original documents in post	1
25	1	0	0	1	Local corporations don't recognise CoA registration	1
Total	35	54	8	97	Total	97

Source: Authors

who delayed the registrations and reasons for delayed registrations are given in Table 8.1.

It is evident from Table 8.1 that most of the respondents (87 respondents) who delayed registrations mostly fall short by one to 5 years. Out of the total respondents with delayed registrations, a major proportion (63.91%: 62 respondents) have the highest qualification either as a PG or Ph.D. The reason for delayed registrations also points towards reasons such as delay in receiving the certificate from the undergraduate university, pursuing master's so that the certificate was with the University/Institution, staying studying or practicing there, working in a different field or academic research, fear of loss of original documents by post, etc. Nevertheless, the COA registration is the sole responsibility of the respondents.

Renewal/Non-Renewal of Registration

The respondents who have done timely/delayed registrations and further maintained or failed to maintain valid registrations by renewal/non-renewal are shown in Table 8.2. The table indicates that, out of the 139 respondents, only 15.82% (22 respondents) did timely registration and maintained validity without fail. Also, one-third of the respondents (33.09%; 46 respondents) have failed to maintain validity and paid a fine.

Valid years of Experience

The year of experience is counted from the date of COA registration and it is strictly followed in academic/research institutions. Contrarily, COA registration does not seem to have any such significance among industry practitioners. In academia, delayed years are not taken into account as experience

Table 8.2 Maintenance of valid COA registration

Registration	Registration status in detail	No of respondents	
Delayed Registrations	Paid fine and renewed	29	97
	Registration currently invalid	6	
	Maintained valid registration	62	
Timely Registrations	Paid fine and renewed	17	42
	Registration currently invalid	3	
	Maintained valid registration	22	

Source: Authors

even though they have been working during those years, especially as researchers, allied professionals such as designers, or pursuing higher education, etc. In contrast, one who has done timely registration but failed to maintain the registration in between, but paid a fine and renewed the registration does not face any issue of loss in experience years.

Conclusion

The analysis of the survey suggests that around two hundred and eighty-three years of professional experience gained from COA-approved intuitional educational experiences of 97 architects is thrown away due to negligence. Considering the critical need for qualified experienced professionals to impart architecture education, a clerical nitpicking approach is impeding the careers of a significant number of formally trained professionals. The information begs a straightforward question. 46 architects have paid a fine to renew the registration, irrespective of this fact they use the 'original year of registration' to count years of experience especially while they are in academia. 'Delayed registration' of the members of our fraternity is a matter of human error but unfortunately in the prevailing scenario, it is one of those very rare cases with no remedy. A principle of natural justice suggests that there cannot be any misdeed/misgiving that does not have a remedy. There are two kinds of negligence i.e., civil and criminal, it is the responsibility of the fraternity to categorize the subject matter appropriately. Civil negligence is pardonable with a fine. Providing relief may not bring disrepute but will inspire confidence in the COA. The young growing profession is already falling short of catering to the exponential growth of the nation. Formal remedies of deprived members of the fraternity will help the cause of architecture education.

Acknowledgment

We duly acknowledge all the respondents who participated in the survey, whose sincere participation has made this study possible.

References

Ansuman, N. (2013). Higher Education in India – A Transition. *The Indian Journal of Political Science, 74(3), 489-500.*

Balaram, S. (2005). Design pedagogy in India: aperspective singanapalli balaram. *Design Issues*, 21(4), 11–22.

COA (2024a). Council of Architecture. Council of Architecture. https://www.coa.gov.in/index1.php?lang=1&level=2&sublinkid=117&lid=130.

COA (2024b). Institution Approval _ Council of Architecture.

Fischer, O. W. (2020). Institutionalized Critique? On the Re(birth) of Architectural Theory after Modernism: ETH and MIT Compared. *The Figure of Knowledge, Chapter 8, 145-159.*

Liberg, L., Blener, P., Collins, M., Dippo, C., and Schwarz, N. (1997). Survey Measurement and Process Quality. John Wiley.

Mazumdar, S. (1993). Cultural values in architectural education: an example from India. *Journal of Architectural Education*, 46(4), 230–238.

Ministry of Human Resource Development, Government of India. (2020). National Education Policy 2020.

Narayanan, V. (2015). The history of the academic study of religion in universities, centers, and institutes in India. *Numen*, 62(1), 7–39. https://doi.org/10.1163/15685276-12341354.

Srivathsan, A., and Bhatt, C. (2023). What do numbers reveal about architecture admissions and the state of education_ - Architecture Live! Last accessed: 21 June 2024. URL: https://www.architecture.live/architecture-admissions-and-the-state-of-education-srivathsan-chirayu-bhatt/

The Architects Act (1972). Pub. L. No. Act No. 20 of 1972 (1972).

Wright, R. C. (1952). Freedom's impact on higher education in India. *The Journal of Higher Education*, 23(4), 198. https://doi.org/10.2307/1976779.

Wujastyk, D. (2004). Medicine and dharma. *Journal of Indian Philosophy*, 32(5), 611–627.

9 Critical analysis of artificial intelligence as a design tool in architectural pedagogy

AR. Saniya Aafreen S[a]

MEASI Academy of Architecture, Affiliated to Anna University, Chennai, India

Abstract

Artificial intelligence (AI) is revolutionizing various industries, and the field of architectural pedagogy is no exception. The advent of AI in architecture is still in its initial stage but offers promising results. This paper is a study of the transformative potential of AI tools in architectural education and the critical analysis of the ways in which AI tools are enhancing the learning experience for aspiring students and architects. The purpose of this study is to discuss in detail how AI tools and platforms are aiding personalized and adaptive learning, catering to individual student needs and learning styles, thereby allowing students to experiment with innovative design concepts and receive real-time feedback on their creations. The paper will advance by addressing the benefits of AI tools that can be used in architectural pedagogy in comparison to conventional learning methodology tools in various stages of design process in architectural studios, including the need for enduring faculty development, particularly in today's educational landscape with further discussion on the ethical considerations and challenges of implementation of AI tools. In conclusion, this paper is to highlight the potential use of AI in revolutionizing architectural education offering new opportunities for creativity and innovation enriching the learning experience.

Keywords: AI tools, architectural learning methodology, architectural pedagogy, artificial intelligence (AI)

Research Aim

The aim of the research is to understand the application of artificial intelligence (AI) as a design tool in architectural education especially in design studio offering new opportunities for creative solutions, design thinking and also as an innovative pedagogy method thereby enriching the learning experience.

Research Objectives

- To simplify the standard design process used in design studios into brief stages of conventional design process.
- To explore and identify various interesting AI tools for each of the above stages with demonstrations of a few tools as example.
- To tabulate and draw out a matrix with benefits and limitations of each of these AI tools explored in comparison to conventional manual and digital tools used in design studios.
- To analyze and evaluate the potential use of AI tools in various stages of studio design process with the help of the matrix.
- In conclusion to demonstrate that learning through creating experiences displays interesting ways to understand and gain knowledge.

Review of Literature

In simple terms AI is the simulation of human intelligence which can perform simple tasks like perceiving data, reasoning,

[a]san09ya@gmail

DOI: 10.1201/9781003571018-9

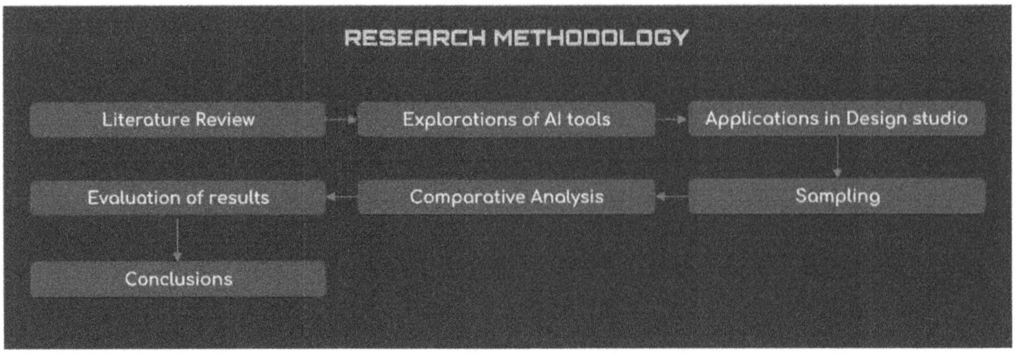

Figure 9.1 Research methodology
Source: Researcher's presentation work

problem solving and even exercising creativity.

AI works on three processes – learning process, reasoning process, and self-correction process.

- Learning process – collects data and converts it to algorithms, creation of algorithms.
- Reasoning process – chooses appropriate algorithms in problem solving, application of algorithms.
- Self-correction process – improves algorithms and creates more responses, patterns of algorithms.

AI is the next revolution in creative design process in Architecture and can enhance creativity, maximize time efficiency, optimize material usage, reduce construction cost, improve sustainability and even as a mood board for design inspiration. Currently AI is in an intermediate stage of development in the field of Architecture and can perform two simple tasks-creates detailed images with brief text prompts and solves simple practical problems with speed and variations.

It has its own limitations and few more aspects in exploration but by and large these tools can be used in certain stages of design process in design studio to create interesting learning experiences.

Research Methodology

The research process as in Figure 9.1 includes:

- literature review and exploration of AI design tools,
- application of the same in design studio- (5th semester)
- sampling of the results
- comparative analysis and evaluation of results
- conclusions.

The research was based on three main tools as explained earlier targeting three types of architecture students:

- **Type – A-** students who can explain ideas and concepts well as descriptive information/ verbally.
- **Type – B-** students who can explain ideas and concepts well as a freehand sketches
- **Type – C-** students who can explain ideas and concepts well with massing model or conceptual models

Analysis – Studio Experimentation

Design process of any design in a design studio can be briefly put into four stages that is studies, concept, planning and presentation of ideas and drawings.

Studies Stage

It is essential to improve reading and writing skills to engage with architectural theory and history, to analyze case studies, draw out inferences of the same and to derive design guidelines, and to communicate ideas effectively. Conventionally these are done by reading widely, practicing writing, seeking feedback and using resources like online resources, writing workshops. Most students seek the above ways to read, write and develop their architectural vocabulary but a few shy themselves away from this process for various reasons. Many students also find report writing a difficult and stressful task and these AI tools can help them do it easily with time efficiency. There are certain tools in AI that can aid to this study stage making this task easier and beneficial. These AI tools have an easy interface with Diverse operations to read, write, edit or rewrite and also summarizes text content from books, articles, epdfs and research papers helping them to write content with definitions, examples, quotes etc. there by helping them learn to structure sentences with the use of appropriate architectural vocabulary like a peer in design studio.

Concept Generation

The most important and critical tasks for any architectural student is to conceive a concept and be able to iterate as many as possible for feedback from academicians and professionals. These concepts are generally developed through many discussions and feedbacks. Conventionally students generate these ideas by site visits, story boards, model making to understand both site conditions and explore massing and form or simple free hand sketches as part of ideation process in design studio. Additionally, digital tools and digital modelling aid this stage of design process to develop unique concepts. However, the recent developments of AI applications or tools have become new trending tools in any architect's toolbox in the professional field. The application of such a tool in one's design studio can become an interesting, educative and creative experience for any student.

This trending and promising tool can be used in three interesting ways when it comes to concept generation and it works based on the concept generative AI. Generative AI is defined as 'a type of artificial intelligence technology that can produce various types of content, including text, imagery, audio and synthetic data:

- **Type-1-TEXT to IMAGE –** descriptive information used to explain ideas or concepts.
- **Type-2-SKETCH to IMAGE –** a freehand sketch as reference with the addition of prompt or without prompt to generate visuals as concepts.
- **Type-3-GENERATIVE FILL –** a photo editing tool that can be used to understand the massing model or concept generated with reference to site conditions.

Predominantly, **Type -1-TEXT to IMAGE** (prompt engineering) is used to generate concepts, as it is easier to use and also can improve students' vocabulary in the process of image generation with creative results as seen in Figure 9.2 below.

Type-2-SKETCH to IMAGE is the use of a freehand sketch with or without the use of prompts that can used to iterate as many possibilities as possible with speed and variations. The application of this tool was tested in one of the design studio exercises to understand its potential use in design studio and its successive outcome as seen in Figure 9.3.

TYPE-3-GENERATIVE FILL which is a photo editing tool that can be used to understand the massing model or concept generated with reference to site conditions as in Figure 9.4.

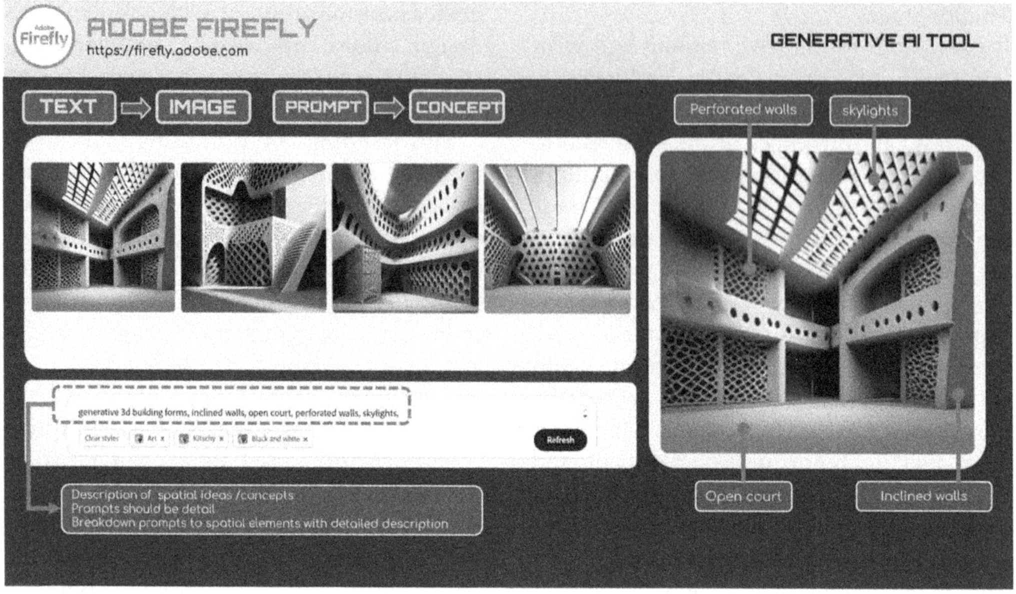

Figure 9.2 Type-1- text to image
Source: Researcher's presentation work

The above image depicts the use keywords as prompts and the image on the right with appropriate concepts based on prompts

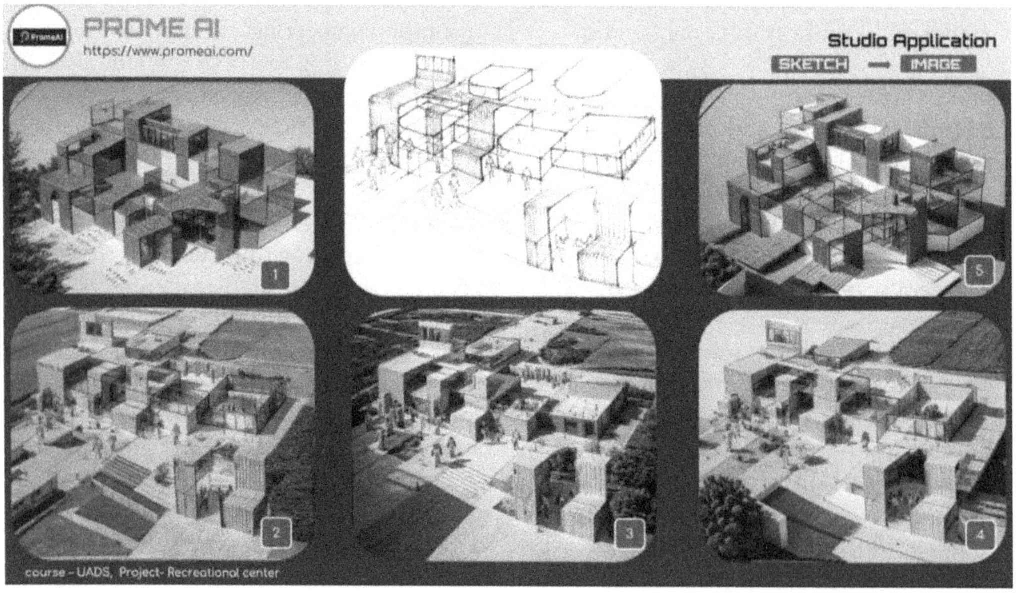

Figure 9.3 Type-2- sketch to image
Source: Researcher's presentation work

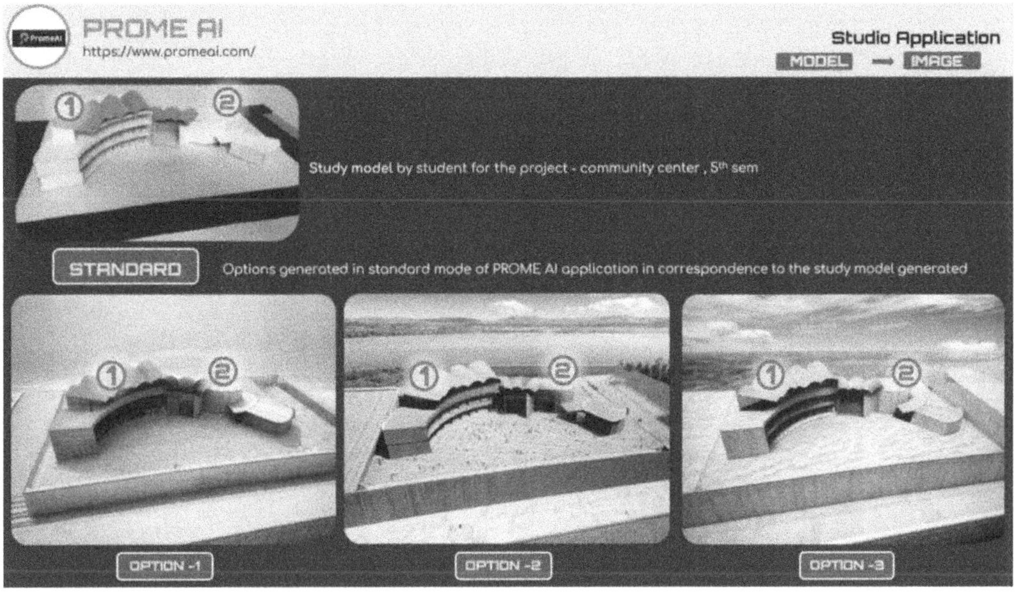

Figure 9.4 Type-3- illustrates the use of a study model to visualize site context and understand form generated

Source: Researcher's presentation work

Planning Stage

Though there are not many applications that can be effectively used or are user friendly there are various tools that can be explored for planning. As much as explored these planning tools are in their formative stages and understand less logistics of planning. However, a lot can be experimented when it comes to zoning options with few limitations as seen in Figure 9.5.

The above image shows the many zoning options experimented with their benefits and limitations.

Presentation Stage

When it comes to presentation students take a lot of effort, time and energy to present their ideas in an interesting and creative method. The biggest challenge in this stage is the rendering of views and drawings. Be it manual or digital, most students exhibit consistent efforts to present the best results when it comes to final presentation of their semester's efforts. There are quite a few innovative and explorative options in terms of material and color palette to render these views especially exterior and interior views with good quality outputs and with less limitations. Students can generate many creative options with time efficiency and good results during design studio hours for effective feedback and promising results as seen in Figure 9.6.

Figure 9.6 shows many options of various material and color palette experimented during studio hours for rendering from original render.

Synthesis of Analysis

A comparative analysis was done for better understanding of the AI tools explored with benefits and limitations draw out in various stages as seen in Figure 9.7.

This was followed by evaluating the same with numerical values to reduce to a bar diagram to conclude the effective use of AI tools in design studios as in as seen in Figure 9.8.

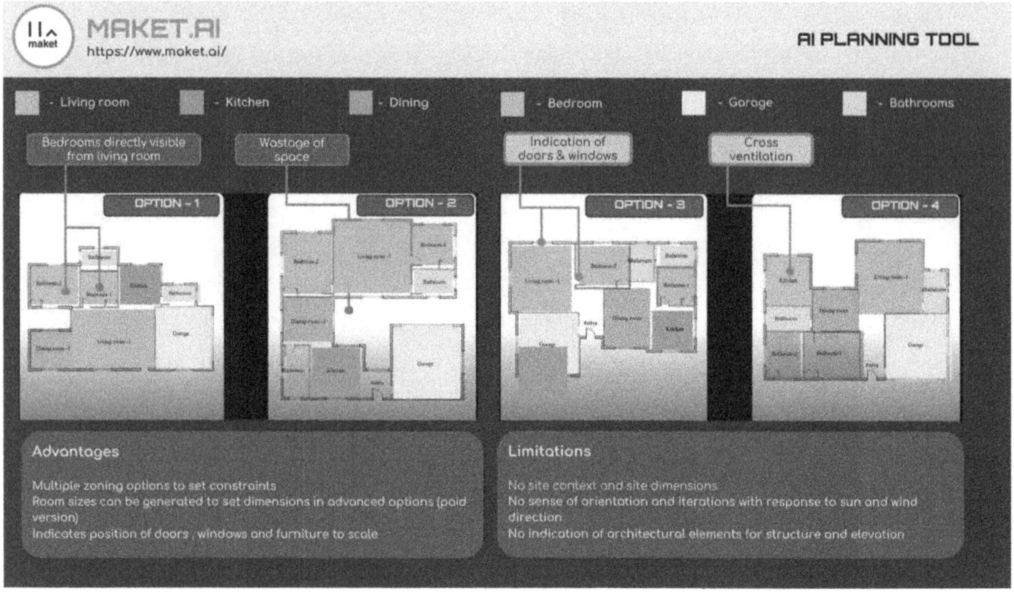

Figure 9.5 Zoning options generated using AI tools
Source: Researcher's presentation work

Figure 9.6 Various rendering options generated using AI tools
Source: Researcher's presentation work

The above image shows the evaluation AI tools done in comparison with manual and digital tools and the resultant values.

Conclusions

The above synthesis of analysis was reduced down to graphs and bar diagrams

Comparative Analysis

STAGES	ANALYSIS	CONVENTIONAL METHOD (MANUAL)	CONVENTIONAL METHOD (DIGITAL)	AI TOOLS
LITERATURE	Benefits	• Forced reading from varied resources • Quick understanding of material from peers / others	• Documentation of work • Cross references and citations	• Time efficiency • Various tones to help writing varied content
	Limitations	• Time consuming and reference to multiple resources	• More of copy paste and less understanding / reasoning of content read	• Plagiorism issues
CONCEPT	Benefits	• Skill development • Improves thought process and reasoning	• Use of multiple software for desired result • Understanding of scale and proportion from outputs designed	• Multiple ideas generated • Photorealistic views
	Limitations	• Time consuming • Sketchy and outputs – no sense of scale and proportion (in many cases)	• Time consuming • Rigid design	• Not to scale • Does not correspond to any geographical location
PLANNING	Benefits	• Functional and logical solutions with reasoning	• Understanding of scale and 3D generated simultaneously • Easy documentation	• Multiple options/ iterations • Some AI applications generate solutions to scale
	Limitations	• Difficulty in documentation and number of iterations	• Rigid design and rare experimentation of form • Less understanding of built-unbuilt / overall form	• No logical solutions especially with respect to geographical location and site conditions
RENDERING	Benefits	• Skill development • Unique and personalized	• Photorealistic quality with detailed inputs	• Photorealistic quality with prompts / application options
	Limitations	• Time consuming – hours to days • Less iterations	• Time consuming – hours to days • Requires high configurations and use of multiple software	• Limitations in architectural details generated
PRESENTATION	Benefits	• Personalized / unique presentation	• Documentation is easier • Organized content / reworking is easy	• Quick generation of graphical content
	Limitations	• Time consuming – hours to days	• Time consuming – hours to days	• Not many options/ variations

Figure 9.7 Comparative analysis of various tools used in design process
Source: Researcher's presentation work

Evaluation

STAGES	CONVENTIONAL METHOD (MANUAL)						CONVENTIONAL METHOD (DIGITAL)						AI TOOLS					
	CREATIVITY	ITERATIONS	TIME EFFICIENCY	QUALITY	DOCUMENTATION	TOTAL	CREATIVITY	ITERATIONS	TIME EFFICIENCY	QUALITY	DOCUMENTATION	TOTAL	CREATIVITY	ITERATIONS	TIME EFFICIENCY	QUALITY	DOCUMENTATION	TOTAL
CONCEPT	3	3	5	3	3	17	1	1	1	5	5	13	3	5	5	5	5	23
PLANNING	5	1	3	3	3	15	3	5	3	5	5	21	1	3	3	3	5	15
RENDERING	5	1	1	3	3	13	3	3	3	5	5	19	5	5	5	3	5	23
PRESENTATION	5	1	3	3	3	15	5	3	3	5	5	21	3	3	5	5	5	21
TOTAL (100)						60						74						82

		1 - POOR	3 - AVERAGE	5 - GOOD
CREATIVITY	Exploration of design contents	Least exploration	Moderate exploration	Good exploration
ITERATIONS	Number of possibilities	1 - 3	3 - 5	5 - 10
TIME EFFICIENCY	Quick and qualitative outputs	Days	Hours	Minutes
QUALITY	Aesthetics + functionality + Output quality	Satisfies any 1	Satisfies 2	Satisfies all 3
DOCUMENTATION	Recording, reworking, quality	Satisfies any 1	Satisfies 2	Satisfies all 3

Figure 9.8 Evaluation of tools
Source: Researcher's presentation work

to understand the effective use of artificial intelligence (AI) tools in design studio and as seen in Figure 9.9. We can understand that Artificial Intelligence tools are in formative stages when it comes to planning but can provide promising results and interesting learning experiences when it comes to concept generation and rendering

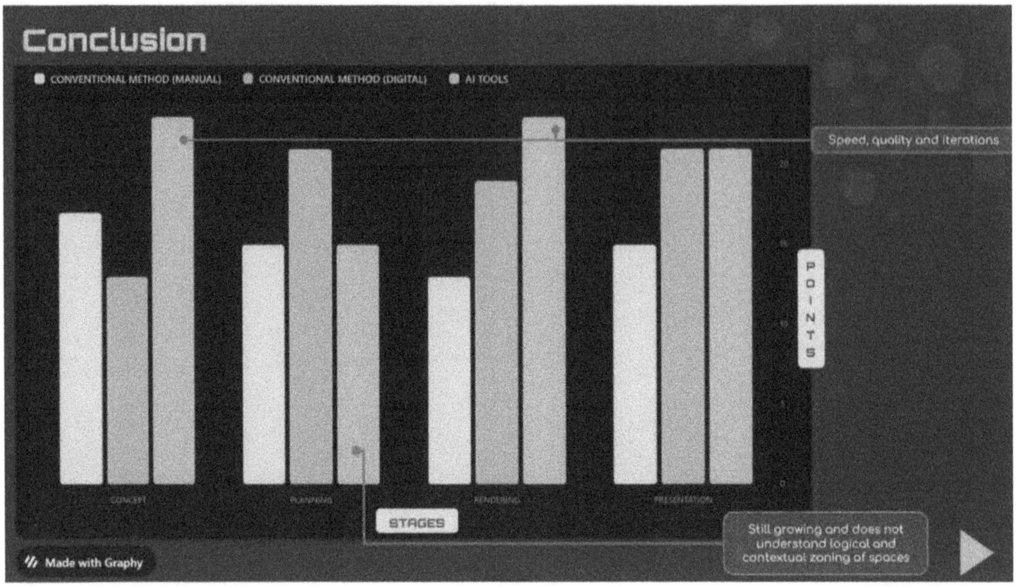

Figure 9.9 Bar diagram
Source: Researcher's presentation work

stages of a design studio with many iterations. Architectural fraternity can make the best use of AI as a design tool to motivate students, sketch and visualize, improve architectural vocabulary with generation of ideas using prompts and improve students' imagination skills. AI tools need to looked and explored to the maximum to suit the trending professional field and prepare students to face practical and creative expectations of today's clients.

Acknowledgements

I extend my sincere thanks to the individuals and institution whose support has been of immense help for my research, especially student RAAJESHWARI - course FDS, and student CAROLA, course – USDS.

Special thanks to Professor Dr. A. NAVARA, HOD (B.Arch.), MEASI cademy of Architecture for her constant support and motivation through my research process.

Finally, to my family, for their understanding and encouragement were a constant source of strength throughout the research process.

References

Artificial Intelligence in Architecture and its Impact on Design Creativity- A Study on how Artificial Intelligence Affect Creativity in the Design Process- CARL BORGLUND

Artificial Intelligence and Architecture Towards A New Paradigm - TAMAS LUKOVICH, Department of Architecture, Ybl Miklos Faculty of Architecture and Civil Engineering Óbuda University, Budapest, Hungary.

CHAILLOU, S. (2019b) AI & Architecture, An Experimental Perspective, www. medium.com/built-horizons

JACK BLACKWELL 9 Best AI Architecture Generator Tools www.purefuture.net/2023/04/26/9-best-ai-architecture-generator-tools-free-and-paid/#google_vignette

www.ijres.org Volume 11 Issue 5 ǁ May 2023 ǁ PP. 398-401. AI in Architecture UTKARSH SRIVASTAVA Student Sunder Deep College of Architecture, Ghaziabad, U.P., India

10 Contemporary concerns in architecture-smart home automation and its architectural integration

M. Narayanan[1,a] *and G. Yogapriya*[2,b]

[1]Research Scholar, Chettinad School of Architecture, Chettinad Academy of Research and Education, Kelambakkam, Chennai, India

[2]Associate Professor, Chettinad School of Architecture, Chettinad Academy of Research and Education, Kelambakkam, Chennai, India

Abstract

Background: The rapid rise in global energy consumption and its detrimental impact on the environment necessitates innovative solutions for optimizing energy use. One intriguing technology that has the potential to drastically lower residential energy consumption is smart home automation (SHA).

Method/Analysis: This paper explores the features and mechanisms of SHA for electricity reduction, showcases successful case studies with quantifiable savings, and analyzes challenges and future research directions. Three diverse case studies are examined, highlighting common success factors and different approaches. The paper concludes by proposing solutions to address adoption barriers and outlining future research areas for maximizing the impact of SHA on sustainable energy consumption.

Findings: SHA offers a variety of functionalities for reducing energy use, including automated routines, enhanced security, optimized energy efficiency, improved accessibility, and personalized entertainment. Specific mechanisms like demand response, predictive control, occupancy-based control, appliance scheduling, and building envelope optimization can achieve significant energy savings. Case studies demonstrate the effectiveness of SHA, with reductions ranging from 30– 40% and even achieving net-zero energy consumption. However, challenges like high initial costs, data privacy concerns, user adoption barriers, and interoperability issues need to be addressed for wider adoption.

Keywords: Case studies, energy reduction, future research, smart home automation, sustainable electricity consumption

Introduction

The rapid rise of global energy consumption poses a significant threat to the planet's ecological equilibrium. The residential sector, responsible for approximately 21% of global energy use (International Energy Agency, 2023), presents a critical area for intervention. Smart home automation (SHA) emerges as a promising solution, offering unparalleled potential to transform how we design, optimize, and manage energy consumption within residential buildings. They are equipped with embedded technologies (sensors, actuators, automated functionalities, etc.) (Bitterman and Shach-Pinsly, 2015; Chanetal. 2008). Hence, they are fundamentally different from the traditional built environments and are equipped to provide functionalities like touchless control, flexible space, energy efficiency, modular construction, home

[a]mnana18@gmail.com, [b]yogapriyag@care.edu.in

DOI: 10.1201/9781003571018-10

healthcare, etc. Smart homes are one of the most studied SBEs, and they are considered the homes of the future. By integrating intelligent technologies within the very fabric of architecture, SHA unlocks innovative avenues for minimizing electricity use and promoting sustainable living.

Methodology

As there is no established framework for the smart environment design process, current design and construction practices overwhelmingly follow a mostly linear design and delivery approach (Ageeetal, 2020; Taysheng, 2009). Hence this paper will dive into smart home automation (SHA) by first reviewing its features and mechanisms for reducing electricity consumption (demand response, predictive control, etc.). Later it explores the roles of consultants in implementing SHA, showcasing successful case studies with quantifiable savings. Three diverse case studies will be analyzed for building type, features, achieved savings per mechanism, and key takeaways. A comparative analysis will identify common success factors and different approaches. Finally, it will tackle adoption challenges (cost, privacy, etc.) and propose solutions like cost-effective models and user-friendly interfaces. Future research directions will focus on affordability, security, user experience, and integration with building design for optimal impact. Through this comprehensive approach, this paper will shed light on SHA's immense potential for achieving sustainable electricity consumption and highlight the key areas for future development.

Background Study

Smart home research started nearly a couple of decades ago, with projects like smart rooms by the MIT Media Lab (Pentland, 1996) being pioneering works (De Silvaet. al.(2012)). Other examples of smart homes are iRoom (Johansonetal, 2002), AwareHome (Abowdetal, 2002), House n (Stephen S. Intille, 2002), GatorHouse (Helaletal, 2005) etc. Yet, the recent explosion of internet connectivity, affordable sensors, and powerful artificial intelligence has breathed new life into this technology, propelling it into the mainstream.

The early days of SHA saw clunky systems built around proprietary protocols and often requiring complex programming. They focused primarily on security and basic automation tasks like turning on lights at sunset or adjusting the thermostat. However, with the advent of open communication protocols like Zigbee and Z-Wave, interoperability grew, and devices from different manufacturers began to talk to each other. This paved the way for a wider range of functionalities and easier user experiences.

The rise of smartphones and voice assistants like Alexa and Google Assistant further ignited the SHA revolution. Suddenly, controlling your home became as simple as speaking a few words. This shift in interface, making technology more accessible and intuitive, opened the door to a broader audience and propelled SHA into everyday life.

Today, SHA offers a sophisticated set of features and functionalities, encompassing:

- Automated routines: Lights adjust based on occupancy and daylight, thermostats learn your preferences, and appliances switch on or off on a schedule, creating a seamless and customized living environment.
- Enhanced security: Smart locks, cameras, and alarms provide real-time monitoring and control, deterring intruders and offering peace of mind.
- Boosted energy efficiency: Sensors and intelligent systems optimize energy consumption by managing lighting, HVAC, and appliance usage based on occupancy and user preferences.

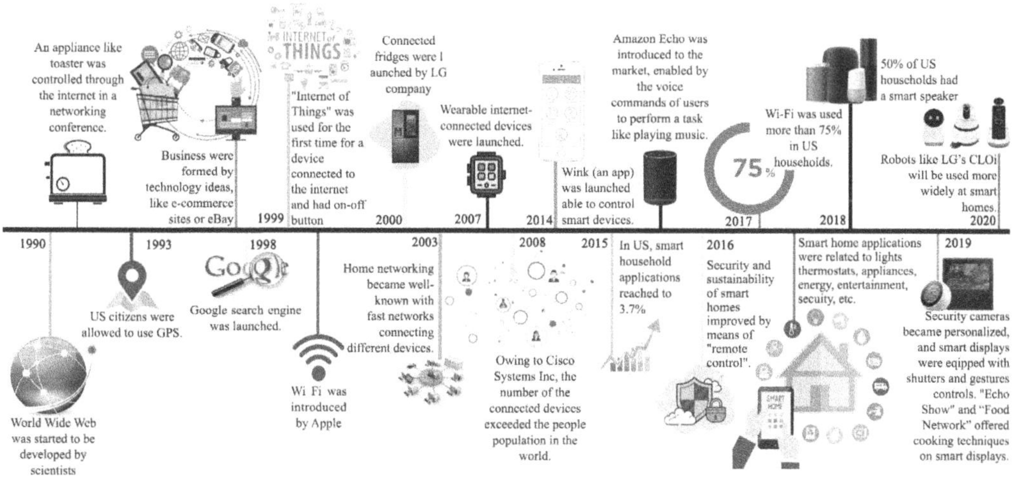

Figure 10.1 Smart home evolution
Source: Karimi et al. 2021

- Improved accessibility: Voice control and automated tasks empower individuals with disabilities to navigate their homes with greater ease and independence.
- Personalized entertainment: Integrated audio and video systems personalize your entertainment experience, letting you control lights, music, and temperature with a single command.

However, despite these advancements, challenges remain. Concerns about cybersecurity and data privacy require robust security measures and user control over data sharing. Interoperability issues continue to plague some setups, highlighting the need for standardized protocols. Additionally, the initial investment in infrastructure and devices can be a barrier for some, particularly in emerging markets.

Looking ahead, the future of SHA is bright. Natural language processing, machine learning, and artificial intelligence have the enormous potential to further personalize experiences, anticipate user demands, and improve energy management. The integration of smart homes with broader smart city initiatives and renewable energy solutions represents another area of exciting potential. Figure 10.1 depicts the timeline that outlines the progression of the Internet of Things (IoT) and smart technologies from the 1990s to 2020, highlighting major milestones such as the advent of Wi-Fi, smart home applications, and the widespread use of smart speakers and robots. Notable events include the development of the World Wide Web, the launch of Google, the emergence of wearable devices, and innovations by Apple, Amazon, and LG.

Mechanisms for Electricity Reduction

SHA boasts a diverse arsenal of mechanisms to curb electricity consumption within buildings. Some of the specific functionalities and concrete examples of how each of these mechanisms translates to energy savings are shown in Table 10.1.

Consultants and Their Scope in SHA

In the ever-evolving world of SHA, various consultants play crucial roles in assisting homeowners, builders, and architects in designing, implementing, and maintaining these sophisticated systems. The main types of consultants involved in SHA and

Table 10.1 Mechanisms for electricity reduction

Mechanism	Description	Concrete examples	Energy savings potential	Source
Demand response	Smart home adapts to grid needs, saving energy during peak times.	Peak saving, price-based optimization, grid-to-home interaction.	10% grid peak demand reduction, 10% homeowner bill reduction.	Aghaei and Alizadeh (2013).
Predictive control	AI and machine learning optimize energy use proactively.	Weather-based forecasting, occupancy patterns, adaptive learning.	Up to 70% energy savings in lighting.	Dongetal.(2019).
Occupancy-based control	Sensors adjust energy based on real-time presence.	Motion sensors, presence detection, zoning.	5.6 kWh/day/household savings.	https://www.energy.ca.gov/sites/default/files/2021-05/CEC-500-2020-015.pdf
Appliance scheduling	Energy-intensive tasks shift to off-peak hours.	Smart scheduling, intelligent modes, renewable energy integration.	15% energy and 40% water savings in laundry.	https://restservice.epri.com/publicdownload/000000003002004652/0/Product
Building envelope optimization	Smart technology optimizes energy use through building integration.	Adaptive facades, energy-efficient materials and building automation systems.	Net-zero energy building achieved.	https://bullittcenter.org/vision/living-building-challenge/

Source: Aghaei, J., & Alizadeh, M. I. (2013); Dong, B., Li, Z., & Yan, Z. (2019); California Energy Commission. (2021); Electric Power Research Institute. (2014); Bullitt Center. (2022)

Table 10.2 Role of various consultants and their scope in SHA

Consultant	Scope	Key activities
Smart home automation consultant	Assessment and project scope definition	Researching and comparing different smart home technologies and solutions.
	System design and selection	Creating detailed system specifications and blueprints.
	Cost estimation and project budgeting	Negotiating with vendors and securing competitive pricing.
	Project management and implementation	Programming automation routines and customizing settings.
	User training and ongoing support	Monitoring system performance and recommending optimization strategies.

Consultant	Scope	Key activities
Architects and interior designers	Integration of smart home systems with building design	Understanding the impact of smart home technology on space planning and architectural details. Coordinating with electricians, HVAC technicians, and other contractors to ensure compatibility with smart home systems.
	Placement of sensors, switches, and other smart home components	Designing user-friendly interfaces and control systems that blend seamlessly with the overall aesthetic.
	Incorporation of sustainable design principles	Incorporating natural light and ventilation strategies to complement smart lighting and climate control systems.
Security and privacy consultant	Risk assessment and vulnerability analysis	Conducting penetration testing and vulnerability assessments.
	Recommendation of security protocols and technologies	Reviewing software updates and patch management procedures.
	Data privacy assessment and compliance	Collaborating with the smart home consultant on secure system design and configuration.
	Security awareness training	Providing ongoing security monitoring and threat detection services.
Electrical and HVAC engineer	Design and installation of electrical infrastructure	Understanding the specific power requirements and communication protocols of smart home devices.
	Integration of smart thermostats, lighting controls, and other HVAC systems	Coordinating with the contractor and smart home consultant on wire routing and panel locations.
	Compliance with electrical safety codes and regulations	Performing load calculations and ensuring adequate circuit capacity for smart home components.
	Troubleshooting electrical and HVAC issues related to the smart home system	Optimizing HVAC system performance to maximize energy efficiency with smart controls.
IT consultant	Network design and installation	Understanding the specific network requirements and communication protocols of smart home devices.
	Data management and security	Configuring routers, firewalls, and other network security measures.
	System integration and troubleshooting	Implementing backups and disaster recovery plans for smart home data.
	Software updates and maintenance	Recommending and implementing software updates to improve system performance and security.

Source: Author

Figure 10.2 One central park, Sydney, Australia
Source: Saieh 2023

their specific contributions are listed in Table 10.2.

Case Studies and Findings

SHA offers more than just convenience; it holds immense potential for reducing electricity consumption and mitigating climate change. To understand the scope of the impact, the following building examples of One central park, Sydney, Australia (Figure 10.2), Klokgebouw, Amsterdam, Netherlands (Figure 10.3) and Bullitt center, Seattle, USA (Figure 10.4) quantify the energy savings achieved through various SHA mechanisms.

Case Study 1: One Central Park, Sydney, Australia (30% Energy Reduction)

- **Smart lighting:** Occupancy sensors and daylight harvesting controls in conjunction with LED bulbs reduced lighting-related energy consumption by 70% compared to conventional systems (Saieh, 2023).
- **Smart thermostats and ventilation:** Predictive algorithms and occupancy detection optimized the HVAC system, leading to a 25% reduction in heating and cooling energy use (Saieh, 2023).

Figure 10.3 Klokgebouw, Amsterdam, Netherlands
Source: Ilse 2018

- **Solar panels:** Integration of a 500 kW rooftop solar photovoltaic system contributed to generating 20% of the building's total electricity demand, further reducing reliance on the grid (Saieh, 2023).

Combined, these SHA implementations resulted in a remarkable 30% reduction in overall electricity consumption compared to a standard building of equivalent size and function.

Case study 2: Klokgebouw, Amsterdam, Netherlands (30% Energy Reduction)

- **Adaptive facades:** Dynamically adjusted based on sunlight and temperature, these facades minimized solar heat gain during summer and heat loss during winter, leading to a 15% reduction in HVAC energy consumption (Ilse, 2018).
- **Smart lighting:** Utilizing daylight sensors and dimming features, the smart lighting system reduced lighting energy consumption by 20% compared to traditional systems (Ilse, 2018).
- **Water-efficient technologies:** 10% less water was used, which resulted in indirect energy savings from less wastewater treatment thanks to low-flow fixtures and rainwater collection systems (Ilse, 2018).

Together, these SHA features delivered a 30% reduction in the building's total energy consumption.

Figure 10.4 Bullitt center, Seattle, USA
Source: International living future institute (2023)

Case Study 3: Bullitt Center, Seattle, USA (Net-Zero Energy)

- **Comprehensive SHA integration:** A central nervous system controls all building systems, optimizing energy use through smart lighting, HVAC, ventilation, and appliances. This resulted in a 90% reduction in energy consumption compared to a typical office building.
- **Rooftop solar panels:** Generating 100% of the building's annual electricity needs, the solar array eliminates reliance on the grid and achieves true net-zero energy status.
- **Water conservation:** Rainwater harvesting and advanced water treatment systems reduce water consumption by 95%, minimizing associated energy usage for water pumping and treatment.

Through its holistic approach to SHA, Bullitt center provides a blueprint for achieving complete energy independence and demonstrates the transformative potential of this technology.

Comparative Analysis

The comparative analysis of the three case studies is listed in Table 10.3. All three case studies showcase the impactful potential of SHA for reducing energy consumption in diverse building types. While One Central Park achieves a remarkable 40% reduction

Table 10.3 Comparative analysis

Case study	Location	Building type	Key SHA features	Energy reduction	Key takeaways	Source
One central park	Sydney, Australia	Office building	Smart lighting, HVAC, solar panels	40%	Holistic integration of diverse SHA features yields high energy savings.	Saieh (2023)
Klokgebouw	Amsterdam, Netherlands	Cultural center	Adaptive facades, smart lighting, water-efficient technologies	30%	Adaptive building envelope and targeted SHA measures offer significant reductions.	Ilse (2018).
BullittCenter	Seattle, USA	Office building	Comprehensive SHA integration, rooftop solar panels, water conservation	Net-Zero Energy	Net-zero goal achievable through extensive SHA and renewable energy deployment.	https://living-future.org/case-studies/bullitt-center-2/, (2023)

Source: Saieh, N. (2023); Ilse, L. (2018); International Living Future Institute. (2023)

through smart lighting, HVAC, and solar integration, Klokgebouw emphasizes innovative adaptive facades and water conservation with a 30% reduction. BullittCenter stands out with its holistic approach, achieving net-zero energy through comprehensive SHA, solar power, and water conservation. Key takeaways include:

- **Diverse technologies can achieve significant energy savings:** From simple smart lighting to advanced adaptive facades, each building demonstrates effective solutions depending on context and needs.
- **Renewable energy integration amplifies reductions:** Solar power plays a crucial role in reducing reliance on grid electricity, with One Central Park and BullittCenter showcasing its impact.
- **Water conservation contributes indirectly:** By reducing water consumption, buildings like Klokgebouw and BullittCenter minimize associated energy expenditure for pumping and treatment.
- **Challenges remain:** High initial investment, user adoption, and data security require proactive solutions for wider adoption, especially in developing contexts.

Discussion and Challenges

Despite the encouraging potential of SHA to curb electricity consumption and foster sustainable practices, significant challenges impede its widespread adoption and optimization. Firstly, the high initial costs of installation and equipment can be a major barrier, particularly for those in developing economies. These upfront expenses can outweigh the perceived future benefits for many individuals. Secondly, inherent vulnerabilities in data privacy and cybersecurity associated with smart technologies necessitate robust security measures and user education initiatives. Without trust and understanding, homeowners may be hesitant to embrace SHA. Thirdly, even with advanced systems, user adoption can be hindered by a lack of technical know-how or user-friendliness. Unintuitive interfaces and complex controls can discourage engagement and limit the effectiveness of automation features. Finally, the absence of standardized communication protocols and device compatibility across different brands creates interoperability issues, frustrating seamless integration and hindering broader consumer adoption. Addressing these key challenges through cost-effective solutions, enhanced security protocols, user-centric design, and industry-wide standardization will be crucial to unlocking the full potential of SHA for achieving sustainable electricity consumption in the future.

Future Research Directions

To overcome these challenges and maximize the potential of SHA for sustainable energy use, future research efforts should focus on:

- **Developing Cost-Effective Solutions:** Exploring innovative financing models, open-source technologies, and modular system designs can make SHA more accessible to a broader range of consumers.
- **Enhancing Cybersecurity:** Implementing robust data encryption protocols, secure communication channels, and user education programs can mitigate security risks and build trust in smart technologies.
- **Improving user experience:** Developing intuitive interfaces, user-friendly apps, and voice-activated controls can enhance the usability of SHA systems and encourage wider adoption.
- **Standardizing communication protocols:** Establishing universal communication standards for devices and systems will enable seamless integration and facilitate compatibility across assorted brands.
- **Integrating SHA with building design tools:** Incorporating SHA functionalities into architectural design software will allow architects to optimize system placement, predict energy performance, and design buildings with sustainable energy use in mind.

Conclusion

The architectural integration of smart home automation (SHA) presents a compelling and comprehensive strategy for achieving sustainable electricity consumption within the residential sector. By leveraging its diverse functionalities and mechanisms, SHA offers significant potential to reduce energy waste, optimize grid management, and contribute to mitigating climate change. Addressing the existing challenges through continued research and development can pave the way for widespread adoption and maximize the impact of SHA on building a more sustainable future.

References

Abowd, G. D., Bobick, A. F., Essa, I. A., Mynatt, E. D., and Rogers, W. A. (2002). The aware home: a living laboratory for technologies for successful aging. In Proceedings of the AAAI-02 Workshop Automation as Caregiver, (pp. 1–7).

Agee, P., Gao, X., Paige, F., McCoy, A., & Kleiner, B. (2020). A human-centred approach to smart housing. Building Research & Information, 49(1), 84–99. https://doi.org/10.1080/09613218.2020.1808946

Aghaei, J., and Alizadeh, M. I. (2013). Demand response in smart electricity grids equipped with renewable energy sources: a review. *Renewable and Sustainable Energy Reviews*, 18.64–72. 10.1016/j.rser.2012.09.019.

Alohan, E. O., Oyetunji, A. K., Amaechi, C. V., Dike, E. C., and Chima, P. N. (2023).

An agreement analysis on the perception of property stakeholders for the acceptability of smart buildings in the Nigerian built environment. *Buildings*, 13(7), 1620.https://doi.org/10.3390/buildings13071620.

Ateliers Jean Nouvel (2018). One Central Park. Retrieved from http://www.jeannouvel.com/en/projects/one-central-park/

Bitterman, N., and Shach-Pinsly, D. (2015). Smart home–a challenge for architects and designers. *Architectural Science Review*, 58(3), 266–274.

Bullitt, Center. International Living Future Institute (2022). Bullitt Center. https://living-future.org/case-studies/bullitt-center-2/

CE Delft (2020). Klokgebouw: An Energy-Efficient Office Building. https://ce.nl/

Chan, M., Estève, D., Escriba, C., and Campo, E. (2008). A review of smart homes—present state and future challenges. *Computer Methods and Programs in Biomedicine*, 91(1), 55–81.

Dasgupta, A. (2021). Rethinking Smart Home Design: Integrating Architectural Perspectives and Technologically-driven Design Thinking within a Framework. https://vtechworks.lib.vt.edu/handle/10919/105654.

De Silva, L. C.,Morikawa, C., and Petra, I. M. (2012).State of the art of smart homes. *Engineering Applications of Artificial Intelligence*, 25(7), 1313–1321.

DEI Creative in Seattle, W. (2014). Living Building Challenge. Bullitt Center Living Building Challenge Comments. Retrieved from https://bullittcenter.org/vision/living-building-challenge/

Dong, B., Prakash, V.,Feng, F., and O'Neill, Z. (2019).A review of smart building sensing system for better indoor environment control. *Energy and Buildings*, 199, 29–46. ISSN 0378-7788, https://doi.org/10.1016/j.enbuild.2019.06.025. (https://www.sciencedirect.com/science/article/pii/S0378778819309302)

EPRI, Smart Grid Demonstration initiative final update. (2014). Retrieved from https://restservice.epri.com/publicdownload/000000003002004652/0/Product

Helal, S., Mann, W., El-Zabadani, H., King, J., Kaddoura, Y., and Jansen, E. (2005). The gator tech smart house: a programmable pervasive space. *Computer*, 38(3), 50–60.

Ilse, (2018).Klokgebouw Eindhoven. Allesvoorevents.nl. https://allesvoorevents.nl/klokgebouw-eindhoven/

Integrated Retrofit solutions for the multitenant light commercial market (2020). California Energy Commission. https://www.energy.ca.gov/sites/default/files/2021-05/CEC-500-2020-015.pdf

Integrated Retrofit Solutions for the Multitenant Light-Commercial Market. (2021). Retrieved from https://www.energy.ca.gov/sites/default/files/2021-05/CEC-500-2020-015.pdf

International Energy Agency (2023).World Energy Outlook 2023.https://www.iea.org/reports/world-energy-outlook-2023.

Johanson, B., Fox, A., and Winograd, T. (2002). The interactive workspaces project: experiences with ubiquitous computing rooms. *IEEE Pervasive Computing*, 1(2), 67–74.

Karimi, R., Farahzadi, L., Sepasgozar, S. M., Sargolzaei, S., Sepasgozar, S. M. E., Zareian, M., & Nasrolahi, A. (2021). Smart built environment including Smart Home, Smart Building and Smart City: Definitions and Applied Technologies. In IntechOpen eBooks. https://doi.org/10.5772/intechopen.95104

National Renewable Energy Laboratory (2020). Energy Savings from Programmable Thermostats. https://www.nrel.gov/

Saieh, N. (2023). One Central Park/Ateliers Jean Nouvel. Arch Daily. https://www.archdaily.com/551329/one-central-park-jean-nouvel-patrick-blanc.

Stephen, S., Intille. (2002). Designing a home of the future. IEEE Pervasive Computing, 1(2):76-82. doi: 10.1109/MPRV.2002.1012340

Taysheng, J. (2009). Toward a ubiquitous smart space design framework. *Journal of Information Science and Engineering*, 25(3).

U.S. Department of Energy (2023).Smart Home Automation. https://www.epa.gov/newsreleases/us-epas-energy-star-program-advances-energy-management-through-smart-homes.

11 Cellular automata based approach for re-imagining residential highrise design

AR. Neha Sahay[a]

Associate Professor, Acharya's NRV School of Architecture, Bengaluru, India

Abstract

Ever since industrial revolution made standardization and mass production possible, high-rise structures became the go to solution for densely packed development, specially to cater to housing needs in the ever-growing cities. But the approach towards designing of densely planned vertical housing has not evolved much in the last century. When dealing with large no. of units, architects are restricted by time constraints and complexity of managing different floor plates to explore multiple solutions. To what extent could a computational approach in the design of high-density housing help to enable certain qualities in: spatial optimization, spatial perception, low contextual volume impact, user quality in flexibility and customization?

In initial stages of mass housing design, a computational decision-support model can be used to its advantage to present alternatives and to evaluate users' preferences. As a computational tool, Structure of Cellular Automata (CA) structure bears close resemblance to the logic usually applied for mass housing designs. Characteristics of CA like self- organization, neighborhoods, context sensitivity and grid-based design can be utilized to arrive at design decisions like relations between clusters and spatial organizations.

The paper looks at exploring a cellular automata based generative design process to redefine the high rise residential tower typology in a mixed use context. The generative process will aid in breaking the stereotyping and repetitiveness of a residential tower.

Simple grid-based logic and climatic analysis is used to formulate base rules for Cellular Automata which generated 2D and 3D simulation when run using grasshopper scripting.

The generated simulations are quickly able to present multiple alternatives for planning and massing of high-rise residential towers which would otherwise be time consuming and impractical in the traditional approach.

Keywords: Cellular automata, computation codel, generative design, mass customization, mass housing

Introduction

For centuries, diversity and inclusiveness have been the key elements in the fabric of Asian cities where low rise urban villages have nurtured communities in informal by lanes and low rise architecture. The advent of industrial revolution brought forth modular and standardized architecture. In a relentless 'Block Attack', massive towers, slabs and blocks with repetitive housing units, floor plans and facades are invading – scraping away the urban villages that evolved over hundreds of years ("MVRDV - Vertical Village Taipei," n.d.). In a country like India especially, that has a history of low-rise communities formed around a courtyard living typology, connected by bustling streets and alleyways, one might ask: 'Why have we locked ourselves into mono-functional towers (Sisson, 2018).

[a]Neha2602@acharya.ac.in

DOI: 10.1201/9781003571018-11

Rem Koolhaas, in 'Kill the Skyscraper' in content, argues that the skyscraper, originally invented as a social condenser, has slowly devitalized social performances throughout history and ultimately lost its programmatic heterogeneity in the form of a high-rise model made up of the repetition of stacking of the ground floor. While this method is effective in producing density, it results in uniformity in terms of look and experience-an object as devoid of differentiation as possible.

Redefining High Rise Design

While it is widely acknowledged that there is a need to bring in the sense of community in a residential complex, with ever-accelerating planning speed and developers' pressure for quick financial return, architects face a design environment that has little concern for variety and architectural quality in the results of their work (Herr and Kvan, 2005). When dealing with large no. of units, architects are restricted by time constraints and complexity of managing different floor plates to explore multiple solutions. There have been research and attempts to break away from the monotony of a high rise design. Sky village by MVRDV/ADEPT architects is a 116 m tall mixed use tower, based on a 60 sqm module arranged around the central core of the building. The grid-size is 7.8 7.8 m which combines size requirements for parking with those for a small housing unit or office type (Sky Village in Rødovre/MVRDV, 2008). The tower allows for different configurations responding to unstable markets, flexibility achieved by re-designating these 60 sqm units.

Computational Approach

Generative algorithms like Cellular automata, L systems, Shape grammars, Genetic algorithms are increasingly being used by researchers to arrive at solutions which were not really possible through traditional design approach. A computational decision-support model is able to present rich alternatives and to evaluate users' preferences for early stages of mass housing design (Dinçer et al., 2014). Once the initial parameters are set, computational tools can act as aids to a designer or an architect to create quick iterations. Interaction between multiple parameters like mass customization, non-repetitive, flexible unit and grid size, vertical and horizontal spatial quality, can be considered and optimized in a highly efficient manner.

Cellular Automata

As a computational tool, structure of cellular automata (CA) structure bears close resemblance to the logic usually applied for mass housing designs. Characteristics of CA like self- organization, neighborhoods, context sensitivity and grid-based design can be utilized to arrive at design decisions like relations of housing block placements, spatial and façade organizations (Dinçer et al., 2014). At Site level CA can be used to evaluate placement of building blocks on site and studying fitness and incompatibility between multiple blocks. While at spatial level it is useful for floor plan generation based on proximity matrix, orientation, placement of cores.

Various research shows that CA have been used as a generative design strategy that creates a design framework for a contemporary housing. The study done by Lee and Kim uses CA as a generative design strategy that creates a design framework for a contemporary house based on spatial configuration of a Traditional Korean House Han-Ok. Based on the rules extracted through spatial analysis of a Hanok (Figure 11.1), their study used CA simulation to offer numerous possibilities that accommodated today's life- style while keeping the core relationships as defined in the rule set (Figure 11.2) (Lee and Kim, 2016).

	Internal Cells
	*Bang** Typical rooms. Function to be determined in later design stage
	*Maru*** Corridor or a balcony or a veranda
	*Madang**** Interior/exterior open space with or without roof. Must be open to sky
	External Cells
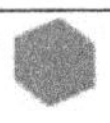	*Entry* Entry to building. Same level as *Sarang-chae*
	Null-A Open space that is subject to be adjacent to *Maru*** or *Madang****
	*Null-B* Empty space by building setback requirement
	*Obstacle* Physical obstacles to avoid

Figure 11.1 Cell rules for CA algorithm
Source: Lee and Kim, 2016

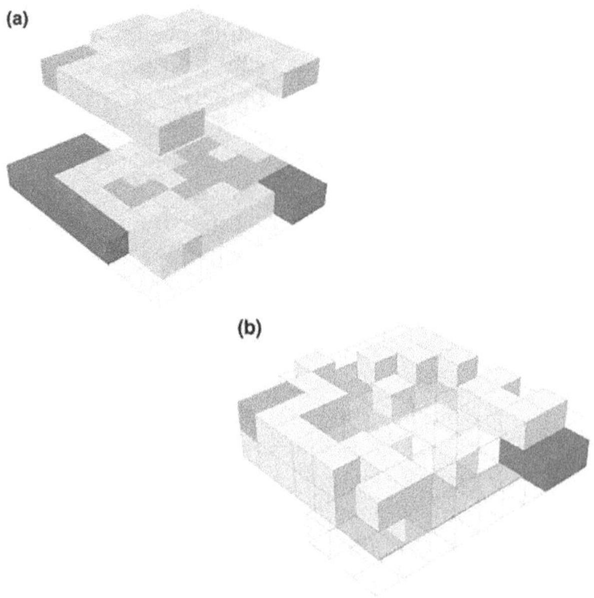

Figure 11.2 (a) Simulation result for 1st and 2nd floor. (b) Morphed 3D massing model
Source: Lee and Kim, 2016

Methodology

The study aims at defining a generative methodology based on CA and testing it by taking core parameters (site area, no. of units, no. of floors etc) of an existing residential project. Grasshopper script was designed to generated iterations which act as decision support models in initial stages of design.

Selection of Test Project

The selected project is Brigade Signature tower which is a 18 acre mixed use development on old madras road in Bangalore (Figure 11.3). By virtue of its location, the medium size, high density development has already ensured its place as one of the successful mixed use projects. Also, since it is one of the recent mixed-use developments, it becomes a perfect project for testing computational design methodology which could then be applied to later projects.

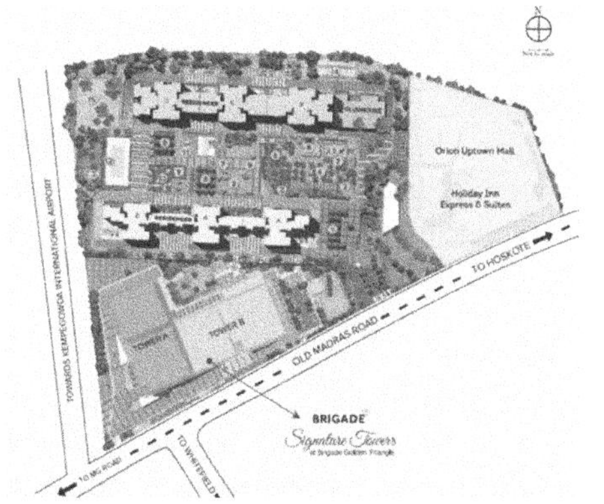

Figure 11.3 Master plan-brigade signature towers
Source: https://brigadesignaturetowers.in/

Master Plan Zoning

For the purpose of design, the project requirement, allocated land areas and achievable built- up area were kept same as the built project. This allows us to focus more on the designing part rather than the area calculation and market feasibility which is outside the scope of the study. But the zoning was reworked based on the issues identified in the project study. (Figure 11.4).

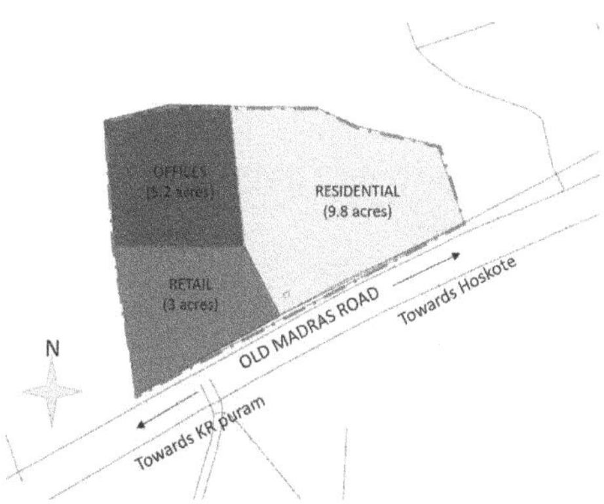

Figure 11.4 Reworked zoning
Source: Author

Defining Cellular Automata Model

A cellular automata Model has it base in two starting parameters: Neighborhood and rulestypes of cellular neighborhood:

Von Neumann Neighborhood - Where definition of neighbors is based on the concept of shared edges. So, a single cell will have four neighbors (Figure 11.5).

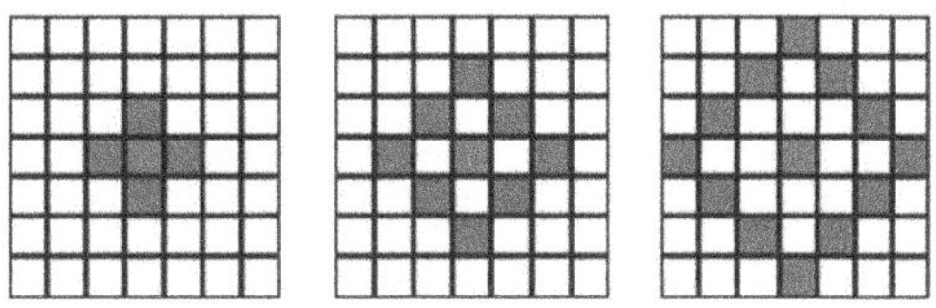

Figure 11.5 Von Neumann neighborhood
Source: Giabbanelli et al., 2019

Moore's neighborhood - Where the definition of neighbors is based on the concept of shared vertices, thus a single cell will have eight neighbors (Figure 11.6).

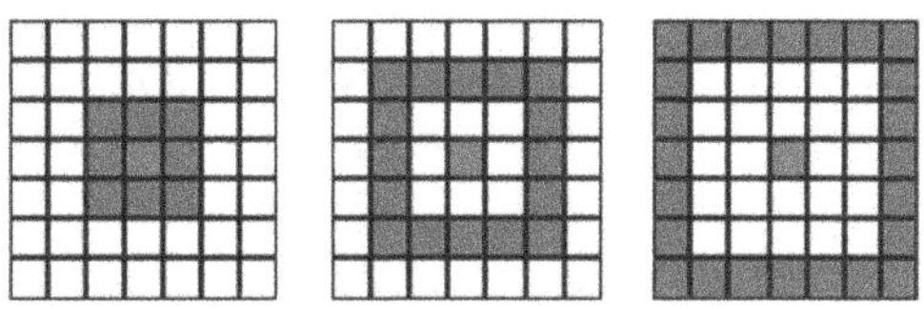

Figure 11.6 Moore neighborhood
Source: Giabbanelli et al., 2019

We can also see how the neighboring cell pattern changes with generation.

The concept of Von Neumann neighborhood was chosen for the study since it is based on shared edges instead of vertices, which gives better control over the adjacency conditions- whether a cell has an external edge for ventilation and light or all edges are shared by other cells.

Defining Rules

As we see in Conway's Game of life simulation, simple rules can give rise to complex models. Also keeping the rules simple and clearly defined helps in negating any anomalies that the simulation may throw up due to multiple dependencies. So just 3 types of cells were defined for 2D simulations based on external edges and lighting conditions required for spaces like cores, corridors and habitable areas (Table 11.1).

Test Simulations-Site Grid with Different Pre-Conditions

Keeping the grid size 6 m 6 m, two approach were tested for population of cells on the site.

1. Identifying the buildable parts of the site and populating cells on the defined region (Figure 11.7).
2. Identifying the tentative building center line (curve) as per designers input and site orientation. Then based on curve attractor logic, the simulations gives the building footprint, clusters and zoning within the building (location of cores and common green areas) (Figure 11.8).

Approach 2 gives greater flexibility to designers in terms of building orientation. Any changes in the curve automatically generates new cluster options.

Generation of Clusters-2D Simulation

Step 1-Running CA algorithm with the defined curve as the attractor curve. The Distance parameter set as 18 meters (three grids) on either side. Running 2D simulation based on defined CA rules. Ten seeds of random nos. were tested generating different cluster patterns guided by the rules (Figure 11.9).

Step 2-Three of the cluster patterns were selected and a ladybug direct sun hours analysis was run on each of them to understand the sunlight penetration in the central common area during the months of October to march. Cluster Pattern 3 showed maximum sun penetration to the central common area. So Pattern 3 was selected for further analysis (Figure 11.10).

Step 3-Assuming 20 storey high towers, Sunrays analysis using Ladybug was done on Pattern 3 massing to determine the height of the south towers so that winter sun can reach the towers behind at different time of the year based on sun angle (Figure 11.11).

Step 4-3D CA algorithm was run on cluster pattern 3 for internal zoning. Identification of core locations based on vertical repetition and clustering of red cells (Figure 11.12).

Step 5-Once the cores were identified, the populated cluster tallying with maximum no. of core location was chosen for 3D population of cells creating the tower.

Generation of Tower Form-3D Simulation (Figure 11.13)

Step 6-Once the core location were identified and fixed using the concentration of red cells, The rules were further tweaked for 3D population of habitable cells. Any live cell with three live neighbors turned green otherwise they turned yellow.

Step 7-Populating a 3D CA of yellow cells for 10 generations with a cell size of 6m 6m 6m gives rise to a 20 storey tower where each floor plan is repeated for two consecutive floors thereby creating double height terraces and balconies.

Step 8-Once the tower cells is populated the height of the tower on the southern side is reduced based on sun rays analysis done in step 5.

In the 3D simulation it becomes clear that the pattern being generated is a repetitive pattern to an extent which works in favor of having fixed cores, structural members

Table 11.1 Rules defined for 2D simulations-population of cells at site level

	DEAD CELLS/CELLS WITH STATE 0		LIVE CELLS/CELLS WITH STATE 1-	
Neighbourhood condition	No. of neighbours =4	No. of neighbours=1	No. of neighbours= 2 or 3	
Generation n			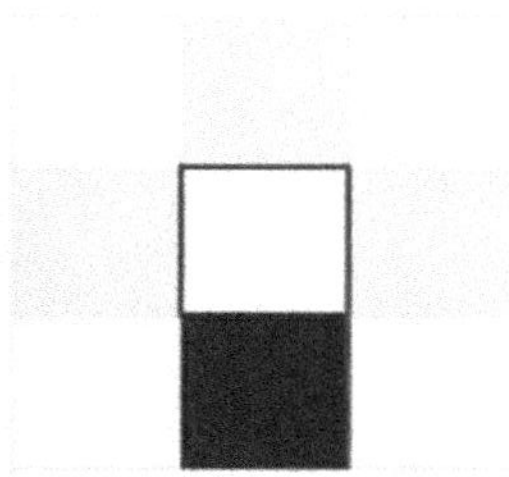	
Generation n+1				
Typology	Cells not necessarily requiring a ventilated space, or an external wall.	Cells with minimal built space in the surrounding.	Cells with atleast one external edge for light and ventilation	
Function	Corridor and Cores	External open spaces	Habitable areas/Rooms, Terraces and Balconies	

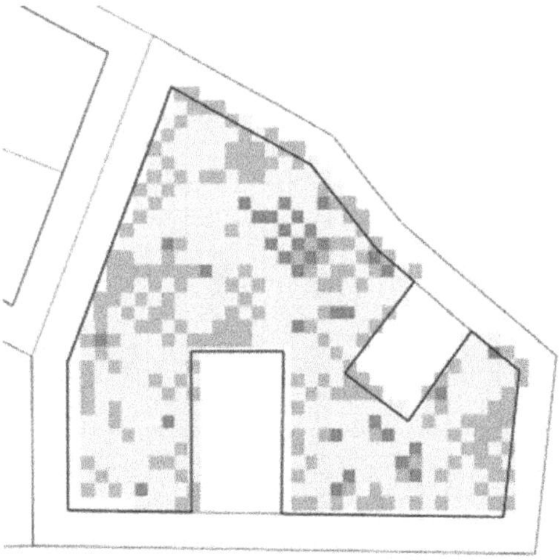

Figure 11.7 Population of cells on buildable areas
Source: Author

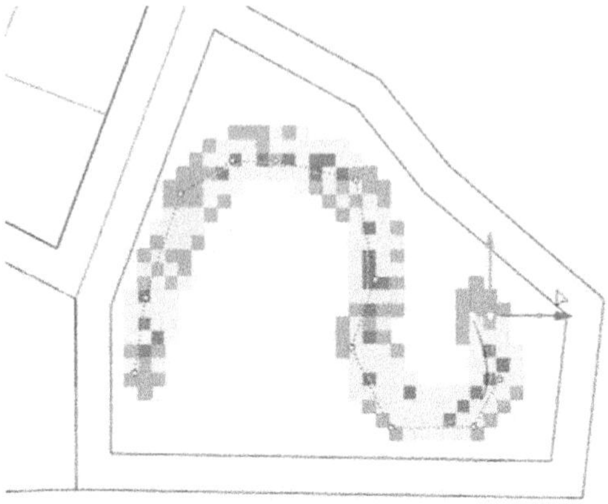

Figure 11.8 Population of cells around curve attractor
Source: Author

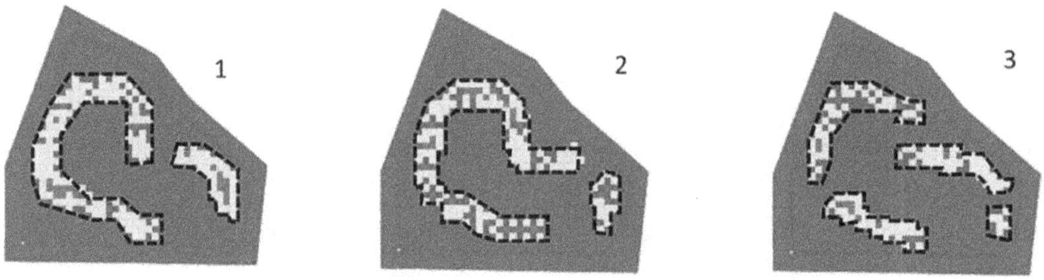

Figure 11.9 Step 1-Generation of clusters
Source: Author

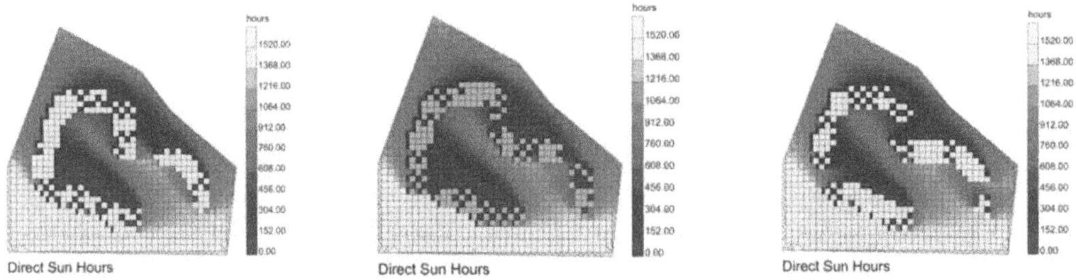

Figure 11.10 Step 2-Ladybug analysis on three identified clusters
Source: Author

Figure 11.11 Step 3-Sunrays analysis on cluster three to determine south side tower height
Source: Author

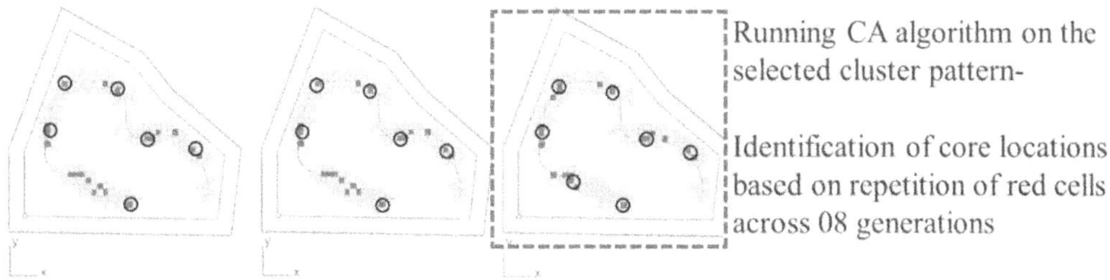

Figure 11.12 Identification of cores
Source: Author

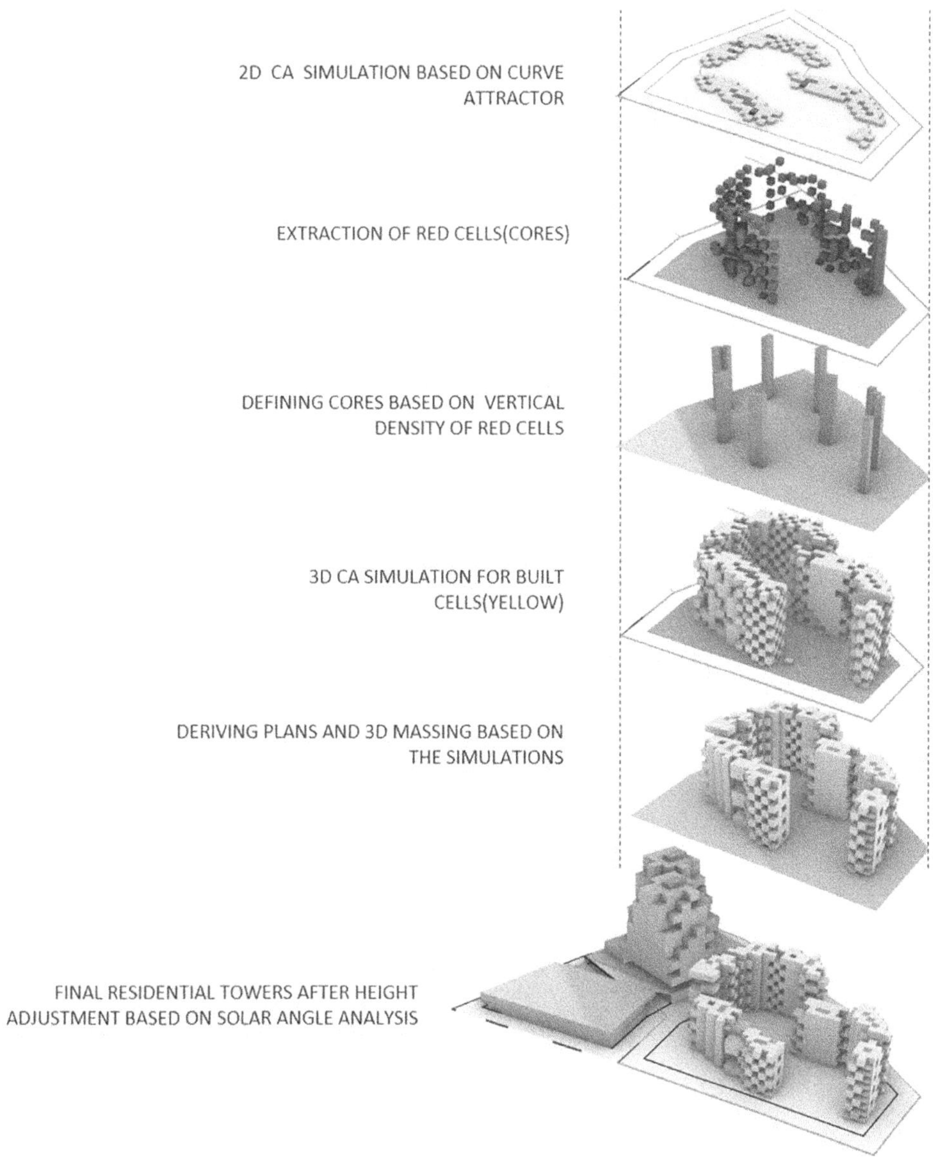

Figure 11.13 3D simulation of residential tower illustrating steps 6, 7, and 8
Source: Author

and plumbing ducts. The 3D output provides a base from which the plans can be derived from. Since each grid is a 6 m x 6 m grid, multiple sized units can be carved out in various combinations.

Conclusion

The study concludes with establishing that computational design algorithms, like Cellular Automata models can successfully be applied to generate design solutions which break the stereotypes of existing design language. Simple starting rules based on adjacency, light, ventilation coupled with climatic analysis softwares can create informed designs for high rise buildings which are visually exciting as well.

The methodology developed can easily be adapted to different sites, area requirements, grid size and tower heights thus making it universally applicable solution for solving similar design problems, still giving unique results.

References

Dinçer, A. E., Çağdaş, G., and Tong, H. (2014). A computational model for mass housing design as a decision-support tool. *Procedia Environmental Sciences*, 22, 270–279. https://doi.org/10.1016/j.proenv.2014.11.026.

Giabbanelli, P. J., Freeman, C., Devita, J. A., Rosso, N., and Brumme, Z. L. (2019). Mechanisms for cell-to-cell and cell-free spread of HIV-1 in cellular automata models. In Proceedings of the 2019 ACM SIGSIM Conference on Principles of Advanced Discrete Simulation. Presented at the SIGSIM-PADS '19: SIGSIM Principles of Advanced Discrete Simulation, ACM, Chicago IL USA (pp. 103–114). https://doi.org/10.1145/3316480.3322886.

Herr, C. M., and Kvan, T. (2005). Using cellular automata to generate high-density building form. In Martens, B., and Brown, A. (Eds.), Computer Aided Architectural Design Futures 2005. Berlin/Heidelberg: Springer-Verlag, pp. 249–258. https://doi.org/10.1007/1-4020-3698-1_23.

Lee, Y., and Kim, S. H. (2016). Algorithmic design paradigm utilizing cellular automata for the han-ok. *Nexus Network Journal*, 18, 481–503. https://doi.org/10.1007/s00004-016-0292-x.

Sisson, P. (2018). Why do all new apartment buildings look the same? [www. Document]. Curbed. URL https://archive.curbed.com/2018/12/4/18125536/real-estate-modern- apartment-architecture (accessed 3.27.22).

Sky Village in Rødovre/MVRDV (2008). [www. Document], ArchDaily. URL https://www.archdaily.com/8649/sky-village-in-rodovre-mvrdv (accessed 3.27.22).

12 Exploring psychological safe design of jury space in architectural design education

Kamakshi Tomar[1,a] *and Suruchi Sharma*[2,b]

[1]Principal Architect, NK Associates, Indore, India

[2]Assistant Professor, REVA University, Bengaluru, Karnataka, India

Abstract

Juries are the instrument used to evaluate design problems solved by students in architecture, enhancing students' understanding and implementation of solutions. These evaluations occur at various stages of the course, culminating in Thesis Juries in the final year. Juries shape student perceptions and are challenging due to factors like delivery, presentation, and physical surroundings. Impartial, proportionate, and fair juries are essential for effective results.

The purpose of this paper is to identify these issues that the students may face during architecture juries and to help create a safer, more favorable environment for them to present their work, during the process of critical evaluation of their work. In the present paper, the authors address the design studio problem which students have worked on from their III Semester of M. Arch in Architecture Education. The Problem under consideration for this research paper, focused on enhancing psychological safety in architectural educational spaces.

Psychological safety being the core idea, and the architectural jury being a space full of anticipation and potential, the authors have attempted to account for all the participants and restructured the layouts of a jury area to facilitate a safe space for presentations, with the central idea of psychological safety for all: the presenters, evaluators, facilitators and supporting personnel, as well as, the spectators, so that all the stakeholders benefit from the jury process.

The major emphasis of the experiment highlights a much-needed shift in the culture of design juries. Such a shift will enable students to be an active part of designing their own jury spaces; it will enable facilitators and evaluators to focus on the project being displayed and the jury's role becomes the mainstay of the goal of outcome-based learning.

Keywords: Architectural design education, architecture design jury, design jury space, psychological safety

Introduction

Educational assessment is a systematic method for enhancing student learning and program refinement, considering knowledge, skills, aptitude, attitude, and conviction. It involves professional evaluations every semester, with impartial external and continuous internal assessments, especially for architecture and engineering courses (Flynn et al., 2023).

Assessment is a crucial process in professional education in aiding the student's career trajectory by encouraging learning, providing feedback, and documenting competency for further skill development. Ongoing and continuous assessment grades, ranks and validates the students' abilities, enabling the validation, certification and licensing processes for professional practice, and the benchmarks to be established for standards (Aderonmu et al., 2017).

[a]ar.kamakshitomar@gmail.com, [b]suruchi.sharma@reva.edu.in

DOI: 10.1201/9781003571018-12

The knowledge of architecture involves observing and contextualizing architectural design solutions within historical periods, communities, and attitudes. Students must face design studio problems to develop effective projects and deliver justifications during jury presentations. This understanding helps them work on final projects and present arguments effectively. The ability to present enhances every year and as the final year approaches, the students with their enhanced and strengthened cognitive structure, add more and more details to their explanations and solutions.

With these enhancements, it is important for the students to be a part of the preparatory process of designing their own jury space. Not only can a good jury space improve their presentation but it can also prove to be a safe space for the presenter. It is important to acknowledge the presence of all the other personnel involved in the jury process- the juror, the mentor, the supporters and the audience. Their interactions may be secondary, but are equally important.

Aim of the Study

With the tenets of psychological safety, the paper aims to enquire how a safe space can be created for all in a jury procedure. The Presenter, the juror, the mentors, the support system for the presenter and the spectators. The Juries are carried out within the framework of the Bachelor of Architecture course and beyond too. The inspiration for this study came about by how students view the jury system. A multi-layered methodology is developed to address these problems. It includes a specific number of design suggestions.

Objectives

The study focusses on restructuring of space in a jury to promote psychological safety for all involved. The presenters and Juror, the facilitator/mentor, and the support system of the presenter, and lastly the audience/spectators viewing the jury. A major objective is to include the students in the design of their jury area so they can personalize their presentation as per their thought process and design process.

Limitation of the Study

Research study, analysis and proposals are restricted to a certain number of designs in order to limit the time frame. In this paper, after a thorough analysis of the Jury system in the final year of B.Arch. and the issues surrounding it, a series of recommendations in the form of design proposals is developed with the aim of enhancing the Jury's functioning and increasing architecture students' acceptance of it.

Architectural Jury Setting

The jury system was initially created as a part of arts education and training, and it was later used in 1795 as a model for assessing architecture students. This was one of the customs that the French system at the Paris-based Ecole des Beaux-Arts (School of Fine Arts) created (Salama, 2010).

Since the mid-1980s, numerous articles have been published on architectural design juries, studio teaching methodologies, and education, exploring communication mechanisms, benefits, drawbacks, and potential solutions to improve current practices in architecture and related fields (Salama, 2010).

Characters of Play

The jury is the act. There is a part for every artist. Flynn et al. (2023) emphasizes that presenters vary in terms of their upbringing, backgrounds, thought processes, and educational attainment. Nevertheless, even though each person has a unique fingerprint, they may all be grouped together (Kanso & Alyyoussef, 2016). Everything about students' presentations, from their origins to their design concepts, varies. To guarantee his presence, a member should

Juror	Priority 1	Priority 2	Priority 3
J01	Oral Presentation	Presentation Boards	Model
J02	Model	Presentation Boards	Oral Presentation
J03	Presentation Boards	Model	Oral Presentation
J04	Presentation Boards	Model	Oral Presentation
J05	Presentation Boards	Oral Presentation	Model
J06	Model	Oral Presentation	Presentation Boards

Figure 12.1 Juror's priorities for evaluation of student's work
Source: Santos 2020

consistently possess unique qualities. It is considered that every performer is familiar with these roles. However, unlike, say, design studios where instructors teach students how to draw—there are no classes where students learn how to play their role in a jury. There is even less training available to instructors on how to appropriately perform their function on juries. It takes trial and error for the students to figure out how to handle jury presentations. There is a need for students to be taught that the environment of a jury room is just as important as the juries themselves.

Characteristics of Spatial Arrangement

The exhibition spaces in architecture schools significantly influence the jury's evaluation process. A strong foundation of qualities, including large areas, clear display areas, natural light, efficient ventilation, standardization, and unification of scales, presentation styles, board types, and suspension equipment, is necessary (Kanso & Alyyoussef, 2016).

Santos (2020) found that that students should prioritize their presentation boards and models as shown in Figure 12.1—the visual component of the required output—in order to effectively communicate information about their designs.

Psychological Safety

Figure 12.2 shows the four stages of Psychological safety: Learner safety, contributor safety, challenger safety, and inclusion safety. A distinct human need that is present in all geographic, demographic, and psychographic categories is highlighted in each of these stages. A condition in which one feels (a) included, (b) safe to learn, (c) safe to contribute, and (d) safe to challenge the status quo, without fear of being embarrassed, marginalized, or punished in some way is what is meant by psychological safety (Kim et al., 2020).

Psychological Safe Space in Learning Environments

Elsamman (2021) in their study highlights the positive impact of respecting psychological language in creating learning

Figure 12.2 Four stages of psychological safety
Source: Clark 2020

environments on students' mental health and overall school effectiveness. Here are few ideas from the study that are worth emphasizing:

1. Spaces designed by their users are more accepted and welcomed; schools that are viewed as social spaces do better than those that are viewed as instructional spaces alone.
2. The school's performance is significantly influenced by its behavioral and aesthetic features, which are enhanced by respecting the diverse backgrounds and cultures of its users.

Figure 12.3 emphasizes on the questions which can promote psychological safety for students. Questions focussing on Hyper visibility of students and how it hinders their confidence, the need and provision of breakout rooms and reset rooms is addressed too. The access to the learning spaces and the scope of movement being a key aspect to be detailed and sensory stimulation within a learning space are questioned with the psychological safety and comfort of students in mind.

Methodology

Examining firsthand the individual experiences and insights from our graduation final year project jury. The primary data for the study were gathered by questionnaires, observations, focus groups, and oral interviews; a survey research design strategy was used in the methodology. The literature, archives, official reports, and records were the sources of secondary data. The design studios, where participants included teachers and students from different schools, made up the sampling frame for the survey as well.

Aspects of ILE Design that have implications for the psychological safety of students	Questions to consider
Hyper=Visibility	Are there spaces for students to be educated where they are not visible to many other students? Are students able to safely engage in self-regulatory behaviours without being being visible to many other students?
Breakout and reset rooms	Are there secluded 'reset' spaces that students can elect to go to when they need them? Is there provision for a range of quiet indoor space that serve different purposes? Are there areas that can be filled with materials that support sensory requirements?
Access to and movement within the learning spaces	Is there a way to enter/exit each of the learning spaces without having to walk through other classes or in front of group of students? Is the accessible toilet private enough for students to have their needs met with a carer without being visible to all? Do leaders have an office/meeting space that is private for when they are addressing personal issues with students, teachers and parents?
Distractibility and sensory stimulation	How are objects and materials used that minimise distraction, discomfort, and distress? How can consideration be given to acoustics, colour choice, lighting, reflective surfaces, wall surfaces and flooring?

Figure 12.3 Questions to promote psychological safety for students
Source: Jennifer, 2021

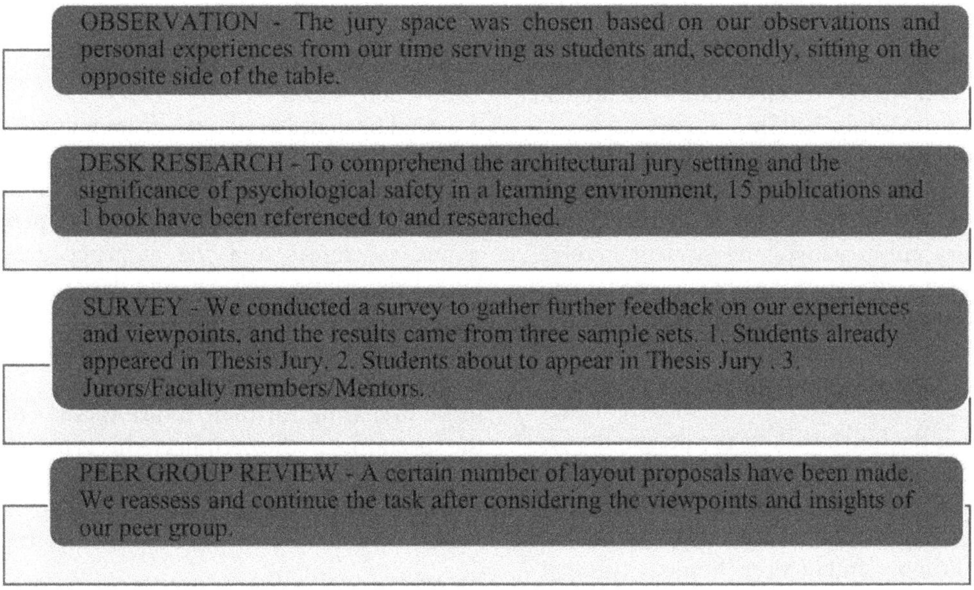

Figure 12.4 Observations and research path followed
Source: Author

The Design Jury's Acknowledged Concepts

Students suggest that jury format modifications involving student participation are most effective. Design studios with disrupted power dynamics are more prosperous. The University of York, Sheffield, and Leicester's CUDE project explored

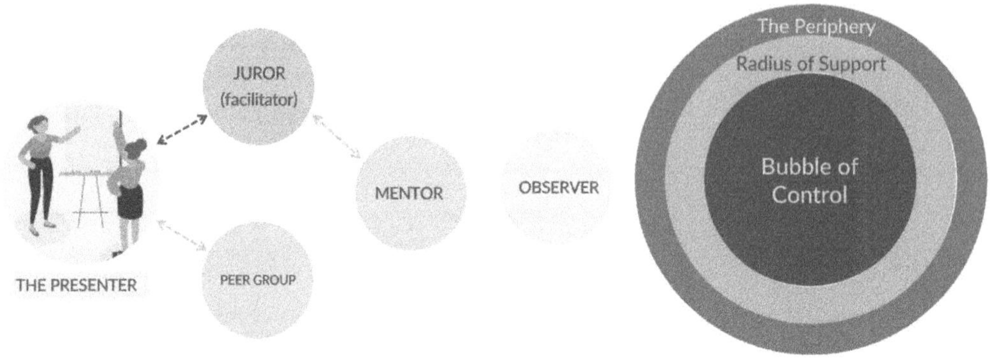

Figure 12.5 Three distinct interactions and concept circle
Source: Author

student-led peer reviews. Sheffield University conducted role-playing reviews to enhance student involvement and knowledge, establishing a review panel to discuss student designs after interviewing real-life examples. After meeting and interviewing 'real-life' examples of these to better understand their perspectives, they established a 'review panel' to talk about the student designs (Milton, 2003).

Alternate review processes enhance student participation and learning capacity. This study emphasizes the need for a change in jury culture, involving student involvement in setting jury settings, project focus, and learning as the final product.

Research Based Analysis and Proposals

Within the framework of the design jury setting, there are three distinct but mutually reinforcing interactions, as shown in Figure 12.5.

- Primary interaction: Student (presenter) → Juror (facilitator)
- Secondary interactions:
  - Student → Peers/friends/support system
  - Juror → Faculty coordinator/student mentor
- Tertiary interaction: The participants of the primary circle of interaction with the observers/audience.
- Inspired by the concept of circle of control, influence and concern by Stephen Covey, the jury space and the three distinct interactions can turn to segregated zones just adjacent to one another.
  - Bubble of control: Primary interaction
  - Radius of support: Secondary interaction and
  - The periphery: Tertiary interaction

Even though a thesis jury involves a close interaction between the juror and the presenter, spectators and the audience gain knowledge about each person's design process. The process being shared is always educational, whether it is with students beginning their architecture journey or those preparing for their future thesis. For the spectators in the periphery, this arrangement therefore counts as inclusion safety and learning opportunities.

Proposals – Possible Outcomes

Studio Arrangements

The idea of providing multiple options for studio arrangements for the jury ensures that the jury's progression is as per the student's comfort and their design journey is portrayed effectively. Figure 12.6 emphasizes on the arrangement for primary, secondary and tertiary interactions.

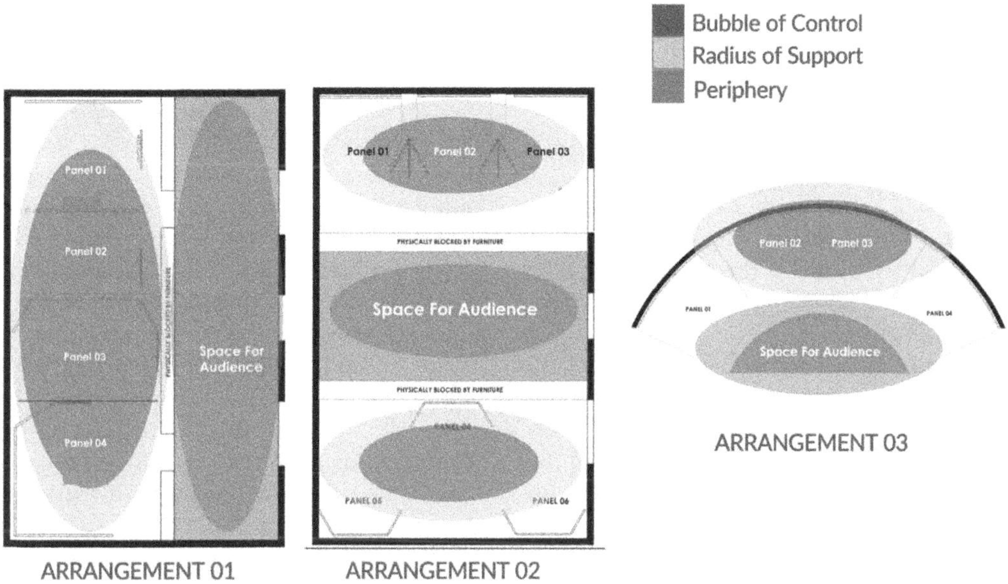

Figure 12.6 Types of possible studio arrangements
Source: Author

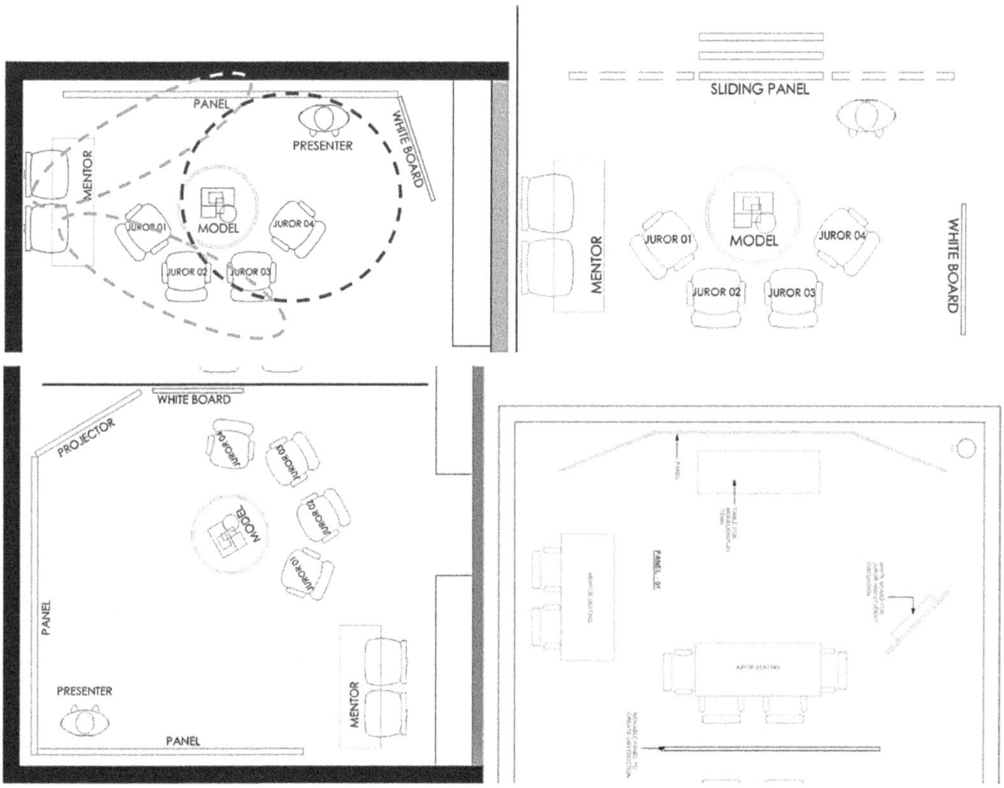

Figure 12.7 (a), (b), (c), (d) Types of possible panel designs
Source: Author

Panel Design

Jury process can be made into a preferred one through the above The Panel where the individual juries are to be conducted are elaborated in Figures 12.7. To provide a panel, a white board, seating for the jurors, the individuals involved in secondary interaction as well as a dedicated space for the model is the idea to make the space comfortable and sensible for a jury proceeding.

Figure 12.7(a) shows the option of having sliding panels where the students are able to dictate the way they share their design journey. The model is placed in the center of the jurors to be able to explain better. It also gives them a way to emphasize their narrative with a unique expression.

In Figure 12.7(b), the panel is fixed, with ample space in front of the seating area for the primary interaction and the model in the middle facilitates the conversation.

Figure 12.7(c) shows a combination for a hybrid presentation with arrangement for sheets on the panel, a white board for discussion, projector for digital presentation and space for mentor/support for the student, jurors, centrally placed model and ample space for the presenter to move and explain their ideas.

Further, Figure 12.7(d) shows a digital heavy arrangement where the panel and model are placed adjacent with scope to move around and discuss, it also showcases a screen to be able to obstruct the jury area from potential noise points.

The above mentioned are options for the jury areas which can be arranged in several ways as per the comfort and preference for the evaluation process.

Conclusion

The design jury is a crucial part of architecture schools' curriculum, focusing on evaluating design projects after the semester. Faculty members participate in final juries, with students facing jurors, studio tutors observing, and an audience of other students.

An admirable approach is one that recognizes the involvement of the presenter in designing their own jury arena with variety of panel arrangements, medium of presentation, the type and shape of panel, etc. for different project contexts, rather than using a single model to address all scenarios. It is advised that the jury space be designed to allow specific space for the mentor, the juror, and the presenter together with their presenting tools. Most significantly, the positioning of the audience should be created so as to allow them to participate visually without interfering with the main engagement.

This research paper's central idea leads to the conclusion that, even though there are numerous ways to display projects in which students participate, it is crucial to provide a psychologically safe environment.

References

Aderonmu, P. A., Awoyera, P. O., Sholanke, A. B., and Erebor, M. E. (2017). Professional methods of assessments in architectural design projects: a focus on the relevant parametric measures in selected Nigerian universities. *Cogent Social Sciences*, 3(1), 1328793.

Clark, T. R. (2020). The 4 stages of psychological safety: Defining the path to inclusion and innovation. Berrett-Koehler Publishers.

Flynn, P., O'Connor, M., Price, M., and Dunn, M. (Eds.). (2023). Rethinking the Crit: New Pedagogies in Design Education. Taylor & Francis.

Jennifer, C. (2021). Psychological safety in innovative learning environments: planning for inclusive spaces. *International Journal of Inclusive Education*, 28(5), 688-704.

Kanso, L., and Alyyoussef, M. (2016). Problems of design juries in schools of architecture. *Architecture and Planning Journal (APJ)*, 23(2), 4.

Kim, S., Lee, H., and Connerton, T. P. (2020). How psychological safety affects team performance: mediating role of efficacy and learning behavior. *Frontiers in Psychology*, 11, 1581.

May Elsamman. (2021). *Psychological language in architecture and its impact on the design standards and criterion of educational buildings*. [M. Arch Thesis]. Helwan University.

Milton, C. (2003). The Jury's Out: a Critique of the Design Review in Architectural Education, (pp. 4–5).

Salama, A. M. (2010). Student perceptions of the architectural design jury. *International Journal of Architectural Research Archnet-IJAR*, 176.

Santos, F. C. (2020). Examining the value of jury critique for architectural design studio courses. *MUHON: A Journal of Architecture, Landscape Architecture and the Designed Environment*, (7).

13 Reimagining the SDGs: A pedagogical perspective through curriculum model

Roopal Deshpande[1,a] *and Lakshmi Rao*[2,b]

[1]Principal, Smt Manoramabai Mundle College of Architecture, Nagpur University, Nagpur, India

[2]Professor, Smt Manoramabai Mundle College of Architecture, Nagpur University, Nagpur, India

Abstract

Sustainable Development is a key term addressed by every sector of the contemporary world. The definition of the term has been ever-evolving since its inception in the previous century. Thus, establishing the fact that the approach to understanding the SDGs needs to be open-ended. Amongst the United Nations's seventeen SDGs, ten number of goals depict direct responsibility and the primary role of architects and the role of the architect for the remaining seven goals needs to be identified through inter-disciplinary research. Architecture Curricula across the world bring to the table a fine blend of multi-disciplines, empowering the architect to work seamlessly across academic boundaries. Hence, the profession shoulders a bigger responsibility to transcend into other disciplines for the achievement of SDGs. Such research will open new avenues and paradigms for architecture professionals in the ever-changing global perspectives. This paper deliberates on the role of the regulating body, practitioners, academicians, and researchers in the built environment in the interpretation of the SDGs. The paper discusses Sustainable Development & SDGs, architecture pedagogy, and outline of curriculum development for B. Arch. program. The paper builds a 'curriculum model' to incorporate the theories of SDGs into architecture pedagogy. The curriculum model aims to facilitate a thought process to stimulate research and generate new theoretical concepts in architecture. The authors have analyzed the guidelines of Council of Architecture, New Delhi for framing the syllabus of B. Arch in India concerning the content on sustainability and its integration into architecture teaching and learnings. The paper concludes by discussing examples of interpretation of SDGs conducted in the master's Urban design studio. The future research can be the implementation of the model in the curriculum through a pilot study and its findings can strengthen the effectiveness of the proposed model.

Keywords: Architecture pedagogy, curriculum model, sustainable development

Introduction

The year 2020 was a watershed year for the planet Earth when COVID hit the world. It was already grappling with poverty, recession, and much more when the virus hit mankind, causing catastrophe. The United Nations initiative to implement the Sustainable Development Goals (SDGs) was entering its fifth year. Some nations were ahead in their progress, while others were still trying to understand the nuances of the target set by the UN. Yet, mankind stood as one, and helping hands were everywhere. Three years later, we witness more determination to march quickly towards 2030 and the achievement of the SDGs.

All the academic domains are contributing in their own way to achieving the goals and targets set by the UN. In this scenario, this paper addresses the role of one particular domain on which many of the SDGs can pivot, and that is the domain of architecture. The architect today is the fulcrum of a sustainable world. Architects' brief extends beyond the realms of the built environment.

[a]roopal76@gmail.com, [b]venkylaks@gmail.com

DOI: 10.1201/9781003571018-13

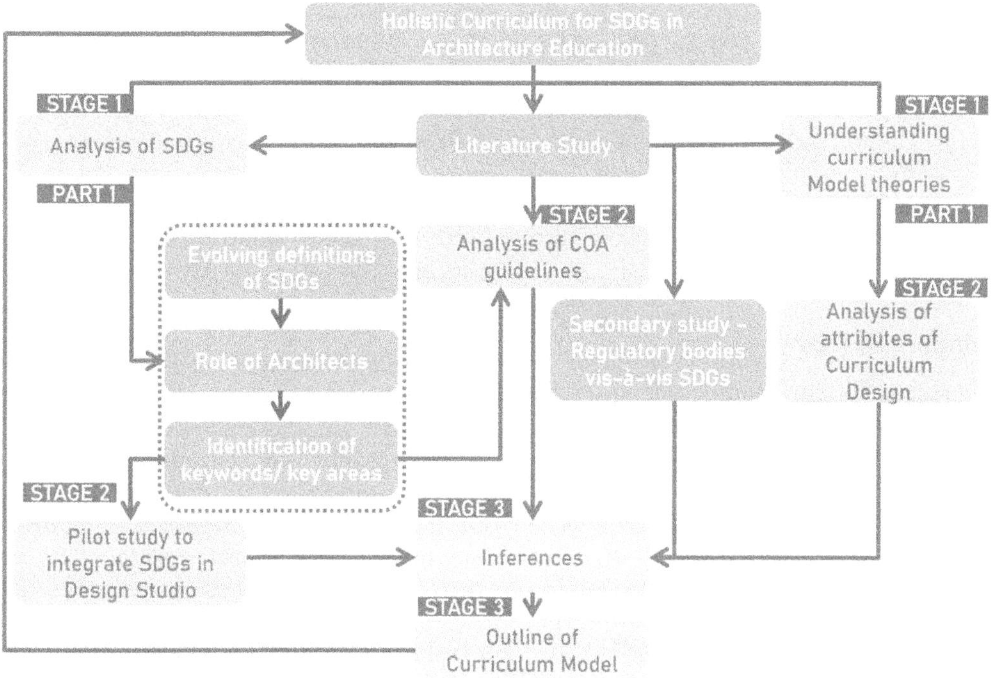

Figure 13.1 Research methodology
Source: Author

Architects, through their academic domain, can significantly contribute to most of the targets of the SDGs, either at a 'conceptual level' or at a more 'tangible level'. This brings us to the major crossroads—the SDGs, the architect-teacher, the student of architecture, and the curriculum transected in the architecture schools. If we believe that the implementation of the SDGs is an urgent need to save the planet Earth from further destruction, the answer definitely lies in architecture.

This paper analyses the role of architects in all the SDGs apart from the distinct goal numbers 6, 7, 9, and 11, where the architects have a primarily responsible role to play. The SDG's as mentioned by United Nations's Transforming Our World: The 2030 Agenda for Sustainable Development, 2015. The authors investigate council of architecture guidelines for preparing the syllabus of all the institutes of the nation. The paper conducts a literature study of the syllabus vis-à-vis sustainable development to understand the status of implementation of SDGs in the theory or design studies. The paper concludes with a proposal for a holistic curriculum model for architecture education towards the fulfilment of the SDGs.

Research Methodology

The methodology consists of three stages (Figure 13.1). Stage I consists of a literature review, conducted in two parts: Part 1 of the literature review focuses on the identification of key areas of sustainable development goals where the architecture profession can contribute.

Part-2 of the literature review deals with understanding of a curriculum model

Stage II deals with the study of the COA (Council of Architecture, 2020) (Minimum Standards of Architectural Education) Regulations 2020 document to examine the inclusion of the identified key areas of the

SDGs. At this stage, a pilot study to demonstrate the integration of SGDs into design studios is discussed. This stage further deals with the analysis of the attributes of the curriculum model. Stage III is oriented towards drawing inferences and proposing an outline for a holistic curriculum model for fulfilling the SGDs through architecture pedagogy.

Literature Review

This section briefly studies the historical overview of the formation of the SDGs. The literature review studies the United Nations document 'Transforming Our World: The 2030 Agenda for Sustainable Development' (United Nations, 2015). The authors have also studied papers pertaining to the role of architecture curricula in sensitizing students to environmental issues. Finally, the study of attributes of a 'curriculum model' is conducted through a literature review.

Historical Overview of Evolution of SDGs

Doric (2008), discusses the background of concept of sustainable development. He mentions that it traces back to 1798 when Malthus raised the alarm to the way exponential population growth of cities were happening for economic reasons. The migration of people to cities was resulting in deterioration of the environment. In 1968 Carson, an ecologist pointed out the concern of use of pesticides for agriculture will not only harm the human health, but shall have an sever adverse impact on the environment. During the same year, European nations met at Rome to discuss the importance of being united to deal with environmental issues through the book 'Limits of Growth'. By the year 1979, it was confirmed that environmental issues are not confined to certain regions and each profession had a role to play to combat the problem. 1987 was the watershed year when the term sustainable development was coined (Doric, 2008).

The goals for achieving Sustainable Development were drafted in Agenda 21 in four sections. In 2000, the millennium goals were drafted into eight points. Finally in 2015, the 17 sustainable development goals were formulated (Klarin, 2018). Thus, for the first time, the discussions pertinently directing to architects, planners and designers with setting goals for Clean water and sanitation, Affordable and clean energy, industry innovation and infrastructure, and sustainable cities as goal number 6,7,9 and 11 respectively were penned down.

It is important to critically evaluate the preparedness of architects to contribute towards the global mission. One of the objectives of the paper is to study the existing work carried out in this direction. (Tyana, 2020), has analyzed the syllabi of ten globally prominent architecture schools to conclude that the technical subjects like building sciences and construction technology have assimilated the aspects of sustainability. Santini points that a fragmented approach is witnessed in practice as sustainable development is interpreted largely to construct energy efficient structures (Tyana, 2020).

A similar study of top 25 architecture institutes in India with NIRF rating is conducted by Sharma et al. (2023). The authors conclude that concepts and principles of sustainable development are included in the syllabus, yet it falls short of holistic integration of the SDGs (Sharma et al., 2023).

The analysis of the literature indicates that there is a strong push need to build a holistic curriculum model to incorporate sustainability and its development goals in the present architecture curriculum.

All studies and researches conducted so far have focused on incorporating the concept of sustainability in design, building technology and humanities (Tyana, 2020). However, the inclusion of the all the aspects of SDGs has not yet been studied in depth. Having established the premise that the architect plays an important role in each of the goals, the authors, in this paper, have crafted a basic, simplistic curriculum model which will engulf the present curriculum and thus will lead to more sensitivity amongst architecture students to the real meaning of sustainability.

One of the most popular models of curriculum development has been the one proposed by Ralph Tyler in 1940 (Tyler, 1949). A simple four step model which is relevant even in today's educational practice, the model in simplistic terms describes the purpose, identifying learning experiences related to the purpose, organizing the experiences, and evaluating the purpose. Considered by many to be one of the strongest models of curriculum development (Ornstein, 1993), the authors have taken some of the guidelines suggested by the model. The purpose is clearly to re-imagine the SDGs through pedagogy, identify suitable learning experiences at every stage of the architecture curriculum leading to organisation of the experiences, evaluation, and a robust repository (Sethi, 2014; Tyler, 1949).

Analysis

The analysis is bifurcated in two sections. In Section 1, authors conducted detail study of the 17 SDGs and total 169 targets spread across the SDGs. The study identifies the contribution of architects as 'tangible' i.e. with evident role and as 'conceptual' which have potential areas of future research, particularly interdisciplinary research (Table 13.1).

Table 13.1 Analysis of SGDs

SDG goal	Key-term/key area	Key-term reframed for searching in COA regulation 2020	Role
1-No poverty	Social protection system	Inclusive public spaces	Conceptual
	Basic services	Affordable building service	Tangible
	Ownership and land control	Affordable housing	Tangible
	Appropriate new technology	Appropriate technology	Tangible
	Climate related extreme events	Disaster resilient structures	Tangible
	Resilience to environmental shocks and disastrous	Disaster resilient structures	Tangible
	Mobilizing natural resources	Energy efficiency in building design	Tangible
2-Zero hunger	Knowing indigenous people and families	Heritage documentation	Conceptual
	Help maintain ecosystem	Environment friendly building	Tangible
	Storage for seeds and plants	Agro-architecture	Conceptual
	Investment in rural infrastructure	Rural architecture	Tangible
	Food commodity markets	Local market design	Conceptual
	Access to markets	Local market design	Conceptual

SDG goal	Key-term/key area	Key-term reframed for searching in COA regulation 2020	Role
3-Good health and wellbeing	Controlling communicable diseases	Strictures during pandemics	Conceptual
	Safe buildings	Safe buildings	Conceptual
	Safe roads	Safe roads	Conceptual
4-Quality education	Built/upgrade education facilities	Teaching-learning spaces	Tangible
	Create inclusive learning environment	Teaching-learning spaces	Conceptual
5-Gender equality	Gender equal public spaces/ services	Inclusive public spaces	Conceptual
	Development of infrastructure	Public Infrastructure	Conceptual
	Social protection	Safe spaces	Conceptual
6-Clean water & sanitation	Equitable sanitation and hygiene	Environmentally friendly building services	Tangible
	Reducing proportion of untreated wastewater	Environmentally friendly building services	Tangible
	Efficient water usage	Environmentally friendly building services	Tangible
	Water and Sanitation management	Environmentally friendly building services	Tangible
7-Affordable & clean energy	Clean energy research	Renewable energy	Conceptual
	Renewable energy	Renewable energy	Conceptual
	Expand infrastructure and upgrade tech	Renewable energy	Conceptual
8-Decent work & economic growth	Inclusive work place	Inclusive work place	Tangible
	Secure working environments (specially women migrants)	Women safety	Conceptual
	Sustainable tourism promotes local culture	Heritage conservation	Tangible
9-Industry, innovation & infrastructure	Built resilient infrastructure	Disaster resilient structures	Tangible
	Sustainable and resilient infrastructure	Disaster resilient structures	Tangible
	Upgrade infrastructure and retrofit industries	Sustainable industrial designs	Tangible
	Clean environment	Eco-friendly design	Conceptual
10- Reduce inequalities	Inclusive infrastructure	Inclusive public spaces	Conceptual
	Facilitate safe migration	Affordable housing	Tangible

SDG goal	Key-term/key area	Key-term reframed for searching in COA regulation 2020	Role
11-Sustainable cities	Affordable and safe housing	Affordable and safe housing	Tangible
	safe transport systems	Safety on road	Tangible
	Inclusive transport systems	Efficient accessibility	Tangible
	Sustainable human settlement/planning	Sustainable human settlement/planning	Tangible
	Safeguard heritage and cultural sites	Heritage conservation	Tangible
	Disaster management	Disaster management	Tangible
	Air & water pollution-waste management	Air and water pollution-waste management	Tangible
	Inclusive green, safe public spaces	Inclusive public spaces	Tangible
	Regional development planning	Regional development planning	Tangible
	Climate change issues	Disaster management	Tangible
	Using local material	Vernacular architecture	Tangible
12-Responsive consumption & production	Sustainable development and lifestyles in harmony with nature	Eco-friendly architecture	Conceptual
	Promote local culture	Documentation of local culture (Study tour)	Conceptual
13-Climate action	Climate related hazards and disasters	Disaster management	Tangible
	Climate change related planning and management	Disaster management	Tangible
14-Life below water	NIL	NIL	NIL
15-Life on land	Conservation of land ecosystem	Conservation of ecosystem	Conceptual
	Reduce degradation of natural habitant		Conceptual
16-Peace, justice & Strong Institution	NIL	NIL	NIL
17-Partnership for goals	Promote environmentally sound technology	Environment friendly building	Conceptual
	Measures for mitigation of climate change	Disaster management	Tangible

Source: Author

Apart from the goal 6, 7, 9, 11 which have distinct contribution of architects and planners, there are some more goals like 1, 4, 8, 10, 13 which indicate a tangible contribution of the professionals. Goal numbers 2, 3, 5, 12, 15, 17 have key-areas which opens new avenues for the architects (Table 13.1)

From the academic perspective, the conceptual key terms can be picked as thrust area for formulating the design studio brief. For example, Inclusive Public Spaces, Heritage Documentation, Local Markets, Infrastructure during Epidemics, Building Safety, Eco-friendly Buildings are areas around which the design projects can be developed (Table 13.1). Such design studios need to have a 'pre-design' phase to orient the students to the fulfillment of the respective SDGs. This will help the students to take informed decisions during different stages of their designs process. While dealing with 'conceptual level' key areas, the institutions need to encourage the inter-disciplinary research for finetuning the design outputs.

An Experiment to understad the challenges of implementation of the 'conceptual' key term was conducted at M. Arch (Urban Design) third semester 'Landscape Design & Sustainable Development & EIA' studio at Smt. Manoramabai Mundle College of Architecture (SMMCA), Nagpur. The project brief was to redevelop ***Kasturchand Park,*** a prominent regional level public space of Nagpur. A detail analysis of the existing park and its precinct were carried out through physical and social surveys. As part of pre-design, a detail study of the Sustainable Development Goals was conducted and analyzed by the students to identify the most appropriate goal or target which could be achieved through the urban design proposal. Apart from the Goal no. 11 (Sustainable Cities & Communities), the other goals like Goal-9 (Industry, Innovation and Infrastructure), Goal 8 (Decent work and Economic Growth) Goal-3 (Ensure Healthy lives and Promote well-being) and, Goal 10 (Reduced Inequality) became the thrust area during the proposals (Figure 13.2). This process had opened up to new thought process inclusive of the aspects like social, economic, cultural and environmental issues.

The key-areas referred as 'tangible' deals with more technical aspect of building sciences and human settlements. 'Tangible' aspects can be integrated into core or elective theory subjects with deliberate reference to the SDG. The key areas like, Disaster Resilient Structures, Affordable Building Services, Rural Architecture, Inclusive Educational Spaces, Vernacular Architecture, Urban & Regional Planning can be organized as mainframe subjects. The primary idea is that through these subjects, a clear mention of the contribution of architects toward fulfilling the SDGs is imparted to the students thus sensitizing the students to a deeper level regarding sustainability. In the Section 2 of analysis, the authors examine the presence of the key areas as mentioned in Table 13.1 in the (Council of Architecture, 2020) (Minimum Standards of Architectural Education) Regulations, 2020 document.

The key terms searched are as mentioned below (Table 13.2).

The analysis of Council of Architecture (2020) (Minimum Standards of Architectural Education) Regulations, 2020, clearly signifies that the 'tangible' key-areas are already included as guidance for the institutes of the nation to frame their syllabi (Table 13.2). Some of the key areas are positioned as core professional course and compulsory subject. Although, some key-areas are covered under electives, they provide an opportunity for the institute to develop further and sensitize the students. Some of the key areas like affordable building services, sustainable industrial design, road safety needs to be integrated into existing subjects or at the master level courses.

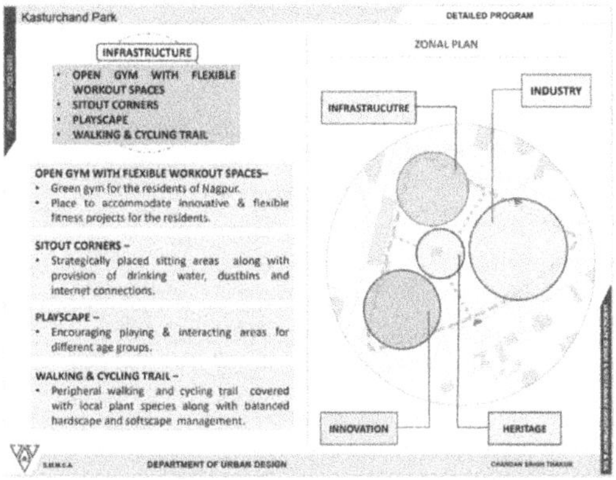

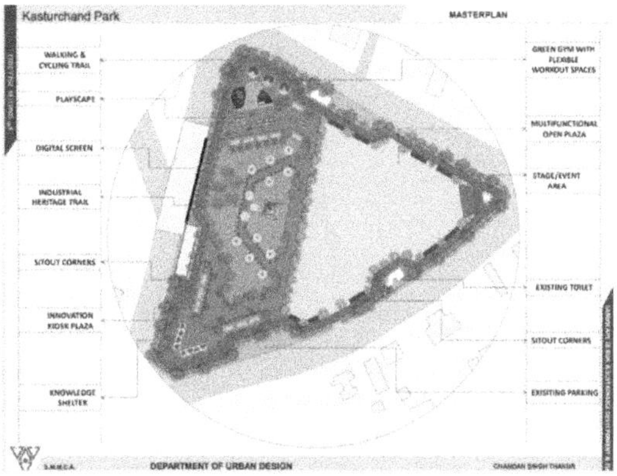

Goal 9: (Industry, innovation, and infrastructure)

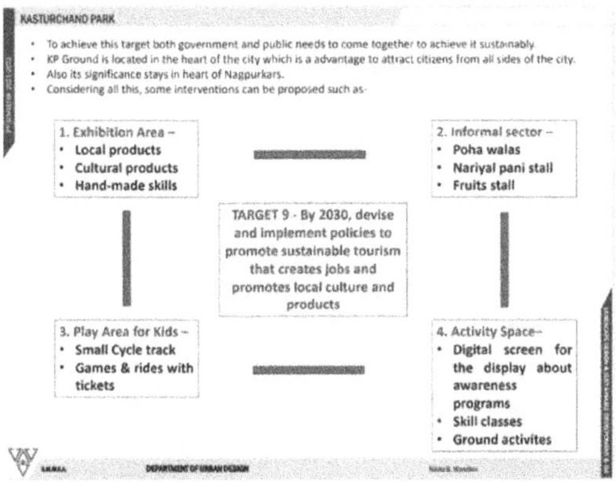

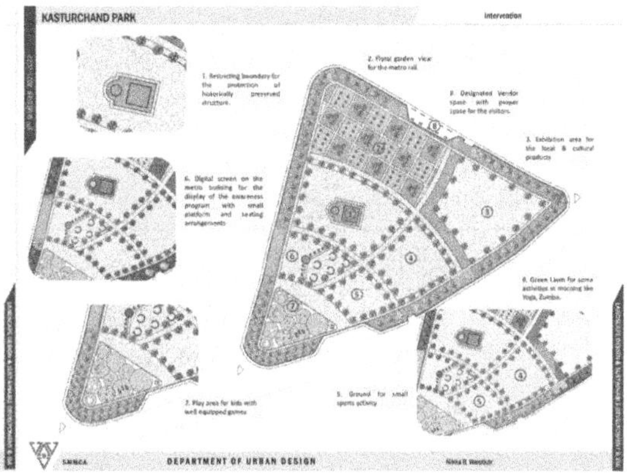

Goal 8: Decent work & economic growth (Target- sustainable tourism that creates jobs and promotes local culture and products

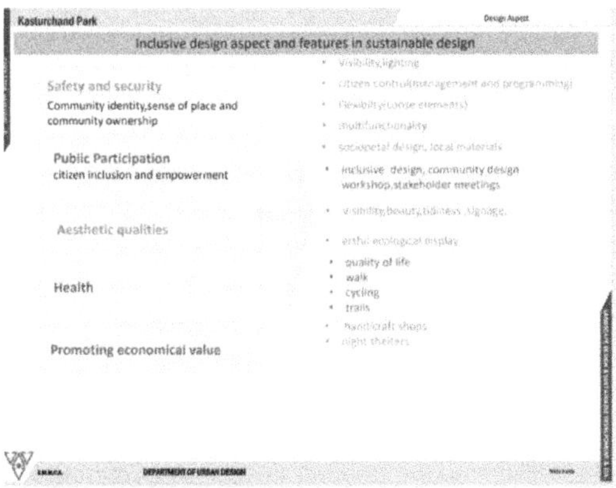

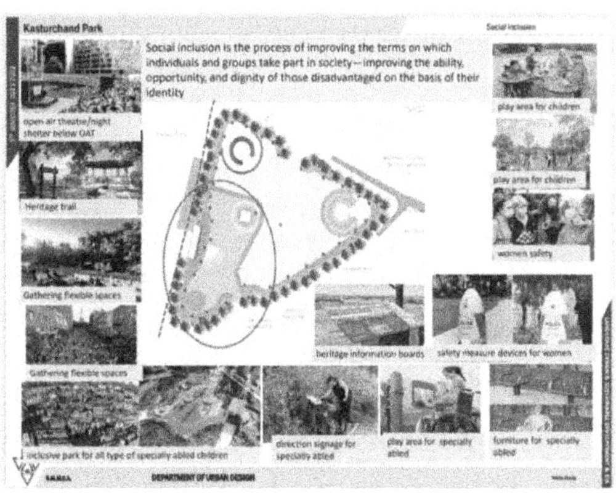

Goal 10: Reduce inequality target 10.2 empower and promote the social, economic and political inclusion of all

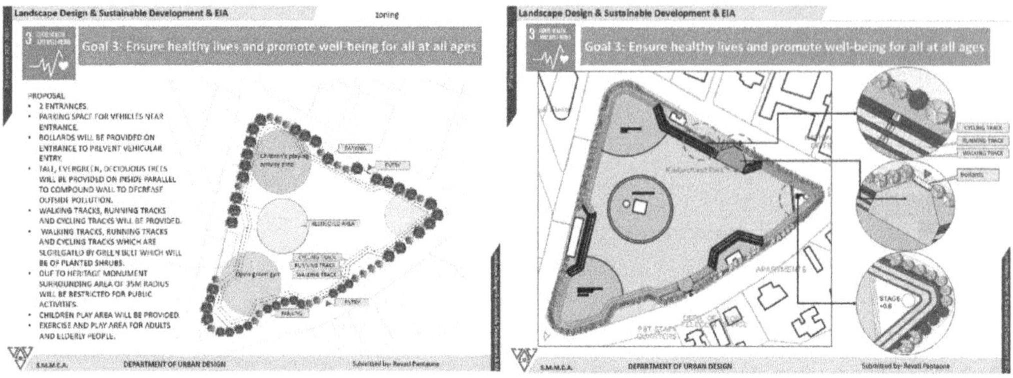

Goal 3: Ensure healthy lives and promote well-being for all at all ages

Figure 13.2. SDGs based studio outcome of 3rd Semester M. Arch urban design students
Source: SMMCA library archive

Table 13.2 Key-terms of SDGs in COA guideline document

Goal no	Tangible key areas	Subject title	Type	Remarks
1	Sustainable cities	Sustainable cities and communities	Professional elective	Addresses to SDG-11
2	Affordable housing	Housing	Professional core	Factors that influence housing affordability
3	Appropriate technology	Appropriate building technologies	Professional elective	
4	Disaster resilient structures	Building services earthquake resistant architecture	Professional elective building sciences and applied engineering professional core	Building Services
5	Energy efficiency in building design	Green buildings and rating systems Building performance and compliance	Professional elective	
6	Environment friendly building	Environmental science for architecture	Professional core	Lab based subject
7	Rural architecture	Human settlements planning	Professional elective	Mention as passing reference
8	Heritage conservation	Architectural conservation	Professional elective	
9	Air & water pollution-waste management	Environmental science for architecture	Building sciences and applied engineering	

Goal no	Tangible key areas	Subject title	Type	Remarks
10	Regional development planning	Human settlements planning	Professional core	Mention as passing reference
11	Vernacular architecture	Vernacular architecture	Professional elective	
12	Affordable building service			
13	Sustainable industrial designs			
14	Safety on road			

Source: Author

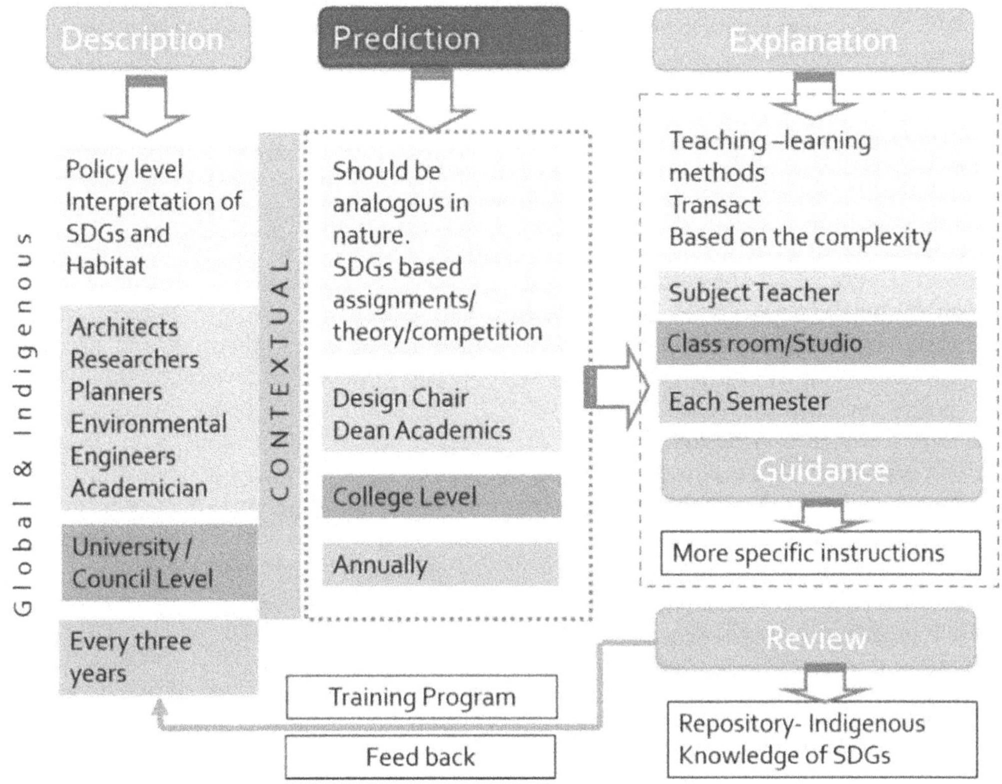

Figure 13.3 Proposed curriculum model
Source: Author

Inference

The universities and the regulatory body need to take methodical steps to contribute to the global mission by compiling the experiences and experiments conducted by the institutes. This will facilitate formulating a repository and reference material for not just architects but for the experts from other fields. This repository will guide the revisions required in the Council of Architecture (2020) (Minimum Standards of Architectural Education) Regulations.

One of the important approaches for a holistic integration of the SDGs into architecture curriculum is to apply the model of curriculum design across academics, research, and practice. The model broadly has five important sections; description, prediction, explanation, guidance, and review. The model facilitates bringing the theory into practice. It works as a tool to think about the SDGs from a broader perspective. It addresses issues beyond the syllabus and encourages research. It not only ensures the perception of all the facets of a subject but also helps generate new theoretical concepts (Figure 13.3).

The first phase 'description', proposes to look at the policy level contribution and interpretation of the SDGs and built environment (Figure 13.3). The authors propose to bring on board architects, researchers, planners, environmental engineers, academicians, bureaucrats, and experts to draft the policy description at national level with due consideration of global as well as regional scenarios. This task is conducted at the apex level i.e. COA and or at university level. This section can be reviewed at the interval of three years (Figure 13.3).

The second section is 'prediction.' The section emphasis to be analogues in nature. This section needs to be implemented at institutional level with theories, studios, assignments as 'mission' towards fulfillment of policy drafted in the previous section. At the institute level, Dean academics/Design Chair can play a pivotal role. The evaluation can be done through academic audits/faculty assessment on annual basis. Further, the progress can be recorded by COA through the Form B of Extension of Approval.

The third section of Curriculum model zooms down to the classroom level referred as 'Explanation'. It deals with studio/classroom level teaching-learning methods, transacts at different levels, undergraduate, post-graduate and diploma levels depending on the complexity. The role of the subject teacher is important at this stage, as everything planned at 'description 'and 'prediction' level is put to practice by the subject teacher. (Figure 13.3).

The fourth section is 'guidance' dealing with more specific instructions, guidelines, assignments formats, evaluation system etc. This stage is at one-to-one interaction with students to help them implement the concepts of SDGs. At this stage the teacher can document her/his experiences. Also giving feedback to the Dean/Chair the modifications required for the upcoming semesters.

The last and the most crucial stage is 'review' which works at the micro-level through feedback systems. Feedback from all the stakeholders. Each institute can develop their own format considering the vision and mission of perceiving SDGs. This stage can create new areas of research, new projects, new theories, and practices. A large content for developing the FDPs.

The assignments/theory can form a mega repository of indigenous knowledge or region-based knowledge regarding SDGs. This will set tone while revisiting the policies every 3 years. This provides a methodical means to addressing the issue.

Conclusion

The comprehensive study of SDGs vis-à-vis architecture education provides a primary insight that the institutes and the regulatory body need to look beyond the primary SDGs i.e. (6, 7, 9, 11) pertaining to architecture. The analysis of the historical background reveal that the goals are evolving and getting descriptive and more specific as we approach towards 2030. The 17 goals are fluid in nature and have cross-references. For instance, 'Disaster Management' is addressed in goal no 1, 9, 11, 13, 15, and 17. Thus, institutes, while targeting study of any key-area need to critically understand all the aspects of SDGs pertaining to it. Lastly, while dealing with any tangible or conceptual key areas through design studio, assignments, subjects, or a unit in a subject, deliberation on the targeted SDG will help in sensitizing the students. The findings of this paper can be taken ahead to understand how research, particularly interdisciplinary research can be conducted.

In addition to the focus on SDGs, there is a dire need today to introduce students of Architecture to the various government schemes in India which address or are linked to the achievement of some of the goals. Some of these issues are also addressed in the National Education Policy (NEP) which in the next few years will be implemented in totality.

References

Council of Architecture (2020). Minimum Standards of Architectural Regulations 2020. New Delhi: Council of Architecture.

Doric, P. B. (2008). A history of the concept of sustainable development: literature review. *Annals of the University of Oradea,* 17(2), 581.

Klarin, T. (2018). The concept of sustainable development from its beginning to the contemporary issues. *Zagreb International Review of Economics and Business,* 21(1), 67–94. doi:10.2478/Zireb-2018-0005.

Ornstein, A. C. (1993). Curriculum Foundations, Principles and Issues. Boston: Allyn and Bacon.

Sethi, S. (2014). Principles of Education and Curriculum Development. Delhi: K.S.K Publishers and Distributors.

Sharma, T., Bhatnagar, T., and Kalra, D. (2023). Mainstreaming sustainablity education: a case from architecture education in India. *Sambhasan,* 3(1), 196–222.

Tyana, S. (2020). Guilty by association: addressing sustainability in architecture education. *International Journal on Environmental Science and Sustainable Development,* 5(2): 60.

Tyler, R. W. (1949). Basic Principles of Curriculum and Instruction. Chicago: University of Chicago Press.

United Nations (2015). Transforming Our World: The 2030 Agenda for Sustainable Development. United Nations.

14 Effective communication in remote work environment - lessons learnt during pandemic

AR. Tanya[1,a] *and AR. Prerana Hazarika*[2,b]

[1]Assistant Professor, BMS College of Architecture, Bull Temple Road, Bengaluru, India

[2]Assistant Professor, School of Art and Architecture, Sushant University, New Delhi, India

Abstract

The role of efficient communication within a workplace is vital to the success and resulting productivity of any organization. It contributes directly to facilitating employee engagement, fueling motivation and innovations, and promoting retention and crisis management skills. Up until the advent of the COVID-19 pandemic, organizations typically relied on verbal communication as the primary tool of contact or correspondence aided by appropriate technology. A worldwide lockdown, however, brought about a radical shift and necessitated an overnight transition from physical to virtual means of communication. Conventional organizations, although reticent initially, gradually adapted to ICT and explored innovative and viable working models. Remote learning was one of the most challenging sectors as it primarily involved students – a dynamic cohort largely equipped with limited attention spans. In order to cater to this particular idiosyncrasy, a hoard of technological tools were experimented with, with the goal of providing the ideal interface for enhanced involvement and mutual exchange of ideas. The paper attempts to dissect the wide-ranging approaches adopted during this period and its consequent impact on the efficiency and psychological well-being of remote learners, educators and administrators alike. The paper further examines the feasibility of diverse models of remote collaboration in the current educational climate as well as the scope for their evolution in the future.

Keywords: Employee engagement, remote collaboration, remote learning, technological tools, working models

Introduction

Effective communication in the workplace is vital for productive management and efficiency. In-person communication can elicit more positive responses compared to nonverbal communication, as body language and tone are conveyed clearly, avoiding potential misinterpretation. The education sector was significantly impacted by the abrupt shift from in-class to remote learning. Academic institutions underwent an unprecedented change when COVID-19 lockdowns were imposed globally. The pandemic forced public school teachers to quickly transition to online instruction (Coker, 2021). This technology-enabled remote model challenged conventional student-teacher relationships. Although not a new idea, extended lockdowns necessitated structured, lasting solutions for remote learning. A major concern was learning and getting comfortable with applications like Microsoft Teams, Blackboard, and Zoom so that virtual classes could proceed smoothly (Almpanis and Joseph-Richard, 2022). This paper seeks to understand the

[a]tanya@bmsca.org, [b]prernahazarika@sushantuniversity.edu.in

DOI: 10.1201/9781003571018-14

various remote learning models implemented and how teachers and students overcame technological hurdles, informed by a critical review of global teaching and learning practices.

Theoretical Background

Over the years, remote learning has been a common way to deliver educational content. Benefits of remote learning include expert instruction, lower costs, flexible scheduling, access to more study materials, live demonstrations and discussions. However, online distance learning typically follows a set structure with pre-selected content and timelines. With the global pandemic, the concept of emergency remote teaching (ERT) was adopted as a temporary shift to online content delivery due to the critical situation (Hodges et al., 2022). This sudden change required faculty to quickly adapt lesson plans for online delivery. Major technological gaps, lack of access to proper digital tools and occasional network or application issues posed significant challenges for educators, students and administrators. Moreover, the rushed pace of these changes was overwhelming for many. As mentioned in Table 14.1, a number of studies were reviewed for this paper.

Description of Included Studies

The article titled 'Structure of Remote Work in Public Education: A Rapid Scoping Review' was published in 2021 in the International Research in Education journal. It was authored by David C. Coker. The purpose of Coker's study was to examine the structure of remote work in public education. To conduct the research, Coker used the method of a rapid scoping review to gather and analyze relevant literature on the topic. The study identified three main components that comprise the structure of remote work in public education: technology, educator roles and responsibilities, and organizational support from institutions. Regarding technology, Coker found that

Table 14.1 Research papers/articles referred to for the study

Title of research paper/article	Author(s)
'Structure of Remote Work in Public Education: A Rapid Scoping Review.' International Research in Education	David C. Coker
'The Difference Between Emergency Remote Teaching and E-Learning.' Frontiers in Education	Moore Hodges
'Lecturing from Home: Exploring Academics' Experiences of Remote Teaching during a Pandemic.' International Journal of Educational Research Open	Timos Almpanis and Paul Joseph-Richardb
'Emergency Remote Teaching in Higher Education: Mapping the First Global Online Semester.' International Journal of Educational Technology in Higher Education	Melissa Bond, Svenja Bedenlier, Victoria I. Marín and Marion Händel
'Remote Learning During COVID-19: Lessons from Today, Principles for Tomorrow.'	Alberto Muñoz-Najar, Alison Gilberto, Amer Hasan, Cristóbal Cobo, João Pedro Azevedo and Maryam Akmal

Source: Authors

reliable internet connectivity and suitable digital tools are crucial for effective remote work. The study also stressed the importance of professional development and training to help educators adapt to remote teaching approaches. In terms of roles and responsibilities, the study noted that remote work requires educators to take on multiple roles like instructor, facilitator, and evaluator. Clear guidelines and expectations are needed to ensure successful teaching and learning remotely. For organizational support, the review revealed that administrative backing and collaboration between educators and stakeholders are vital for successful remote work. This includes establishing communication channels, providing resources, and addressing challenges faced by educators. In summary, Coker's rapid scoping review offers insights into the structure of remote work in public education. It emphasizes the significance of technology, educator roles and responsibilities, and organizational support to effectively implement remote teaching and learning.

The article titled 'The Difference Between Emergency Remote Teaching and E-Learning' published in 2022 in the journal Frontiers in Education by Hodges, Moore, and colleagues examined the contrast between emergency remote teaching and e-learning, specifically in light of the COVID-19 pandemic. The purpose of the study was to differentiate these two modes of education. The authors analyzed the key features and differences between them. They contended that emergency remote teaching refers to the temporary shift from in-person instruction to remote teaching owing to unanticipated events like the pandemic. In comparison, e-learning is an established and deliberate approach to teaching and learning utilizing online technologies and pedagogical methods. The study highlighted several major differences between emergency remote teaching and e-learning. First, emergency remote teaching is often rapidly implemented without the extensive planning and preparation of e-learning. It is a temporary solution to maintain educational continuity during crises. Second, the pedagogical approaches used in emergency remote teaching may be more limited versus e-learning. In emergency situations, educators often use familiar tools and techniques that may not fully leverage the potential of online learning technologies. Furthermore, the study stressed that emergency remote teaching can involve a narrower range of educational activities and assessments compared to e-learning. Due to time constraints and the need for immediate solutions, educators may focus on content delivery rather than engaging students in interactive and collaborative learning experiences. In summary, the article by Hodges, Moore, et al. delineates the difference between emergency remote teaching and e-learning. It emphasizes that emergency remote teaching is a temporary and reactive response to crises, while e-learning is a planned and comprehensive approach to online education. Recognizing these differences is vital for educators and institutions to effectively navigate and adapt to remote teaching and learning circumstances.

The article titled 'Lecturing from Home: Exploring Academics' Experiences of Remote Teaching during a Pandemic' authored by Almpanis and Joseph-Richard was published in 2022 in the International Journal of Educational Research Open. The objective of the study was to examine the experiences of university professors and lecturers who had to abruptly transition to online teaching amid the COVID-19 pandemic. The authors used qualitative research methods such as interviews and surveys to gather data from academics regarding their experiences with remote instruction. The focus of the study was specifically on the difficulties faced, advantages identified, and techniques used by academics while delivering lectures from their homes.

The key findings of the study were as follows: First, academics encountered various challenges with the sudden shift to online teaching. These included technological problems, trouble adapting teaching approaches to virtual platforms, and issues with engaging students effectively in a remote setting. Second, some benefits of remote teaching were recognized, like increased flexibility in scheduling and the ability to pre-record and re-use lectures. Remote teaching also created opportunities to explore innovative instructional methods and materials. Third, the study outlined strategies used by academics to improve remote teaching, including adopting new digital tools and platforms, incorporating interactive activities to engage students, and providing clear instructions and communication channels.

The study also stressed the importance of institutional support in enabling successful remote teaching experiences, noting that adequate technical assistance, training, and resources were crucial for academics to effectively deal with the difficulties of online instruction. Overall, the article by Almpanis and Joseph-Richard offers insights into academics' experiences transitioning to emergency remote teaching during the pandemic. It highlights the challenges, benefits, and strategies related to lecturing from home, underscoring the need for backing and resources to ensure effective remote teaching practices.

The study titled 'Emergency Remote Teaching in Higher Education: Mapping the First Global Online Semester' published in 2021 in the International Journal of Educational Technology in Higher Education by Bond, Melissa and colleagues examined the experiences and practices of emergency remote teaching during the initial worldwide transition to online education in colleges and universities. The researchers conducted a large survey to gather data from educators globally who had switched to emergency remote teaching because of the COVID-19 pandemic. The study aimed to map out the difficulties faced, tactics used, and assistance received during this changeover. The findings revealed several key insights. First, educators encountered various challenges when transitioning to emergency remote teaching. These included technological problems, lack of training and preparation, issues with engaging students and assessing them, and the emotional toll of the rapid transition.

Second, the study highlighted the strategies educators used to adapt to emergency remote teaching. These included using video conferencing platforms, asynchronous learning systems, and online resources. Educators also employed various pedagogical approaches like flipped classrooms and collaborative learning to improve student engagement. Third, the study emphasized the importance of institutional support and collaboration. Educators reported that institutional support in the form of technical help, training, and resources played a vital role in enabling the shift to emergency remote teaching. Collaborating with colleagues and sharing best practices were also valuable sources of support. Overall, the article by Bond et al. (2021) provides valuable insights into educators' experiences during the global transition to emergency remote teaching in higher education. The findings highlight the challenges encountered, strategies used, and the significance of institutional support in facilitating effective teaching and learning in emergency remote contexts.

The report 'Remote Learning During COVID-19: Lessons from Today, Principles for Tomorrow' published in 2021 by Muñoz-Najar et al., 2021 and colleagues aimed to provide insights and key takeaways from the experience of distance learning during the COVID-19 pandemic. It also outlined principles to guide the future of remote education. The report highlighted several important findings and suggestions. First, it stressed the significance of fairness in remote learning. Access to technology and internet connectivity were crucial for ensuring equal opportunities for all learners. The authors recommended addressing the digital divide and providing necessary resources to underserved groups. Second, the report emphasized the need for pedagogical innovation in distance learning. Educators were encouraged to explore diverse teaching strategies and technologies that promote active and engaging learning. The report also highlighted the value of fostering social interaction and teamwork among students in remote settings. Third, the report discussed the importance of teacher training and support. Professional development programs and ongoing training were suggested to equip educators with the skills and knowledge to effectively facilitate remote learning. Furthermore, the report stressed the significance of continuous assessment and feedback in distance learning. It proposed leveraging digital tools and platforms to evaluate student progress and provide timely feedback. The report also underlined the importance of student wellbeing and mental health in remote environments. It recommended approaches to address social isolation, promote emotional support, and cultivate a sense of community among students. Overall, the report by Muñoz-Najar et al., 2021 and colleagues offers insights and principles for remote learning based on lessons from the COVID-19 pandemic. It emphasizes equity, pedagogical innovation, teacher support, assessment, and student wellbeing. These recommendations can guide future remote education practices and policies.

During the pandemic, the main model adopted was synchronous as institutions preferred to retain scheduled campus classes to minimize the impact of distance learning and ensure compulsory attendance. This delivery mode is effective for live workshops, demonstrations and real-time discussions. Platforms like Zoom, Google Meet, Cisco Webex, Microsoft Teams were used by educators to conduct timetabled classes and timely assessments.

The alternative common method of instruction is asynchronous, which provides more flexibility regarding learning time. Lecture material can be pre-recorded or saved, allowing users to access it when convenient. Assessments may have time limits. This asynchronous approach has been used for years on various online platforms like Coursera, Byju's, upGrad, and others that offer diverse courses in all subjects. Both modes of course delivery have been studied and compared as mentioned in Table 14.2 to develop an understanding of the pedagogy.

Methodology

In order to comprehend the impact of ERT during the pandemic, a qualitative survey was conducted for educators in addition to referring to a vast number of studies documented on the same. Following criteria were assessed accordingly:

C1 - Realigning Course Structure

1. Altering teaching methods better suited to remote learning
2. Redesign of teaching materials
3. Modification of lesson plans if any
4. Inclusion of video demonstrations/ workshops in lecture content

C2 - Methods and Mode of Content Delivery

1. Standard span of a lecture conducted remotely
2. Relevance of attendance during remote learning
3. Importance of Content delivery vs Knowledge retention
4. Adjustments to cater to students' diverse learning needs
5. Learner engagement strategies used in remote teaching
6. Educational institutes coping up with remote working mode

Table 14.2 Comparison between synchronous and asynchronous modes of course delivery

Synchronous	Asynchronous
Course delivery in real time	Pre–recorded course content can be accessed at any time
Offers no scope of flexibility in terms of timings	Offers scope of flexibility, can be accessed at one's own time availability
Live attendance can be monitored	Live attendance cannot be monitored
Offers lesser access to users from different time zones	More open to global users
Double channeled communication is possible. Open discussions or interactions are possible in this mode of delivery	Single channeled communication. Only relevant topics are discussed, no scope of face to face interactions with instructor
Lecture span guided by timetable, uninterrupted course delivery	Entire lecture span can be split up and content may be consumed as per flexible timings
Real time quizzes and assignments may be assessed	Graded or non-graded time bound assignments may be uploaded
Requires stable network in the duration of the lecture	Content can be accessed whenever stable network connection is available

Source: Authors

C3 - Evaluation and Feedback

1. Altering assessment methods
2. Quality assurance processes followed
3. Changed way to provide feedback to students

C4 - Additional Student Support

1. Modified approaches to student support and guidance
2. Additional modes of communication for industry support (like discussion forums/expert groups)

C5 - Faculty Well-Being and Development Initiatives

1. Additional training and development opportunities during remote working period
2. Becoming aware of newer tools of digital learning
3. More equipped to transition to digital learning after the emergence of the pandemic
4. Effect on work-life balance
5. As an educator, preferred mode to run courses

Findings

A qualitative survey was designed for educators to comprehend the experience of educators teaching at university level and the following was deduced from a sample survey of 56 respondents.

C1 - Realigning Course Structure

A total of 48% of respondents stated that their previous method of teaching was ineffective during the lockdown period and hence they had to redesign their teaching materials.

As indicated by the data in Figure 14.1, a stark 84% respondents believed in redesigning their course structure and teaching methodologies. They also added that they incorporated video content into their lecture content to increase student engagement.

As indicated by the data in Figure 14.2, 70% respondents believed in focusing both on reducing the course content as well as modification of lesson plans to suit the emergent needs of students.

C2 - Methods and Mode of Content Delivery

As indicated by the data in Figure 14.3, 84% respondents claimed that the standard

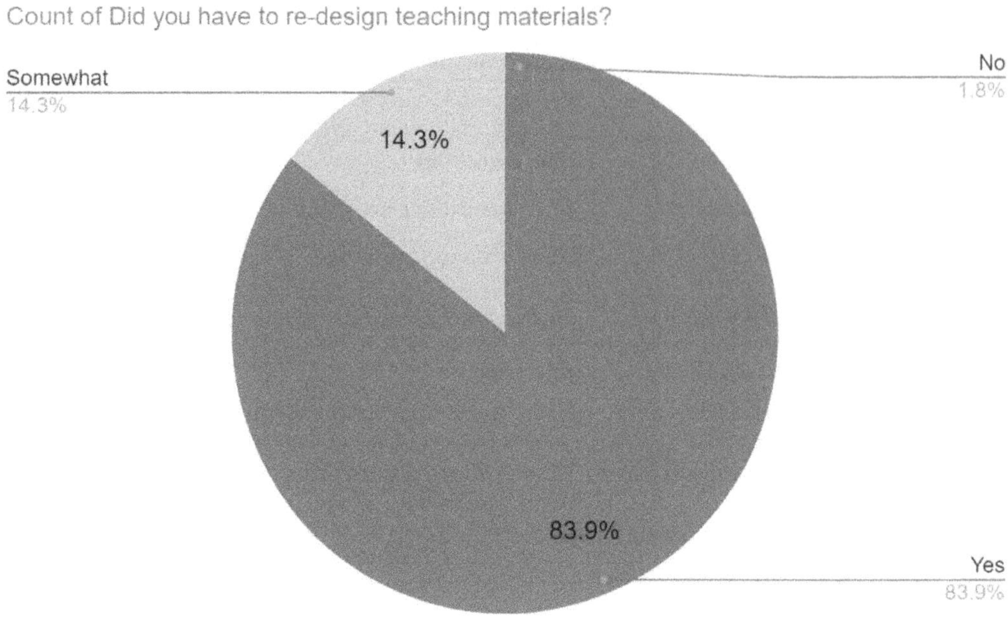

Figure 14.1 Depiction of data reflecting percentage of respondents who redesigned their teaching materials
Source: Authors

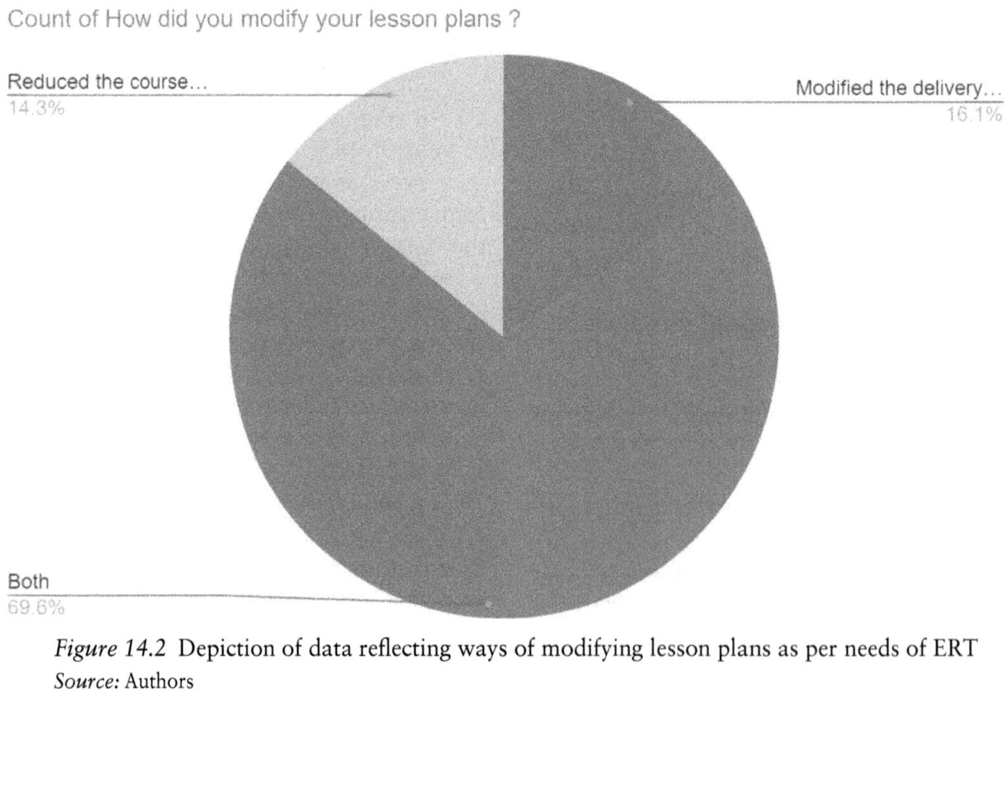

Figure 14.2 Depiction of data reflecting ways of modifying lesson plans as per needs of ERT
Source: Authors

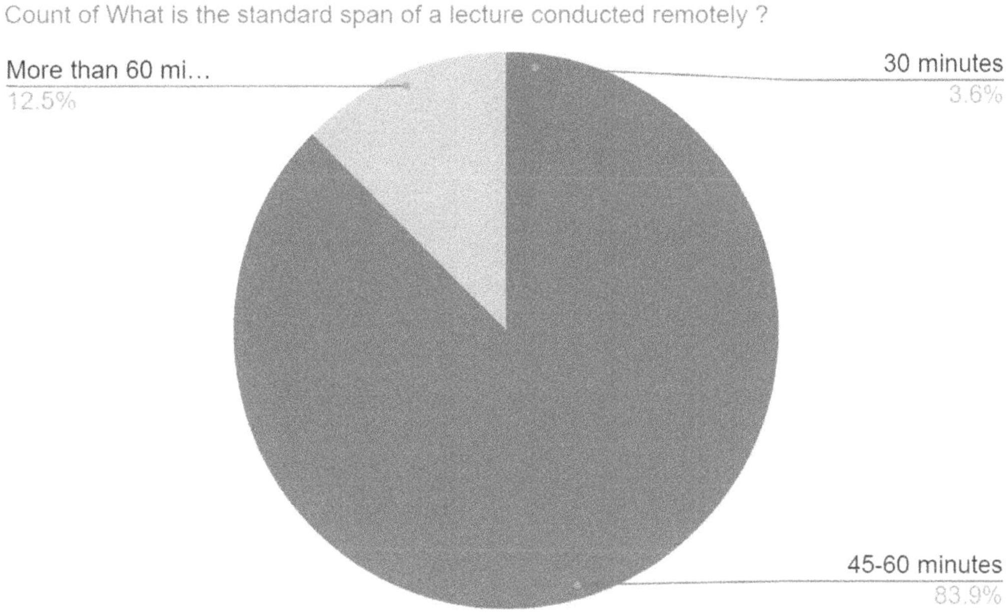

Figure 14.3 Depiction of data reflecting standard span of a lecture
Source: Authors

span of a lecture was 45-60 minutes. 94% respondents believed that attendance during this period was very crucial to the learning process.

As indicated by the data in Figure 14.4, 84 of respondents focused more on content delivery against a mere 16% who put more effort into retention of knowledge. Further they also mentioned using different learner management strategies like 1:1 sessions with students, altering existing student-teacher ratio of a class, introducing appropriate LMS (Learner Management System) like TCS, Google Classroom etc, and peer learning. They expressed the need for educational institutes to cope better with remote working mode.

C3 - Evaluation and Feedback

As indicated by the data in Figure 14.5, 91% respondents resorted to assigning quizzes at the end of lecture delivery in order to maintain student engagement throughout and to ensure compulsory attendance in the entire duration of the class. Quality assurance measures like peer review process, external examiner/jury, self assessment and proctored online exams were adopted during this period. A radical shift in student methods of providing student feedback was also noted. Educators took to one on one discussions, frequent follow ups and overall took an individual approach to handling student issues.

C4 - Additional Student Support

As indicated by the data in Figure 14.6, 97% respondents mentioned the use of supporting industry forums and expert groups for broader exposure. Students were provided additional guidance by means of group discussions, regular counseling sessions and individual discussions.

C5 - Faculty Well-Being and Development Initiatives

As indicated by the data in Figure 14.7, 95% respondents claimed to have undertaken different training activities during this time which helped them add on to their

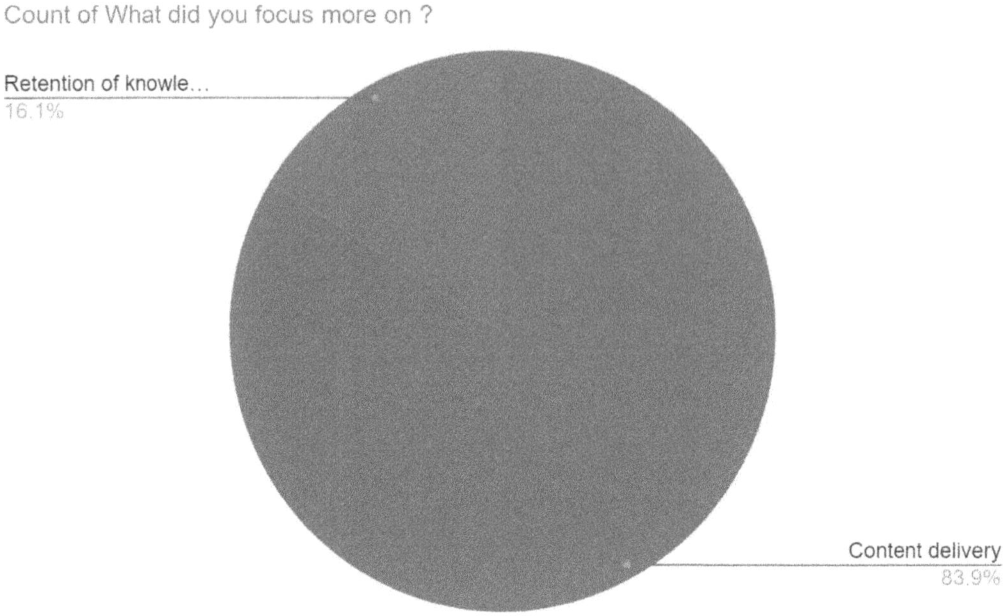

Figure 14.4 Depiction of data reflecting focus on knowledge retention vs content delivery
Source: Authors

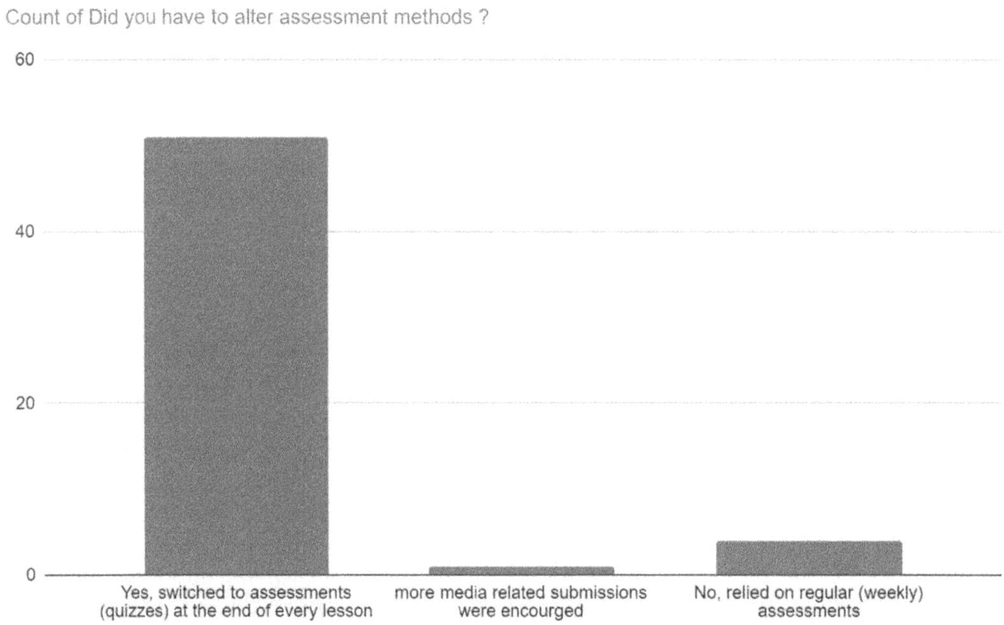

Figure 14.5 Depiction of data reflecting methods of assessment followed during remote teaching
Source: Authors

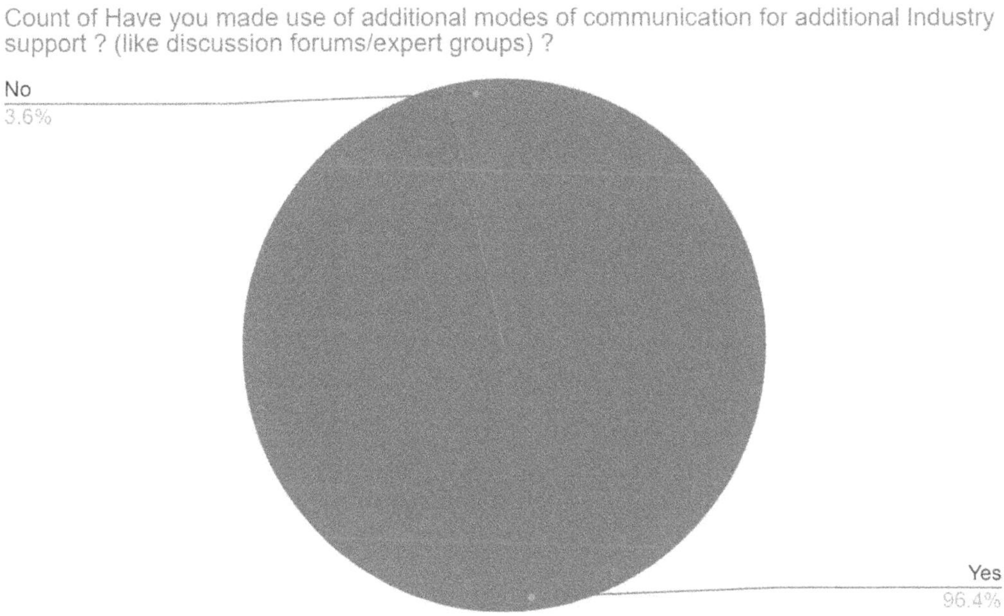

Figure 14.6 Depiction of data reflecting use of additional forums for communication
Source: Authors

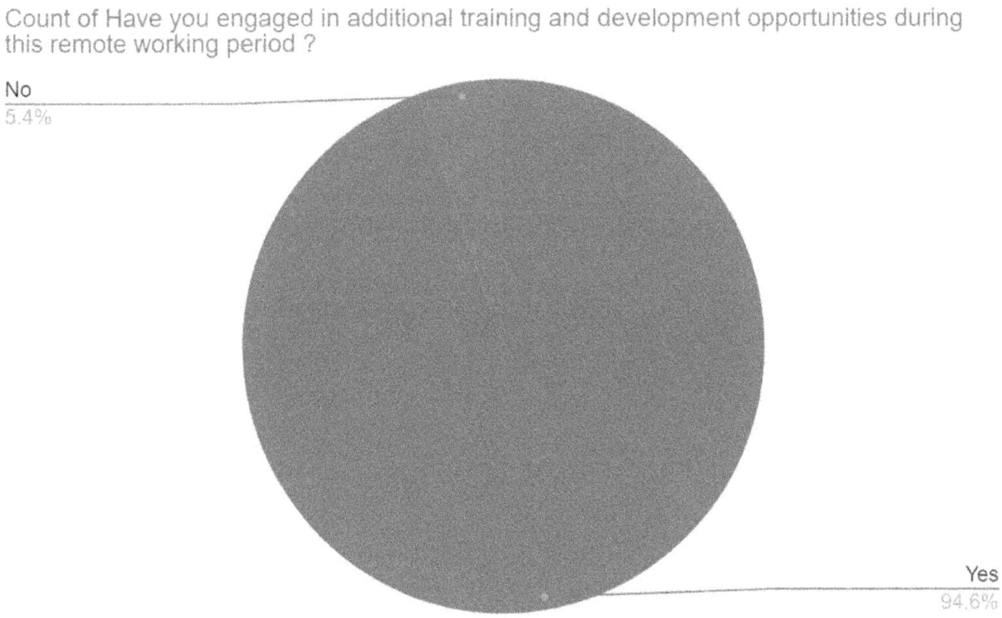

Figure 14.7 Depiction of data reflecting additional training and development activities undertaken by faculty during the pandemic
Source: Authors

academic profile. They have further stated that the lockdown period compelled them to learn new age digital tools to aid in ERT with the need for more training.

As per the data in Figure 14.8, 90% of the respondents stated that their work-life balance deteriorated due to lack of boundary between the work and personal front.

As per the data in Figure 14.9, 77% of respondents chose a hybrid mole of teaching rather than sticking to the conventional on campus teaching. 20% believed in the offline model while a meager 5% of respondents wanted to switch to the online model of teaching.

Discussion

The study illuminated the difficulties and obstacles experienced while teaching remotely during the pandemic. The results show that there was mostly a severe lack of readiness and insufficient technical support during a critically urgent time. Over the course of this research, it was also inferred that emergency remote teaching required far more time and preparation than regular in-person instruction. Additionally, educators were expected to be well-versed in digital communication tools and learning management systems (LMS) for a smooth changeover from campus to remote learning platforms within a very short timeframe. Furthermore, access to stable and fast internet connection was the fundamental requirement of remote learning, so the lack of it therefore had serious repercussions for both students and teachers. Network problems also led to significant disruptions during proctored exams.

Recommendations

Notwithstanding the abundant difficulties during lockdowns, various approaches were tested to give the best substitutes for customary face-to-face teaching and

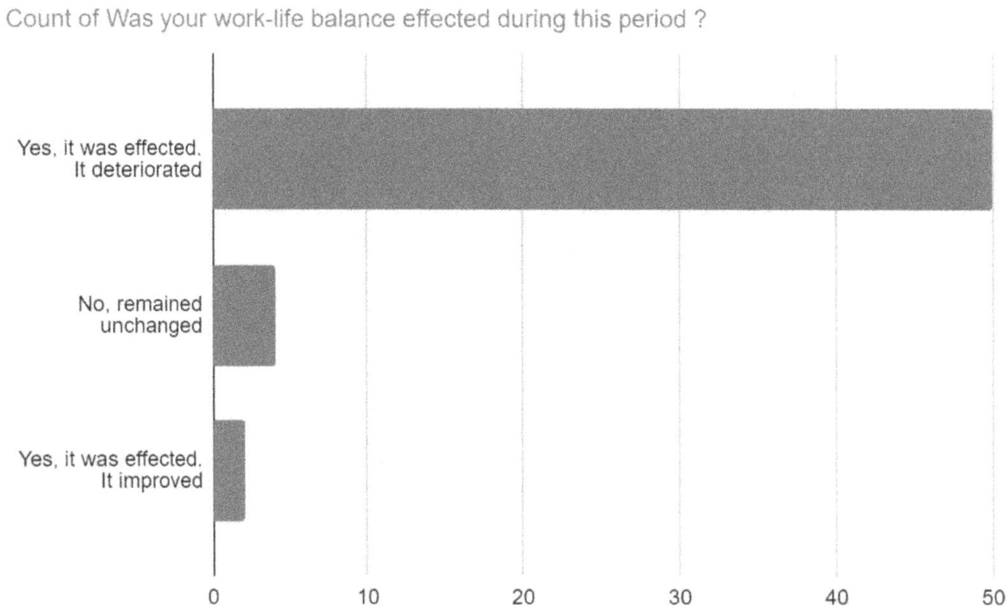

Figure 14.8 Depiction of data reflecting the effect on work-life balance during the pandemic
Source: Authors

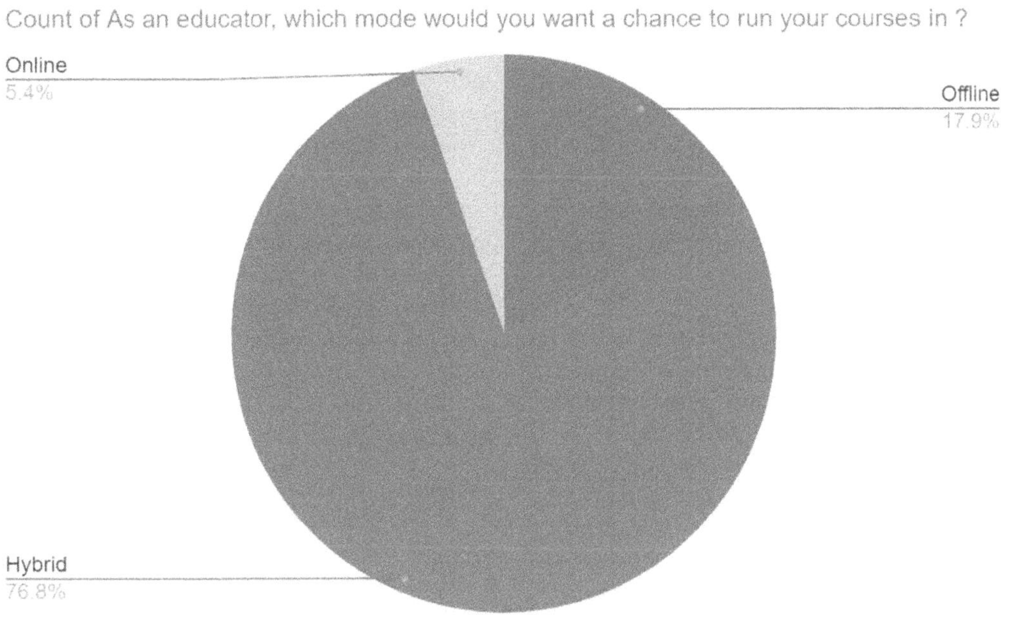

Figure 14.9 Depiction of data reflecting the mode of teaching chosen by educators to follow from hereon
Source: Authors

learning. From this study's findings, some suggestions have been made for the education sector regarding remote learning going forward.

Compulsory extensive training programs on digital learning systems and tools for both students and educators at all levels to facilitate smoother remote learning. Restructuring course content with greater emphasis on delivery catering to students' attention span is vital.

When designing digital modules to suit shorter attention spans and improve retention, several strategies can be utilized. Below are some recommendations to minimize course material while optimizing engagement and outcomes:

Divide information into smaller, manageable chunks or modules, each focused on a specific concept. This enables better digestion of info and reduces overload.

Incorporate diverse multimedia like videos, animations, infographics, and interactive activities. These diversify learning and cater to different styles, making content more engaging and memorable.

Provide clear, concise instructions communicating objectives and expectations for each section/activity. Use straightforward language to avoid overloading learners. Bullets, numbered lists, and visual cues can also provide clear instructions.

Prioritize essential concepts, skills, or knowledge learners must grasp to achieve objectives. Focus on these key elements and avoid redundant info. This helps learners concentrate on what truly matters and prevents overload.

Include interactive quizzes, knowledge checks, and scenario activities throughout modules. Interactive assessments promote engagement and reinforce learning by enabling application of acquired knowledge.

Encourage participation and collaboration via discussions, group projects, or peer feedback. Social interaction fosters community, motivation, and accountability, enhancing attention and retention.

Offer short, bite-sized microlearning resources accessible on-demand. Microlearning can be consumed quickly during breaks, reinforcing understanding without lengthy content feeling overwhelming.

Use storytelling to create narratives or scenarios that contextualize information. Storytelling grabs attention, enables emotional connections, and aids memory retention.

Implement spaced repetition to reinforce learning over time. Reviewing key concepts at intervals helps solidify long-term memory and combat the forgetting curve.

Personalize content and delivery based on individual learner preferences and needs. Provide options for learners to choose paths, explore interests, or set goals. Personalization increases engagement and motivation, improving attention and retention.

Balance positive feedback with constructive criticism. While highlighting strengths, addressing areas needing improvement is equally important. Aim for balanced, fair feedback guiding learners toward goals.

Prompt self-reflection before giving feedback. This builds self-awareness and active participation in learning. Encourage identifying own strengths and areas to improve, using feedback to validate/expand on self-reflection.

Use remote desktop sharing to assist with complex technical issues in real-time. This allows support staff to access screens and efficiently troubleshoot.

Offer training resources or workshops to help learners get familiar with the platform and address common technical challenges. This empowers independent troubleshooting and reduces dependency on tech support.

Regularly evaluate digital learning module effectiveness through feedback, assessments, and analytics. This feedback loop enables continuous improvement and ensures engaging, effective content for learners with shorter attention spans.

References

Almpanis, T. and Joseph-Richard, P. (2022). Lecturing from home: exploring academics' experiences of remote teaching during a pandemic. *International Journal of Educational Research Open*, 3, 100133. doi: 10.1016/j.ijedro.2022.100133.

Bond, M. et al. (2021). Emergency remote teaching in higher education: mapping the first global online semester. *International Journal of Educational Technology in Higher Education*, 18(1), doi: 10.1186/s41239-021-00282-x.

Coker, D.C. (2021). Structure of remote work in public education: a rapid scoping review. *International Research in Education*, 9(2), 27, doi: 10.5296/ire.v9i2.18474.

Hodges, C. B., Moore, S. et al. (2022). The difference between emergency remote teaching and e-learning. *Frontiers in Education*, 7, doi: 10.3389/feduc.2022.921332.

Muñoz-Najar, A, et al. (2021). Remote learning during COVID-19: lessons from today, principles for tomorrow. *Report*, 1(1), 62.

15 Knowledge acquisition: Framework for architectural education

Anand Khatri[a]

Professor, Apeejay Institute of Technology School of Architecture & Planning, Greater Noida, Uttar Pradesh, India

Founder, Director, Urban Village Charitable Trust, Noida, Uttar Pradesh, India

Abstract

Architectural education and practice are immersive. "Built" for the architect is a holistic involvement, a language, an expression and a self-generative process. It develops a visual grammar. For the masses, architecture is merely a shelter, building, creative self-projection, a place that safeguards family, a possession or an address. An architect develops a signature style through an ingenious mind. Building and architecture amalgamate art, techniques, and designs through the conceptual idea, with the living. Buildings are silent schools while historic cities and ruins carry narratives of how ancient societies lived and thought. Buildings live longer than the average human lifespan. Despite such strengths, an architect's knowledge could never percolate to the masses. Since its 19th century re-organization as a profession, architects have been able to serve only a limited few. There is also a progressive isolation of the profession, which is causing its decadence. Technology led professions are replacing architects. Within architecture, there is deeper design thinking and better grammar of form. So the challenge is dual, there is an inner evolution and an external disconnect. A fresh graduate, despite spending five years for education feels unsure and hesitant. The master programs are not extensions of the thrust areas of five years of bachelors. They open new dialogues. A practicing architect hires engineers for maximum jobs, while architects are not able to do all jobs arising out of the practice. This research paper written from a COATRC lecture explores the strength of architectural education and highlights the need for revisiting the syllabus. It tries to make a statement by mapping the inter connections that could give birth to a new definition of an architect. Discussion on knowledge acquisition in the growth of sociology and economics in the last century and a comparative narrative in architecture draws parallels. Grading system in AKTU and syllabus is as an example of pedagogic illustration. Architecture today is becoming just an input education for being absorbed into allied fields. Preparedness of a fresh architect to take projects is discussed. The paper concludes in findings useful for pedagogues to reconstruct the training for upcoming architects, making it better equipped to practice architecture.

Keywords: AKTU syllabus, economics, knowledge acquisition, output tools, pedagogy, sociology

[1]Anand Khatri is a professor at Apeejay Institute of Technology - School of Architecture & Planning. He did his Master in Arch Conservation from SPA Delhi 1996 and a B. Arch. From GCA Lucknow 1988-93. With a 25-year experience in practice of architecture, across the scales of Interior, Below the Line Design, Retail, Architecture and Heritage, he is now an academic and researcher. He is the founding director of Urban Village Charitable Trust and POIESIS.

[a]kcasanand@gmail.com

DOI: 10.1201/9781003571018-15

Introduction

There is a dichotomy in perception of architecture, training of architects, use of built, appreciation of buildings and output of architecture. Buildings and shelters are in use at all scales in the world, but the percentage of ownership of the built with architects is very less. Architects contribute to just or a little less than 2 percent of the buildings in the world (LaBarre, 2008) (Wakeford, 2008) (Smita Khan, 2019). Human civilization progressed in buildings and cities. It remained safe under shelters and experienced light, shade and rains, which could have been a challenge. Architects as champions of built form enjoy design integration of engineering (Beninger, 2014). Great civilizations communicate through their buildings. Think about the way the word architect is used: the architect of economy, for example, or the architect of a whole variety of things. It's not about being a builder, it's about being somebody who can synthesize different components and things based on bigger ideas (Rotondi, 2012). The principal architects are limited only to the drawing board and a few site visits. There is a limitation in role definition, within the practice of profession itself. The problem accentuates when builders copy details or make multiple replications of prototypes designed by architects. There are few methods to explore copyrights or connect the built replications to design.

Such a limitation of an architect is because the ownership role assumed by the architect is very authoritatively colonial-clerical that started as a drafting course (Mehta, Towards a new Pedagogy. Architexturez Imprints, 2001). With the evolution of many more professions, the presence of architect across building construction, as compared to that of engineers is very limited. Imagine a situation where Michelangelo, Raja Ravi Varma, Jamini Roy etc. only made sketches or drawings for their paintings/ sculptures and were limited in their work on the periphery. Imagine what would have been art if they hired contractors to shape their painting or sculpture. Imagine Laurie Baker, Revathi Kamath limiting their output to drawings and assigning the work of translating them to engineers and building management experts. We all know we would not have had what we see as architectural details in their buildings today.

Architecture and building is an expression and a powerful tool of a civilization (Mehta, Towards a new Pedagogy. Architexturez Imprints, 2001). Architects are thinker doers, thrust with the responsibility of reconfiguring environment in which they work. No other profession has this mandate (Beninger, 2014). When we look back at historic cities and architectural remains, we immediately know how old societies lived and thought. Architecture is thus a language, and a building-book and this method of research by looking at technology is unique to architects. This is not been recognized as a method within the structured format of research. Etymological explorations of the word 'architect' across several languages reveals that the word addresses an individual who commands the making buildings (Fabian Dejtiar, 2018) and originates from either a carpenter or a sculptor. 'Architect' from architektōn in Greece and Vaastukaar from Vaastushilpa in Hindu systems represent a basic connection of evolutionary trades connected to the profession. All are suggestive of an ability to command building process/method/owner. 'Architect' by its professional name definition, has never taken the complete ownership of the built. Subjects like construction, structures and design remain as separate and are taught in compartments. The deviant education structure of architecture and the limitation of the role of the architect who is design enabled, but implementation dependent and construction shy is what has created a crisis call for research discussions and rethinking into the future of architecture (Mehta, Decolonizing Architectural Education in India. Architexturez Imprints,

2020). Such discussions are a sign that the profession is self-aware and willing to change, but the architect is stuck within a self-referential jargon (Chandavarkar, 2019) (Smita Khan, 2019). It is not the profession or practice that is restricting architectural growth; it is the definition of 'architect' within the profession itself that is collapsing.

Despite the importance of built, and architecture being a forever need, architects could never percolate to the masses (Chauhan, 1999). Since the early 19th century, as an organized profession, has never been able to serve the needy. Because of its low applicability, the 20th-century architects, are progressively getting isolated while its output is in great demand. Because they are no industry ready, fresh graduate feels weak and unsure to take on projects. In addition to this is the low pay scales of jobs. B. Arch. does not suffice as a terminal qualification. When we see the connection of B. Arch. to post graduation, we realize that the spectrum developed away from the mainstream of practice. M. Arch. does not help architects to be better architects, but takes them off on a different route, as can be seen in Figure 15.1. Within the prevalent trends of M. Arch. Courses, few are in the core subjects in B.Arch. e.g. Design, Drawing, Construction or in Structures. There are insufficient directives in the COA, 2006 document Minimum Standard for Architectural Education Guidelines for PG Programs section 10.0 on the Development of Post Graduate Programs.

Pedagogic exploration of the strength of architectural training and the need for revisiting the syllabus needs a better dialogue. The interconnection of architecture with the living world (Chhaya, 1999) and graduation extension into postgraduate courses should stay within the licence-ship of the architecture. Because of the lack in the reintegration of the information from the profession, back into the course, it has no developed a depth. Design as the only key subject is too non-technical to keep the interest of the knowledge-enabled world (Aditi Deep Singh, 2022). In search of more grounded connection with reality, 50% of graduates who seek admissions in post-graduation, opt for planning courses. This is at a junction when planners have an independent vertical and see themselves different from architects.

Material and Methods

The research emerges through discussions held with architect and members of alumni groups – as Founder Architects Association Noida Zone, as Fellow of Indian Institute of

<table>
<tr><th></th><th>Name of core subjects</th><th>M. Arch. as extension to core subject</th><th>Available M. Arch. Courses related to the core subjects</th></tr>
<tr><td>1</td><td>Design</td><td>No M. Arch. in Design. M. Des. is only at IIT Mumbai or IIT Delhi, and leads to product design, not architecture</td><td>Urban Design, Industrial Design, Landscape,Interior, Housing Design,</td></tr>
<tr><td>2</td><td>Construction</td><td>Construction management, Construction systems, Construction system coordination etc. is missing.</td><td>B.E.M. as masters' course is largely on scheduling the engineering management.</td></tr>
<tr><td>3</td><td>Structures</td><td>No M. Arch. related to the subject</td><td rowspan="3">No M. Arch. related to the subject. Not even listed in the suggested list of M. Arch. Courses in the Appendix A of the COA 2006 document on M.Arch.(Architecture, 2006)</td></tr>
<tr><td>4</td><td>Drawing</td><td>No M. Arch. related to the subject</td></tr>
<tr><td>5</td><td>Practice</td><td>No M. Arch. related to the subject</td></tr>
</table>

Figure 15.1 Table of masters in architecture courses in key subjects needed by architects
Source: Author

Architects, as Vice Chairman GCA Alumni North, as founder Urban Village Charitable Trust, as alumni of GCA and SPA Delhi Department of Conservation. The paper brings together my own professional experience of three decades as an architect and one decade with academia. It involves my insights into the syllabus after I had sufficient practice, to incorporate learnings from theory and practice. The structure of the syllabus approved by the Council of Architecture is used for discussions.

Discussions

Positioning Architecture as a Profession Within the Four Industrial Revolutions

Understanding of the four industrial revolutions with architecture gives us a perspective. Architecture as a profession was born with the first industrial revolution towards the end of the 1700s, with the use of coal power. This was one of the most significant period in architectural history, when innovations in masonry, reinforced concrete, and steel utilization started to take hold. Construction possibilities increased. Roles were divided, there was aggregation of labour force. Stronger, higher, longer and durable buildings, railways and bridges were constructed. The Second industrial revolution came with the discovery of gas. Architecture and other branches of education emerged due to plenty and the need to look deeper into the plenty. Economics, Sociology etc. were born in this era to look into the needs of people. Architects were awarded licences and the profession took a shape close to what it is today. The Third industrial revolution began in the 1970s with the electronics and nuclear energy. With it began modernism. After this age when technology funnelled the flow of capability and competence, an architect has become merely an interpreter of techniques. The fourth industrial revolution Education 4.0, which comes with automation and technology, has caused a change in the way people think and today architecture is in question with just design as a standalone strength of the architect.

Knowledge Acquisition

Sociology and Economics as subjects developed around the time of the first and the second industrial revolution, as shown in Figure 15.2. Both the branches evolved in the same time as architecture did. Artists like Michelangelo, builders and painters of the picturesque period, came within the umbrella of Architecture. Sociology and Economics became root knowledge to multiple thoughts, but on the other hand, Architecture gave birth to new branches of knowledge like planning, urban design, build-heritage conservation, landscape and it thus becoming an input qualification. On the contrary, sociology absorbed anthropology (AMU, studocu, 2024); economics integrated trade and accounting, mercantilism with its ambit (LaHaye, 2023). While mercantilism was for building a wealthy and powerful state, a form of economic nationalism while economics is the nature and causes of the wealth of nations. On drawing a comparison, we can say that the former is close-ended; the latter is an enquiry into the process. Understanding the evolution of Sociology as a field of knowledge requires an in-depth understanding of the change in the nature of economics, politics and the evolution of science. The awareness of the world, the acknowledgement of individual roles in the collective, replaced the limited nationalist spirit. Change from agriculture to industry and the inventions, changed feudalism and gave birth to capitalism. Interdependence of countries on each other for material and workforce recognized the united human perspective.

However, in this debate of information acquisition, the only commonality is that both sociology and economics evolved into people's methods, whereas architecture, in the fight for style and exclusivity, limited its definition to a select few. There is no license or practice for people's areas, for shelters, for villages and hence its

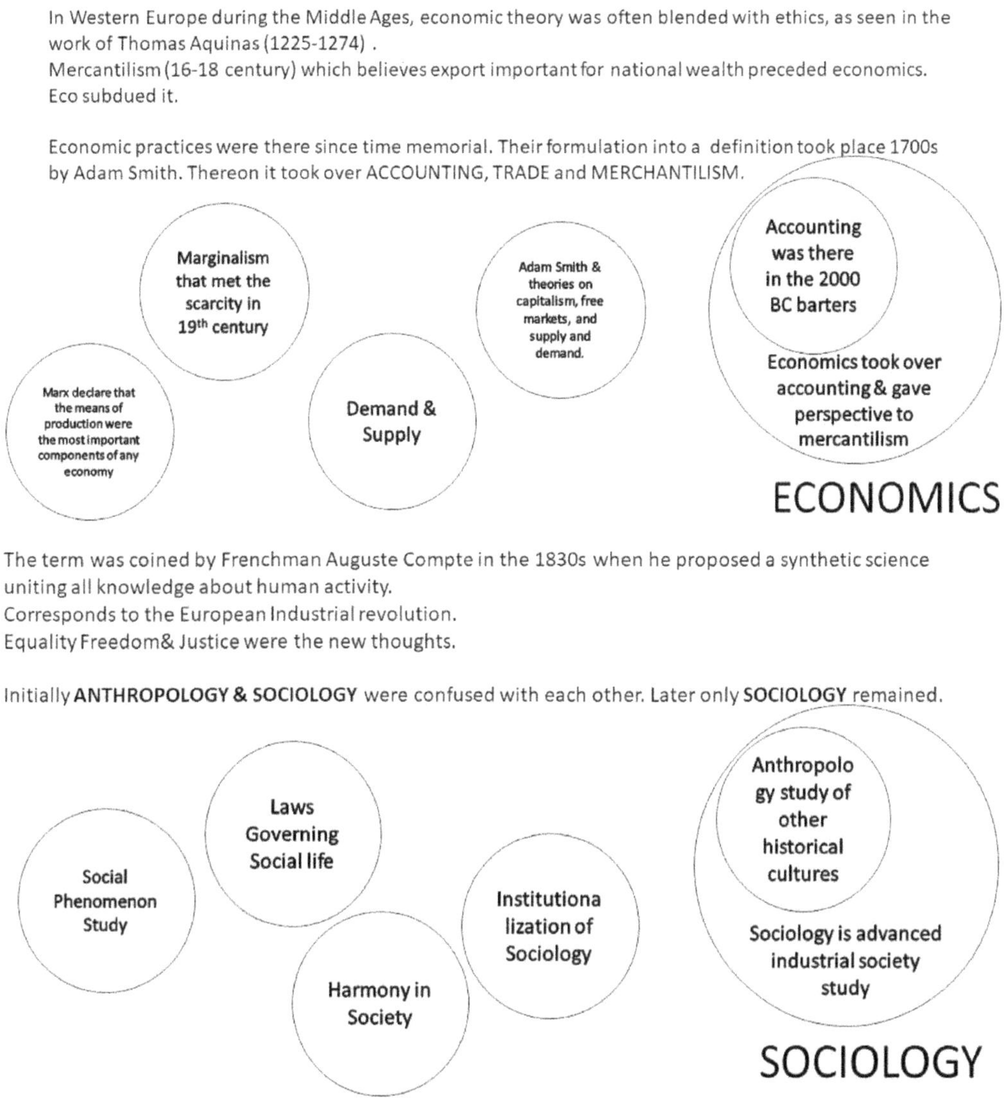

Figure 15.2 Evolution of sociology and economics, as subjects that aggregated knowledge, within their ambit
Source: Author

percolation is restricted. The result is that in the 21st century industrial revolution 4.0 where exclusivity became common information, architect is now merely a manager of many roles within an outline that fits all, an outline called architectural design. In the absence of a wide outreach and a knowledge base, architectural design, the root subject has not yet incorporated the climate dialogue, green buildings, adaptive reuse, and resilience as an integrated approach. They are optional over the geometric form based opulence generally considered as design. Geometry, measurement systems, crafts are no more integral to architectural design. Betterment of design is possible with better materials and better technology, and this too is out of the reach of the core strength of architecture. The intellectual acquisitions of lead thinkers of sociology and economics strengthened the foundation of the subjects and helped in them

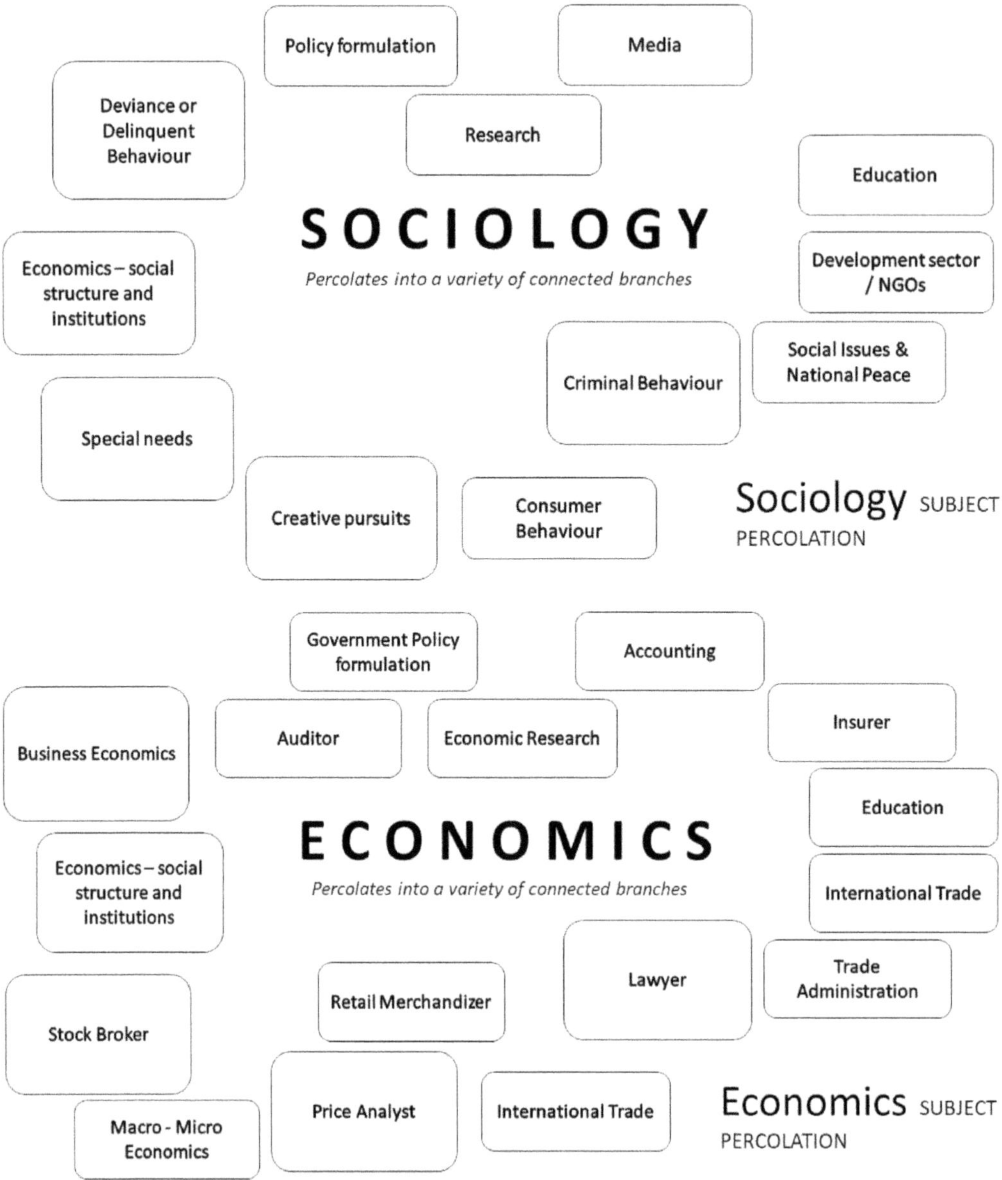

Figure 15.3 Subject percolation diagram explaining how the subjects have percolated into different other branches of knowledge and professions
Source: Author

being of better service to the commons. The process in Figure 15.3 explains the stages of development of knowledge along with the growth of the subject.

There are economists and sociologists who come to lecture architects for the betterment of the understanding of these subjects within design. Not just this, the PhD research in architecture requires social and economic surveys. Social research and economic research become a part of architectural research, but the use of architectural research method is yet to be developed. This Figure 15.3 explains the percolations of these two subjects. We must acknowledge that most of these heads are also architectural domains.

There are two important questions:

Q1. Is design a part of PhD research in architecture?
Q2. Do architects take lectures within the economic and sociology fraternity? Architecture is an indisputable part of all economics and of all societies.

The strength of a subject develops from a strong core. Architecture as a subject is yet to be developed as a field of study, as a subject (Arjit Kumar Gupta, 2020). It is dwindling between being a profession and being a branch of knowledge at the root of new and emerging professions. We can also see that subjects like sociology developed special branches where it studied special needs, criminal behaviour, delinquent and deviant behaviour, but on the contrary, mainstream architecture stayed away from vernacular, disaster affected, villages, urban villages etc. Engineers and planners write policies. They wrote the National Building Codes, which architects now follow.

Emerging Scope of Architecture

The scope of architects outside the Council of Architecture licence to practice is very vast. Figure 15.4 illustrates the areas where architects now earn. The list integrated from personal knowledge and from the practice of architects, is still only a small reflection of the profession. This list is not covered within the bachelor's program as subjects, nor listed in the practice license. There is thus a need for redefinition and acquisition of these subjects within architecture.

Knowledge Intrinsic Only to Architects

The syllabus of the course has recently inculcated a brief introduction of urban design and landscape as electives, but there is no field of knowledge that is intrinsic only to architects. A design process involves a general of nine stages through which implementation takes place. Even within an architect's office, the least paid remains an architect, because they lack training to handle the nine stages. Apart from the peripheral involvement in design, the principal

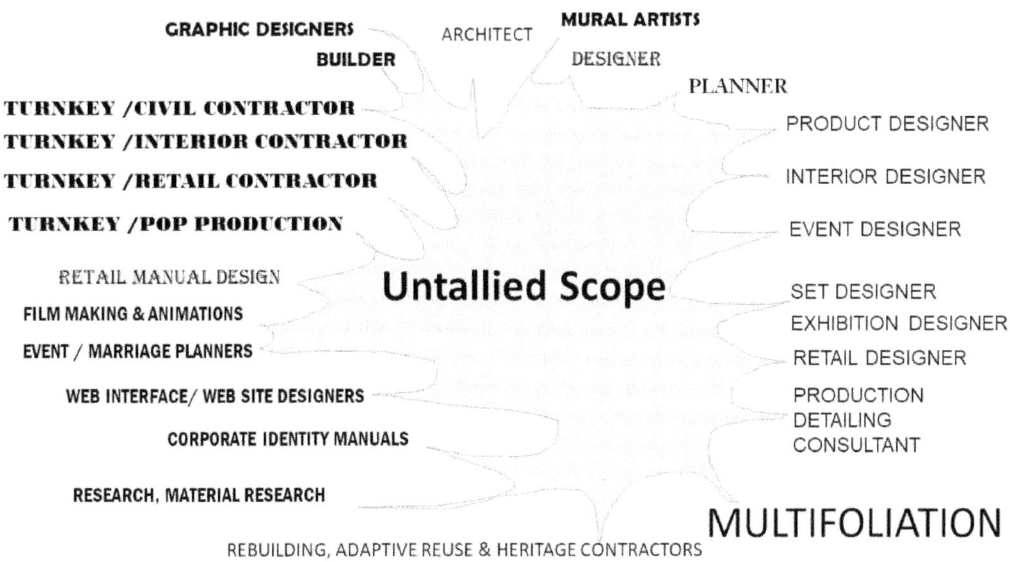

Figure 15.4 Areas where architects find economic generation, employment, work and earn
Source: Author COA TRC lecture *(Khatri, 2024)*

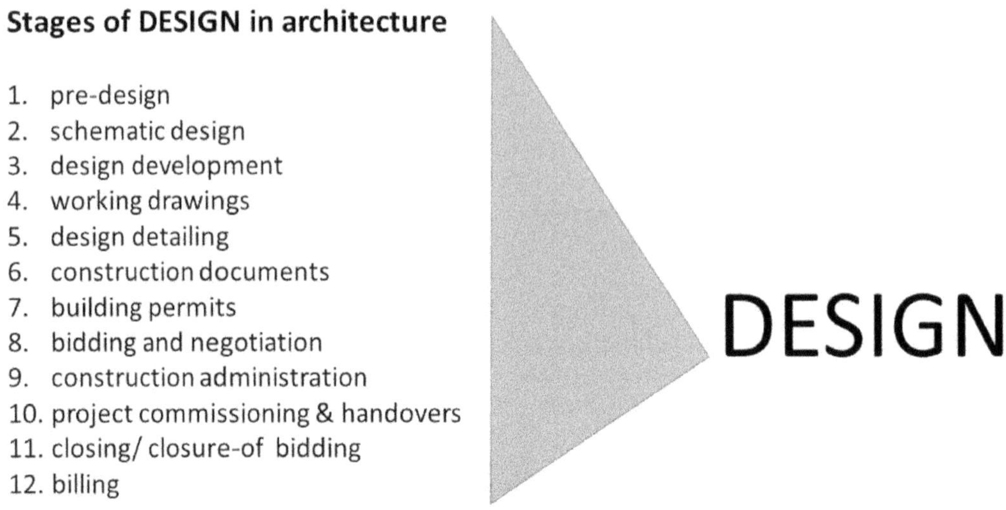

Figure 15.5 Nine stages of design
Source: Author (Khatri, 2024)

architect entrusts all stages to people from other branches of education. No architect can entrust structural design to another, or services design to an architect. Engineers developed even the National Building Code while architects all along their training and practice have been digging only into design. AI or new software and tools are also now doing this in a better way.

Syllabus and Course Credits

Within the course of architecture, the credit focus is on tools and not on knowledge intrinsic to architectural practice. Subjects have become outdated and general and hence the capability to use a subject directly in the profession is not there. Accounting, marketing, client management, bidding etc. is not taught. Fresh graduates are thus not ready for practice. The result is that the practice runs through the involvement of other disciplines and there are new courses replacing the more successful economic trends, to the extent that they fall out of the preview of an architect. Planning is one such example.

Subjects in focus in the credits in COA Appendix A and the syllabus of university shows that there are certain projects which are in focus and they are neither extended into post graduate studies, nor are they part of doctoral studies.

Conclusion

Architecture is till-date the profession responsible for buildings and thus responsible for shaping economics. The practice licence of architecture must take within its ambit the successful outreach of the profession.

Architect must learn all stages in the process of the execution of a built project. The profession can be strengthened when architects are successful at undertaking all stages of structural design, construction, MEP, energy, urban design and all these activities fall within the licence of architectural practice.

Syllabus needs a complete overhaul. Design structures and construction must integrate. There must be points that marry

ARCHITECTURE OUTPUT TOOLS		ARCHITECTURE INPUT INTERCONNECTIO NS		BEARINGS OF ARCHITECTURE	
Design	6	History	2	Arts and Graphics	2
Construction	4	Sociology	3		
Structures	2				
Communicatio n	3				
Drawing	2				
TOTAL	17		5		2

AKTU COURSE CREDITS

Figure 15.6 Course credits from AKTU syllabus
Source: AIT SAP syllabus, A.P.J. Abdul Kalam Technical University

Sl. No.	Subjects of Examination	Minimum Number of periods 50-60 minutes duration
1	Architectural Design	600
2	Building Construction	360
3	Building Materials and Sciences	60
4	Architectural Drawing and Graphics	360
5	History of Architecture	120
6	Workshop Practice	120
7	Landscape Design	60
8	Structural Mechanics and Theory of Structures	300
9	Surveying and Levelling	60
10	Building Services and Equipment	90
11	Humanities	60
12	Estimate & Costing	60
13	Principles of Human Settlements	60
		2310

Figure 15.7 Course credits in appendix A COA minimum standards of architectural education regulations 1983
Source: Council of architecture website

these subjects into one unified understanding, rather than function in compartments.

Education of important subjects like history, sociology and economics within the course is very superficial. History of architecture could be a master's subject while the other two could be integrated into a critical and operational understanding of the profession.

At the masters level, courses that give better specialization into building, must be evolved. There is a complete disconnect between the B. Arch. and the M. Arch. Programmes in India.

Acknowledgement

Most of the studies originate from research undertaken for the COATRC Bhopal program with the SRM University at which the author was a speaker. This paper and the diagrams therein thus acknowledge the lecture by the author on *"Configuring architectural education for better reabsorption and correlation with social sciences"*.

References

Aditi Deep Singh, J. S. (2022). Reflections on the current Architecture Education in India: Realities and reforms. *JETIR - Journal for Emerging Technologies and Innovative Research*, 482-489.

AMU. (2024). *studocu.* Retrieved from Emergence of Sociology and Social Anthropology: https://www.studocu.com/in/document/aligarh-muslim-university/general-sociology-i/emergence-of-sociology-and-social-anthropology/22126603

Arjit Kumar Gupta, D. R. (2020). *www.spav.ac.in/pdf/ae2020.* Retrieved from www.spav.ac.in: chrome-extension://efaidnbmnnnibpcajpcglclefindmkaj/https://www.spav.ac.in/pdf/ae2020.pdf

Beninger, P. C. (2014, February 25). The Future of Architectural Education in India – The Crises and Challenge. *Architecture Live.*

Chandavarkar, P. (2019). A Pedagogy-Centred Curriculum. *New Design Ideas, Vol 3, No.* 2, pp. 124-129.

Chauhan, A. (1999). *Humanise Architectural Education for the 21st Century.* Retrieved from Architexturez Newsletters: https://architexturez.net/doc/az-cf-21656

Chhaya, H. D. (1999). *In Symposium on New Directions of Architectural Education - the necessity of a cultural paradigm responsive to the majority.* Retrieved from Architectural Education a Holistic Framework, Greha: https://architexturez.net/doc/az-cf-21654

COA. (2006). *Minimum Standards for Architectural Education Guidelines for Post-Graduate Program.* Retrieved from www.coa.gov.in: https://www.coa.gov.in/index1.php?lang=1&level=0&lid=11&linkid=7

Dejtiar, F. (2018, July 30). *Etymology in Architecture: Tracing the Language of Design to its Roots.* Retrieved from arch daily.

Fabian Dejtiar, T. C. (2018, July 30). Etymology in Architecture: Tracing the Language of Design to its Roots. *Arch. Daily.*

Khatri, A. (2024). Refiguring Architectural Education for better Correlation. In V. S. Khatri, *Future of Architectural Education in India* (pp. 259-269). New Delhi: AITSAP.

LaBarre, S. (2008, October 1). *www.metropolismag.com.* Retrieved from Metropolis: https://metropolismag.com/programs/truth-in-numbers/#:~:text=Consider%20the%20figure's%20many%20permutations,in%20all%20likelihood%20only%20two

LaHaye, L. (2023). *www.econlib.org.* Retrieved from Econlib: https://www.econlib.org/library/Enc/Mercantilism.html

Mehta, J. (2001). *Towards a new Pedagogy. Architexturez Imprints.* Retrieved from www.architexturez.net: https://architexturez.net/doc/az-cf-21236#off-canvas

Mehta, J. (2020). *Decolonizing Architectural Education in India. Architexturez Imprints.* Retrieved from www.architexturez.net: https://architexturez.net/doc/isbn/81-902816-6-6

Rotondi, M. (2012, November 01). Architectural Education. Arch Daily. (R. Frey, Interviewer)

Smita Khan, H. K. (2019). Nemesis in the Genesis, Reforming Architectural Education in India. *Jomard Publishing*, 148-153.

Wakeford, B. B. (2008). *Expanding Architecture: Design as Activism.* London: Metropolis Books.

16 Origami in architectural education

Devi Renuka[a]
Assistant Professor (Grade 2), School of Architecture, VIT, Vellore, India

Abstract

This research articulates a pedagogy of origami in architectural education. It indicates the relationship between architecture and origami. In the present, rapid construction has made origami more relevant, as it lends to both, prefabricated construction and manufacturing of modules. Digital computational methods complement the design process to identify principles, concepts, elements, geometries, and technologies. This forms the theoretical framework which is clarified by the practical construction of models both in physical and digital forms. The significance of origami in generating holistic approaches to architecture education from theory, concept, design, and construction to meet the needs of people, is systematically articulated through this research. Design concept development and structural principles shall embrace origami demonstrations. A series of models reveal the construction of forms and mathematical geometries of three-dimensional patterns that could be evolved from planar surfaces. This research focuses on anthologising the application of origami principles across various courses of present architectural education and understanding the efficacy of the process among the students. Existing studies and origami practices aid in mapping the folding with the architectural courses. The better learning of the students is gauged through an origami pedagogy workshop. At the end structured survey is conducted among the students to assess the efficacy of the learning through origami. The results are analysed concerning cognitive theories to arrive at conclusions. The study shall be applicable in the current architectural education and serve as a process for design development in practice.

Keywords: Foldable architecture, folded plates, innovative teaching, modular origami architecture, origami architectural design, origami structure

Introduction

Origami as a word has its origin in the Japanese word orukami. Oru means 'fold' and kami means 'paper' (Wang-Iversona and Lang, 2011). Origami is the art of folding a single sheet of paper into a three-dimensional form with various complexities (Lang, 2004). Origami has percolated its way into many engineering, medical, and architectural applications. The foldable, flexible properties of origami are explored in architecture (Marco Meloni, 2021). Origami has been applied in architecture by various architects since the 19th century. Numerous design applications have been made possible by origami, which has proven to be an abundant source of inspiration. New techniques and forms have also been found. It has been determined that origami helps with pointers and ideas on how to imagine various possibilities in architecture classes. The study concludes that origami provides an interface to develop cognitive expertise in spatial configurations and finding shapes. This study investigated whether origami should be part of the architectural curriculum with respect to exploring form, structure, material, and other possibilities.

[a]renukadevi.n@vit.ac.in

DOI: 10.1201/9781003571018-16

Research Questions

Based on the existing practices in teaching and learning, the following questions are contemplated.

1. How do we use origami in the courses across the architecture curriculum?
2. Does origami help students in understanding various concepts and form development?
3. Do the students prefer to have origami-based learning as part of their architecture curriculum?
4. What are the areas in which origami principles could be used in architectural education?

Objectives

- To investigate the use of origami principles to teach architectural design and structural concepts.
- To map the student's perception in including origami in architectural education
- To explore various possibilities of including the origami principles in the curriculum through recommendations.

Literature Review

Origami and Architectural Education

Author Megahed suggests origami as a means to study architectural forms, spatial evolution, and aesthetics for students (Megahed, 2017). The innovation in form creation and methods of construction are explored through various exercises given to the students. The study 'Origamics in Architecture' discusses various applications of origami in space, medicine, and architecture (Hagiwara et al., 2009). Pertaining to architecture, the form creation, origami for computational architecture, and kinetic architecture are discussed in detail (Hagiwara et al., 2009). Students and teachers were involved in an exercise and survey to gauge the effectiveness of origami in the design along with the attempt to integrate STEM carried out by the authors Kennedy et al. (2016). In the paper 'transformable building structures in architectural engineering education', the integrated disciplinary approach driven by flexibility, modularity, deployability, and transportability through origami is studied (Phocas et al., 2022). The study explores a range of underlying principles, one of the principles being origami. Authors Nestrovic et al., explores the understanding of geometric principles of folding in the development of new designs among architecture students (Nestorović et al., 2016). A course was introduced to develop algorithms using origami folds and students created a prototype resulting in innovative design solutions. In the study published by Gur et al., architecture students were given exercises to develop three-dimensional products based on different types of symmetry. In conclusion, it is been suggested to include various elective courses for the students based on origami and visualization (Hülya Gür, 2020).

Origami in Building Structures, Parametric and Applications in Practice

Besides education, the use of origami in practice is studied by various authors. The study by Xiang et al. (2020) reveals the use of origami-based metamaterials and the energy absorption capabilities of the origami-based structures (Xiang et al., 2020). Among engineering and mathematics, the application of origami in architecture is discussed in the study done by authors (Marco Meloni, 2021). The study discusses origami in developing building skins that are not only aesthetically pleasing but also structures/facades that can enhance structural performance, reduce energy demands, and develop adaptive and climate responsive.

Innovative Education Methods in Architecture Curriculum

In recent years, due to the advanced technological development, the architecture education process in India is to be reviewed. In that context, several discussions and Development programs are conducted by the Council of Architecture, India. Many practicing architects have put forward their perception of altering architectural education to align with the existing needs and developments. Architect Vibuti Sachdev has expressed her concerns about keeping up with the present trends and challenges According to architect Sachdev, the architecture curricula are to be more student-centric and experiential learning (Sachdev, 2023). Architect Joanne Elliot in her blog, mentions that architectural education should be generative, eco-friendly, and experimental (Elliot, 2023). In both the mentioned studies the involvement of digital media is emphasized to suit the current trends. Innovative teaching approaches including virtual reality are explored. The study carried out by authors Hozan et al., in Iran has explored the use of collaborative learning methods through virtual reality in architectural education. Various levels of collaboration through VR are discussed (Latif Rauf and Shareef, 2022).

Various innovative education methods have been explored by various academicians and researchers in the last few decades.

Methodology

The research on the necessity of origami principles in architectural education is conducted by experimental method. An origami workshop is conducted for two days for the fresher's program in architecture (B. Arch.) for 60 students. The rubric in the curriculum which covered fractals was mapped with the two-day origami workshop. The concepts of fractals and the Fibonacci series were explored using Tomoko Fuse's origami Navel shell (Fuse, 2012). The mathematical principles behind constructing this model are deeply investigated by a group of students in the USA (Kaur and Soong, 2021). The students were given clear instructions on the folding methods and the various steps involved. Three sizes of models were made by each student to understand the resulting proportions. The use of 1.6 as a measurement guideline to space the curvature lines of the shell represented the golden ratio. However, some of the students had accidentally experimented with less or more than the golden ratio which resulted in different curvatures. Students took two days of 8 hours to learn and complete three models of Fuse's shell as shown in Figures 16.1 and 16.2.

Figure 16.1 Students folding Tomoko Fuse's origami shell during origami workshop as part of foundation design course
Source: Author

Figure 16.2 Shells made by the author to demonstrate the concept of golden ratio and fractals in the workshop
Source: Author

The outcomes of the origami workshop were recorded in the form of photography, and videography. A structured survey was circulated among the students to understand the application of origami in architectural education. Out of 60 students 45 students have recorded their responses. The questions of the survey were meticulously framed to discern the influence of origami with respect to experimental learning, conceptual underpinning, and geometric explorations. Following are the questions circulated to the students to fill in along with the Likert scale of rating. One being strongly disagree, two being disagree, three being neither, four being agree and five being strongly agree. The last question was open-ended for the students to express their opinions without restriction. The results are interpreted using the Likert scale method and %ages. The outcomes are discussed and the recommendations are suggested. The faculty-to-student ratio worked out was 1:20. The following list of questions was floated using Google form after the workshop ended.

1. Did you understand the concepts of fractals through origami in the workshop conducted?
2. Were you aware that origami can be used in design, architecture, and art before this workshop was conducted?
3. Your preference for making an origami model over a golden ratio drawing
4. Does working with origami principles boost your creativity?
5. Do you prefer origami principles to be used in teaching other courses as well? Architectural design studio, structural system evolution, etc.
6. Is learning through origami is more fun and interesting than the regular learning methods?
7. Will you apply origami principles in your architectural design in further semesters?
8. Do you recommend some of the origami principles to be brought as part of the curriculum for learning design possibilities and structural possibilities?
9. Do you agree that origami principles can explain concepts more clearly than theoretical learning of building structures?
10. In a few lines, explain your thoughts on using origami as a teaching method in an architecture course.

Limitations

Most of the students managed to finish the model with the desired outcome. Some of the students needed assistance in collapsing the model and shaping it. The one-to-one engagement with the students was time-consuming at the same time satisfactory. To engage a large number of students more trained faculties are required to guide. The studio faculties have learnt the model, and assisted with folding and shaping for 60 students. A proportion of students had difficulty in comprehending the folds and a few of them showed less interest.

Results and Discussion

First-year students of architecture were asked for their opinions on using origami techniques to understand the concept of fractals. In the results, it is suggested that the golden ratio/fractals are better understood by making the origami shell model. 44% of them rated agreed and 20% of them rated strongly agreed. It is interesting to know that the students are aware of the relevance of origami in art and architecture. Around 47% of the students strongly agreed and 31% agreed to the same. The rest of them were able to relate the applications of origami in architecture after the presentation was delivered. Although students agreed that the inclusion of origami in architectural learning boosts creative thinking, the opinion on making this particular model to understand fractals was neutral. The reasons identified were the lack of prior knowledge, and lack of confidence since the results are dependent on the steps followed. The Fuse's shell model belongs to the intermediate complexity level. Some of the students did well in the first attempt and some of the students got familiar after the second attempt. In the results, students have shown immense interest in learning concepts of structure, form evolution, and flexibility through origami folds in other courses across architectural education. The origami folds can be mapped with courses such as Architectural History, Design of Structures, Architectural design, Ideation, and Architectural theory.

As listed in Table 16.1 40% of students strongly agree and 22% of the students agreed that the use of origami principles in architecture education is interesting and fun. Something that deviates from the existing teaching-learning practices and at the same time anchored to the subject to be taught. The students of architecture have overall agreed that the origami principles in learning various concepts should be inculcated as part of the architectural curriculum. The results of the 9th question are intriguing. Students have given neutral responses to the question posed on origami preferred over learning in the form of theory. It is clear that the students prefer origami as part of architectural education and point out that it is not practical to learn all of the courses and concepts through origami.

For the tenth question on origami-based teaching in architectural education, which was open-ended, students were given a wide variety of responses. The majority of the responses were on encouraging origami to be a part of curriculum in architecture. Very few showed disinterest. One of the students mentioned "Origami helps us learn about forming spaces. it was fun doing it in groups. We fold paper to make little structures. We start with easy shapes and move to harder ones. It's like a fun puzzle that teaches us about real structures". The other students said 'This way is more interesting than the theory classes'. The level of concentration is increased. It triggers creativity. Some of the students critically expressed their opinions. For example, one student has said 'I found it quite tedious to do but origami was a creative way to learn'.

Question no 10 provided students with freedom to express their challenges in origami method of leaning. 10% of students expressed challenges such as difficulty in handling the material, repetition, boredom and larger time consumption.

Conclusion

The experiment on understanding the interests of students, challenges faced by faculties, conceptual understanding and application of origami principles in architectural curriculum was quite enthralling. Overall students have shown immense interest in accepting the use of origami principles in architectural education. The origami principles could be used to teach courses such as design of structures, architectural design, ideation and architectural theory. Students have expressed that the theory and other practical exercises such as model making, computing shall be combined along with origami principles. It is suggested from the survey results that origami alone is not enough but combinations of various methods that includes origami folding methods could be administered.

Various levels of complexities in origami could be introduced based on the courses. The other way of looking at it is to offer the origami in Architecture as an elective course where students who are interested shall engage themselves in the same. However, the results of the study show that 90% of the students are open to experimenting with

Table 16.1 Likert scale interpretation

Questions	Strongly disagree (%)	Disagree (%)	Neither (%)	Agree (%)	Strongly agree (%)	Likert interpretation
Understanding the concept through origami	0	2.2	15	44.4	20	3.8 (Agree)
Awareness of using origami in architecture	4.4	4.4	13.3	31.1	46.7	4.1 (Strongly agree)
Preference of origami model to learn golden ratio	4.4	8.9	46.7	24.4	15.6	3.3 (Neither)
Creativity boosting	4.5	4.5	25	25	40.9	3.9 (Agree)
Preference with respect to origami-based learning in other courses	4.4	6.7	24.4	40	24.4	3.7 (Agree)
Interest and fun in learning	8.9	6.7	22.2	22.2	40	3.7 (Agree)
Application in design – form evolution	2.2	6.7	24.4	46.7	20	3.7 (Agree)
Recommendation in curriculum	2.2	4.4	42.2	26.7	24.4	3.6 (Agree)
Preference over theoretical learning	4.4	13.3	37.8	26.7	17.8	3.4 (Neither)

Source: Author

origami principles to explore various topics in architecture. Origami-based teaching demands proficiency in origami folds from the faculties. Further research by academicians and scholars shall be carried out in specific mapping of models to various topics in each course.

Recommendations

1. Origami-based teaching modules could be introduced in courses such as Design of Structures, Architectural Design, Ideation, and Architectural Theory across various semesters.
2. The origami geometric models could be mapped with specific topics in various courses in the curriculum.
3. An elective course on origami learning of form evolution and structural design could be offered as part of senior semesters.
4. Origami teaching/learning methods and means shall be embedded in the curriculum.
5. Interested faculties shall be professionally trained in origami-based pedagogy to enhance better deliverance.
6. The outcomes of origami–based learning shall be recorded and necessary moderations shall be made to improve the experimental learning capabilities of students.

References

Elliot, J. (2023). The Future of Architecture Education: Trends, Innovations, and Opportunities for the Next Generation. Retrieved from Futurist. https://www.futuristarchitecture.com/108611-the-future-of-architecture-education-trends-innovations-and-opportunities-for-the-next-generation.html.

Fuse, T. (2012). Spiral Origami Art Design. Viereck Verlag.

Hagiwara, I., Sorguç, A., and Selçuk, S. A. (2009). Origamics in architecture: a medium of inquiry or design in architecture. *Middle Eastern Technical University Journal of the Faculty of Architecture*, 26(2) DOI: 10.4305/metu.jfa.2009.2.12.

Hülya Gür, Y. I. (2020). Reflections of origami to architecture. *Social Mentality and Researcher Thinkers Journal*, 6(37), 1879–1887.

Kaur, G., and Soong, C. (2021). Mathematical modeling of the origami navel shell. *Journal of Student Research*, 10(3).

Kennedy, J., Lee, E., and Fontecchio, A. (2016). STEAM approach by integrating the arts and STEM through origami in K-12, institute of electrical and electronics engineers inc. In Proceedings - Frontiers in Education Conference, FIE, (10.1109/FIE.2016.7757415).

Lang, R. J. (2004). Origami: Complexity in Creases *Engineering and Science*, 67(5). pp. 5–19.

Latif Rauf, H., and Shareef, S. S. Najim Nizar (2022). Innovation in architecture education: collaborative learning method through virtual reality. *Journal of Higher ETheory and Practice*, 21(16). DOI:10.33423/jhetp.v21i16.4909.

Marco Meloni, J. Cai, J., Zhang, Q., Daniel, S. H. L., Li, M., Ma, R., Parashkevov, E, T., (2021). Engineering origami: a comprehensive review of recent applications, design methods, and tools. *Advanced Science*.

Megahed, N. A. (2017). Origami folding and its potential for architecture students. *Design Journal*, 20(2), 279–297. doi:10.1080/14606925.2017.1270511.

Nestorović, M., Milošević, J., Nestorovic, P. and Maneski, M. (2016). Instrumentalization of origami in construction of folded plate structures - design, research and education. *Spatium,* 1(35). pp. 22–29.

Phocas, M. C., Matheou, M., and Haase, W. (2022). Transformable building structures in architectural engineering education. *Architecture, Structures and Construction*, 2, pp. 183–198

Sachdev, V. (August 23, 2023). Revolutionising architectural education: innovations, challenges, and impact. *India Today* www.indiatoday.in/education-today/featurefilia/story/revolutionising-architectural-education-challenges-and-impact-2420309

Wang-Iverson, P., and Lang, R. J. (2011). Origami 5: Fifth International Meeting of Origami Science, Mathematics, and Education. MA United States: A. K. Peters Limited.

Xiang, X. M., Lu, G., and You, Z. (2020). Energy absorption of origami inspired structures and materials. *Thin-Walled Structures*, 157. doi: 10.1016/j.tws.2020.107130

Appendix

1. Did you understand the concepts of fractals through origami in the workshop conducted?
45 responses

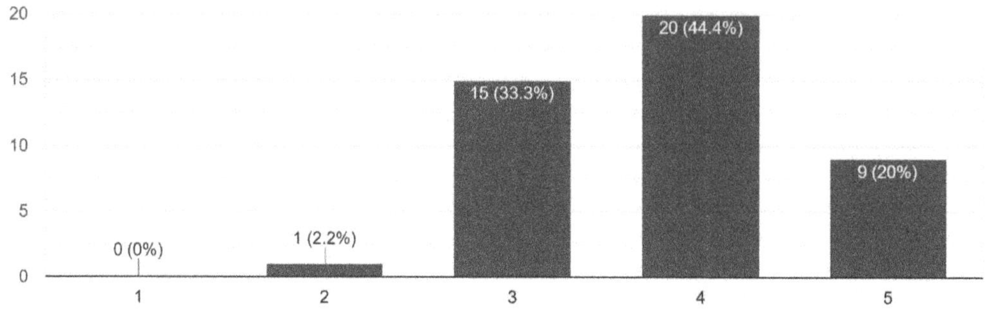

2. Were you aware that origami can be used in design, architecture and art before this workshop was conducted?
45 responses

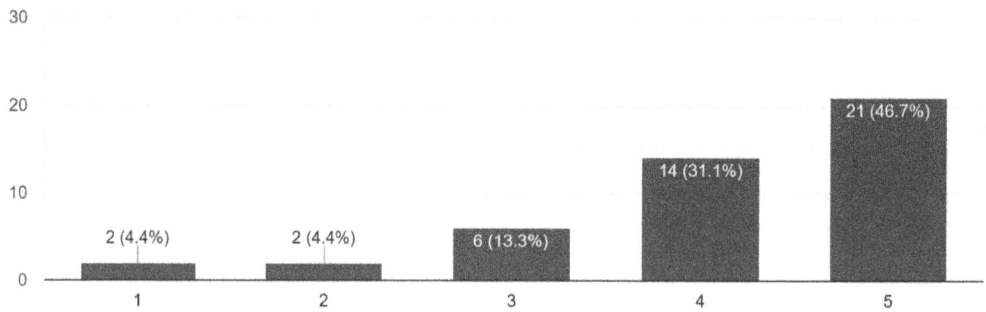

3. Your preference of making an origami model over a golden ratio drawing
45 responses

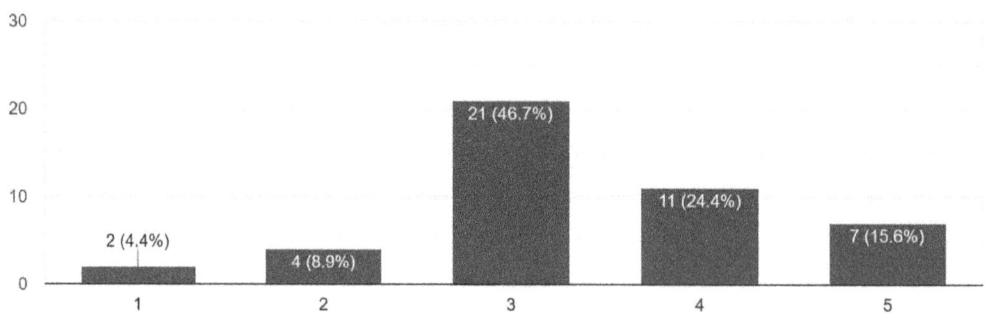

4. Does working with origami principles boost your creativity?
44 responses

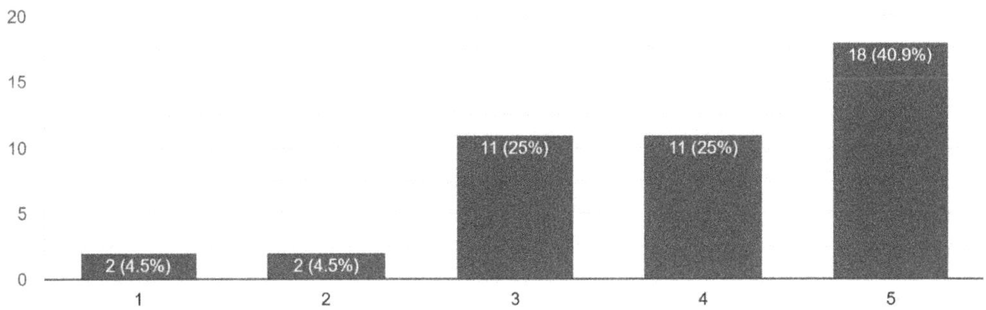

5. Do you prefer origami principles to be used in teaching other courses as well? Architectural design studio, structural system evolution etc.,
45 responses

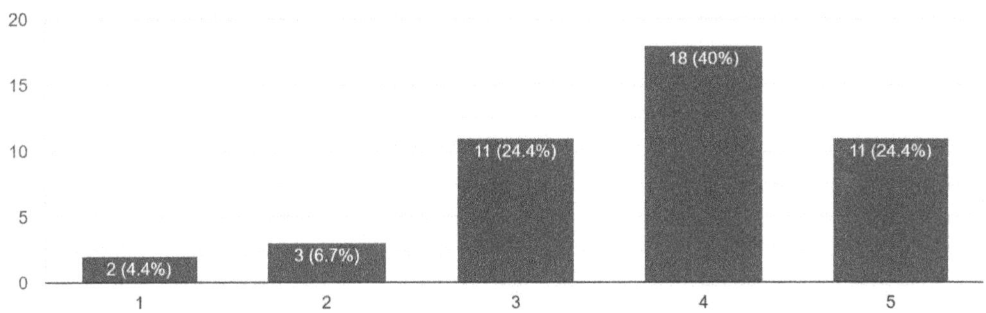

6. Is learning through origami is more fun and interesting than the regular learning methods?
45 responses

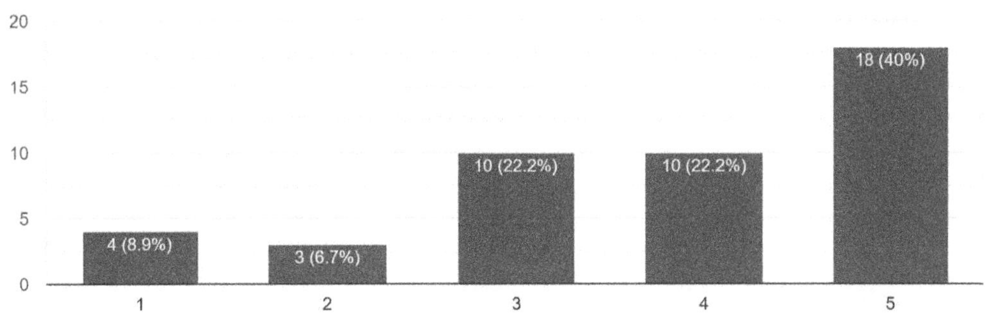

7. Will you apply origami principles in your architectural design in further semesters?
45 responses

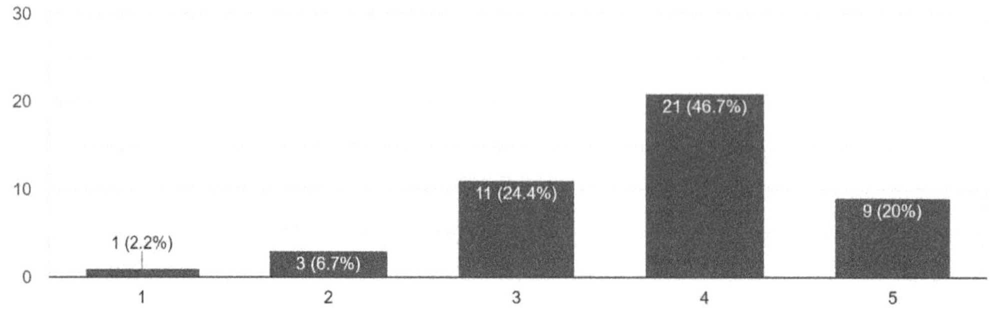

8. Do you recommend some of the origami principles to be brough as part of the curriculum for learning design possibilities and structural possibilities?
45 responses

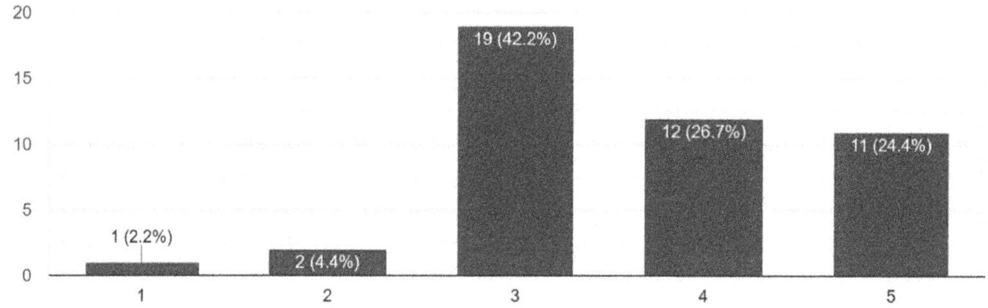

9. Do you agree that origami principles can explain concepts clearly than theoretical learning of building structures?
45 responses

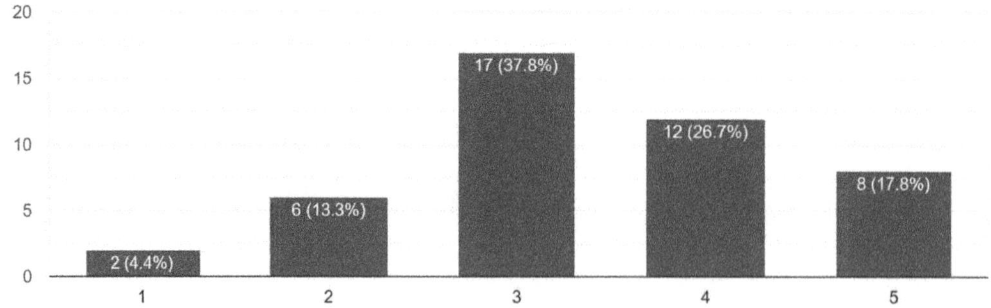

10. In few lines explain your thoughts on using origami as a teaching method in architecture course.

- Good but not much innovative
- It was a fun and very informative and creative way to learn. We could be as creative as we want with origami.
- It can be used because it can tell from how it is made and each step tells how the building is made and where it is balanced and how is the core structure of it is built, etc.
- The folding and cutting of craft paper to make origami is a fun aspect not really seen in academia and the practical applications of it especially in architecture and design makes a lot of sense to have it in the syllabus.
- This way is more interesting than the theory classes. The level of concentration is increased. It triggers the creativity.
- Origami is an both an interesting and intricate form of art. Origami model we made gave us an understanding about the importance of proportion in any design.
- Origami is a useful tool for teaching architectural concepts such as geometry and spatial design. It offers hands-on learning and problem-solving, but it should be integrated with traditional teaching methods for a comprehensive education.
- It actually gives more design possibility to the structure and the concept of fractals gives a structure to it.
- It was fun learning and composing all the pieces together.
- Nice way to explore and understand concepts easily.
- It's a very good teaching method to be applied in architecture course.
- It was a refreshing way to learn architectural elements and concepts.
- Doing the same model again and again was boring and exhausting.
- Origami is mostly done through hands on work. So, it is very useful and helpful for us architects to study and learn about their structure and dimensions.
- It's very practical and hands on work Makes it very interesting to work on.
- Origami helps us learn about forming spaces. It was fun doing in groups. We fold paper to make little structures. We start with easy shapes and move to harder ones. It's like a fun puzzle that teaches us about real structures.
- It's engaging and boosts our thinking process and creativity.
- It helped us understand geometry and ratios more.
- I found it quite tedious to do but origami was a creative way to learn.
- I think it's a fantastic initiative, and I'm eager to see the outcomes.
- It helps us to get knowledge about the materials and it helps to boost our creative ideas more effective than normal kind of teaching.
- It was creative. It was innovative.
- Not that amazing to take over totally One exercise was ok. We should use other materials too.
- It can be time consuming.
- The pattern and the structural rhythm of creating an art into an architectural design can be understood by using origami.
- Origami is something which helps us to creatively involve in making models. These create structures and models which look good and different.
- It helps us recognize some design principles. It also allows us to visualize outcomes of certain forms as we are constructing them. It is very hands-on and while it is fun to work with, it is also an interesting challenge.
- Using origami provides a good hands-on experience and allows us to remember the concepts better with clear understanding. It is very engaging, allowing us to interact with our peers and learn from each other.

17 Possibilities of exploring experiential learning in studios through design-build program in architecture education

Fathima Rishin Razak[1,a], *Chithra K*[2,b] *and Sanil Kumar*[3,c]

[1]Research Scholar at National Institute of Technology, Calicut, Kerala, India

[2]Associate Professor at National Institute of Technology, Calicut, Kerala, India

[3]Assistant Professor at National Institute of Technology, Calicut, Kerala, India

Abstract

Learning is best conceived in an alternative way rather than as an outcome. Kolb's experiential learning theory addresses active learning by increasing student engagement through various activities. Design-Build studio program is an activity having idea generation, graphical communication, experiment and construction as its stages; completing the learning cycle through continuum, perception and processing. This approach also addresses four kinds of competencies that are academically allocated into skill, knowledge, technology and design streams, as mentioned in the interim report of National Education Policy 2020. The paper tries to identify various possibilities of implementing a Design-Build studio brief in the foundation studio, continuing into the second year of the course, as an action research exploring experiential learning. The method may help students to indulge, carry forward and integrate learnings from one stage to the next in these experiments in a cyclic way, more than that occurs during the linear progression, usually followed in design studios. Hence, the study tries to identify the advantages of experiential learning over conventional cognitive learning, in achieving the pedagogical objectives. The brief has a multi-stage structure; where learning goals and desired objectives are framed, activities are assigned, and design development is taken up. The findings can evoke a new content of learning outcomes that is measurable and concise, with meaningful and achievable results. The knowledge gained by students is evaluated through questionnaire surveys. The study concludes by explaining the achievement of its aim, through a design-build system, to shift the architecture education objective away from the emphasis on shape elaboration, towards a concern for usefulness - through decision-making, teamwork and responsibility.

Keywords: Active learning, architecture pedagogy, design-build studio, experiential learning

Introduction

The Architecture design pedagogy in India is rooted in the oral culture and religious mysticisms concentrating on craftsmanship. This was community-based learning which was established through apprenticeship, making an individual acquire skills and knowledge over some time. The temple construction sites in Bhojpur and Ashapuri in Bhopal established this way of learning through a Design-Build process in the 11th century. The teaching-learning happened through oral instructions and construction drawings of details drawn on a 1:1 scale on the rocks on the site, which were to be carved and erected later on (Hardy, 2015). The efficiency of this method of instruction and execution is a proven fact through the built heritage of India. Experiential learning in India can hence be traced back to a centuries-old habitual system.

[a]ar.rishin@gmail.com, [b]chithrak@nitc.ac.in, [c]sanil@nitc.ac.in

DOI: 10.1201/9781003571018-17

Evolution of Architecture Pedagogy in India

The pervasive Education System was introduced in the colonial era by Lord Macaulay, mainly to train middle-level administrative staff to serve the British government (Balaram, 2005). In college education, The British Art Schools established in the early 19th century imposed Western techniques and visual idiom which caused lasting damage to the confidence of Indian craftsmen and craft learning. The concept of 'Swadeshi' ideology and 'NayaTaalim' by Mahatma Gandhi led to the formation of an experimental university for arts, crafts, and design - 'Santinikethan' founded by Rabindranath Tagore, which provided learning environments for freedom of expression and freedom of learning (Balaram, 2005). The learning objective was to understand and reinstate the 'language of hand', oriented towards craft production. Post-Independence, the need for modern design education and technology-based design was envisioned by Pt. Jawaharlal Nehru, which led to the formation of 'The Indian Report'. The report recommended a design institute intending to create an alert and impatient national conscience which has concerns about the quality and ultimate values of the environment, with functions of research, design and service to the society (Eamses, 1958). This led to the formation of Institutions like NID, Sir. J.J. School of Arts, SPA Delhi - which offered formal education in Design in India, with different pedagogical approaches. The design-build studio model is suggested as a pedagogical alternative, but fewer explorations have been done in the Indian context (Vriddhi, 2022).

Purpose and Scope of Research

The research tries to frame a more responsive architectural pedagogy that enables future architects to develop conscious design decisions and create humane environments that emerge in response to societal, cultural, and environmental needs - through the Design-Build studio model. It analyses learnings from the four stages of Kolb's experiential learning theory (Figure 17.1) and evaluates the competencies that are academically allocated into skill, knowledge, technology and design streams, as mentioned in the interim report of National Education Policy 2020.

Pedagogical Objectives should incubate Inquiry, Investigation and Discovery (Salama, 2005). The main research objective is to define the role of experiential learning as a Design-Build studio model, in understanding the design process through Investigation. The particular studio brief tries to address the following pedagogical objectives:

- Knowledge of how to address a public space for the benefit of society
- How to involve people affected by design, in the decision-making process
- Develop contextual knowledge and apply appropriate construction technologies
- Evaluate/justify the design decisions using real-world constraints

The paper tries to identify various possibilities of implementing a Design-Build studio brief in the foundation studio, continuing into the second year of the course, as an action research exploring experiential learning. The brief to develop a public gathering space into a familiar context was introduced as part of AR 22-11 Basic Design, for the first-year students. The design process included public surveys & interviews, workshops (Figure 17.2) and interaction with officials, and the final outcome was presented to the LSGD (Figure 17.3). A survey was conducted in order to know the responses of architectural students towards the current way of teaching architecture design by adopting a design-build program, and to evaluate the students' sensitivity in acquiring the pedagogical objectives. The analysis will help faculty to frame and sequence the process for future improvisations of learning objectives.

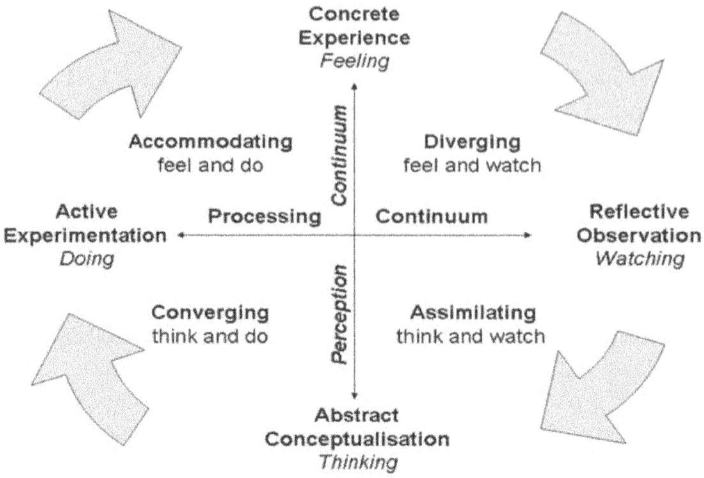

Figure 17.1 Kolb's experiential learning theory
Source: www.structural-learning.com

Figure 17.2 Site documentation, place-making exercise, seating space design
Source: Author

Figure 17.3 Project presentation at Kadavallur panchayat
Source: Author

Experiment with the Design-Build Studio Model

Literature references showcased community-based projects which offer an actual solution being adopted and realised, as an effective way of the Design-Build method (Salama and Wilkinson, 2007). The main theme of a design project in a Basic Design studio in the first year is to have an application of knowledge of the basics of design learned through various creative exercises. The project brief was to design a 'Public Gathering Space'. The area identified is currently acting as a crossing for people to reach to the bus routes connecting three different districts in Kerala. The site had the potential to become an identity of that place with historic significance and hence the intervention of **public art** in the form of sculpture and visual art, was also the focus of the design brief. The program components are initiated by students according to the users' needs and as a result of discussions with them. It mainly includes art installations, resting spaces, a kids' play area, an open gym and landscaped areas. Students participated in taking public surveys and interviews, interaction with government officials, lectures, site visits and documentation, making physical models, hands-on activities and the design-build program.

The challenge for faculties was the shortage of time because the project was not intended for the whole semester, but a culmination of the other creative exercises in the same studio (with a duration of 8 weeks). The methodology was experiential learning, which required the students to finally execute the design on site. The final execution of the project was taken over to the second-year studio for the same students, due to a shortage of time.

Experiential Learning through Design-Build Program

The Basics of Experiential learning involve the process of creating knowledge which gets grasped and transformed by the learner as a result of experience. Grasping occurs during Concrete Experience (CE) and Abstract Conceptualisation (AC), and transformation occurs during Reflective Observation (RO) and Active Experimentation (AE).

Architectural education must prepare students to think and respond constructively. This activity includes successive processes that start with an idea then a drawing, an experiment and construction, and finally an experience of the user (Amer, 2015). These four stages in Design-Build studio correlate to the four stages of Experiential Learning as shown in Table 17.1.

Aim and Activities to Achieve Learning Outcome

The course Basic Design is tailored by the faculty in a way that it initiates with visual communication, progressing to understanding the fundamentals of design and culminating in architecture design projects of a smaller scale, as per University of Calicut Syllabus. The Design-build studio model is applicable to the architecture design project and it consists of thirteen topics and activities as shown in Table 17.2.

Table 17.1 Experiential learning theory & architecture education correlation

	Architecture education	Experiential learning theory
Stage 1	Idea generation	Reflective Observation (RO)
Stage 2	Drawing	Abstract Conceptualisation (AC)
Stage 3	Experiment & construction	Active Experimentation (AE)
Stage 4	Experience	Concrete Experience (CE)

Source: Author

Table 17.2 Aim & activity to achieve the learning outcomes in the specified design-build studio

Step	Aim	Activity
1	**Permission from LSGD** Permission to access the site, context and propose a design, was sought from the LSGD authorities Thrissur District, Kerala, India	A meeting with the governing team was arranged and the idea of involving students in the design process was presented by the faculty team
2	**Project Introduction - 'Public Gathering Space'** The concept of public art and its significance was discussed, The site and existing scenario was introduced to students of first year	Presentation by faculty and project discussion among students
3	**Site and Context Documentation** Site visit to understand the existing condition of the site and its neighborhood	Students visited the site, documented the various activities within the context.
4	**Public Participation** To involve different user groups in understanding the design requirements	Students surveyed and interviewed different user groups, interacted with the government officials
5	**Design Development** Students had undergone through the design stages and produced individual designs as part of academic purpose	Brainstorming using the various terms collected from survey and Individual concept for Design development Review and presentation of the design
6	**Hands-on Workshops** Various hands-on activities were conducted for students to have an experience and also to take up design decisions while designing	Workshop on art and place-making: 1. Using umbrellas as shading devices in an open space in college campus 2. Wall art using 2D compositions 3. Seating spaces, reusing discarded glass bottles 4. Seating spaces using brick and concrete slab
7	**Project Presentation** Students combined the best potentials of their individual designs into one single solution Meeting with LSGD authorities, ward members, Kerala Traders and Industrialists Coordinating Committee (KVVES) was arranged and students presented their works	Entire class listened to the presentation of every student to pick the best solutions with potential. Discussion and debates Presentation of design
8	**Professional Service** Students along with the consultancy cell of the institution prepared approval drawings to be submitted to the PWD and handed over the same	Students studied about the various specifications of materials needed for the project, to prepare the drawings and report
9	**Presentation to the MLA** Students and Faculty presented the work to the MLA, Shri. A.C. Moideen	Students got an opportunity to present their thoughts and design before an elected representative member of the Legislative Assembly
10	**Meeting with the officials** A meeting with officials to discuss the on-site requirements and required extended support from other government departments like KSEB, KWA etc. was discussed	Students got an opportunity to know various government departments involved in the public infrastructure development of a town/city

Step	Aim	Activity
11	**Preparation of Estimate and Fund Allocation** Students prepared a rough estimate of the civil works, landscape and art installation Students prepared a brochure and started with fund collection	Students learned how to prepare specification, estimation and costing Students got a rough idea of the total budget and started searching for sponsorships with the support of the Institution, IIA and LSGD
	Next Phase of the studio to be done	
12	**On-Site Construction** Setting out and construction of the various design components by students	Students doing construction activities to see their design getting executed
13	**Handing Over** The public gathering space designed and executed by the students, to be handed over to the public for functional purpose	Students need to take a feedback after handing over the functional space designed and executed by them, and see how their thoughts have positively impacted a common user

Source: Author

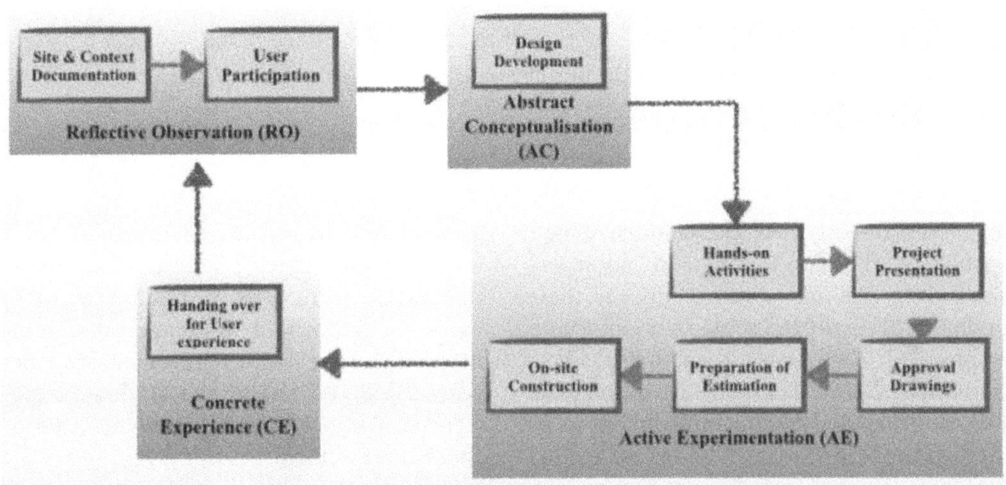

Figure 17.4 Experiential learning & design-.build correlation in the studio
Source: Author

The correlation among the processes followed in the studio and theoretical learning of Kolb's Experiential learning is represented in the form of a flow chart (Figure 17.4).

Pedagogical Objectives and Motivations of Design-Build Program

In the field of architectural education, design-build is a teaching alternative to the theoretical, conventional cognitive learning that is commonly featured in design schools (Canizaro, 2012) It tries to address and analyze the context based on a practical perspective as against the regular hypothetical academic projects. It gives a more complete understanding of the consequences of design choices and provides an experiment-based outcome. The nature of the Design-Build moves beyond traditional intellectual exploration, hovering between the academic (theory) and professional realities (practice) (Aydemir et al., 2020). Students

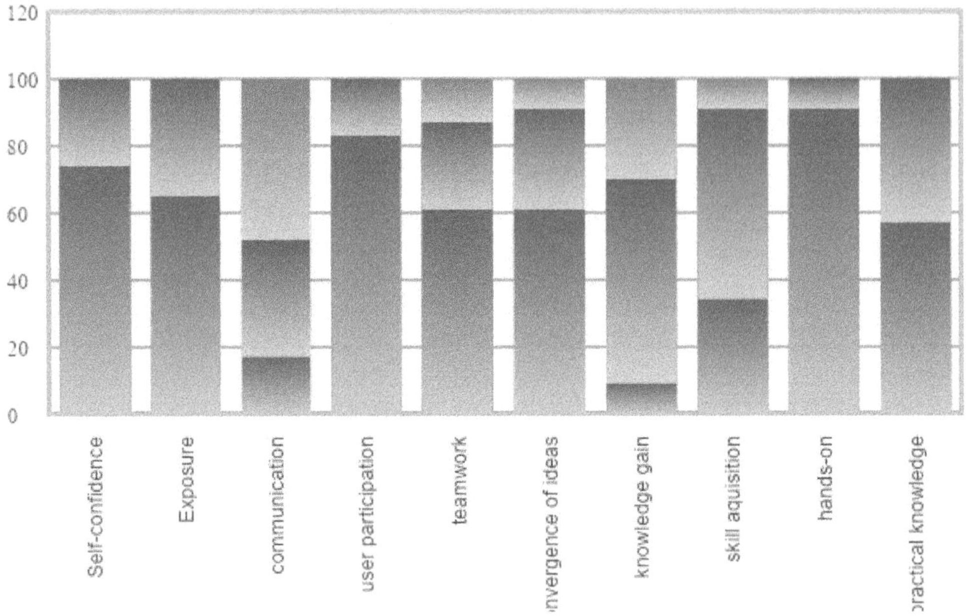

Figure 17.5 Survey results about the learning outcomes of design - build studio
Source: Author

are taken out of their classrooms, which encourage the development of interpersonal skills and inspires confidence in their designs; they gain the tools that can make a physical difference in people's lives. (Stonorov, 2018). There are many reasons for adopting the design-build program in architectural education. The learning outcomes of the studio were surveyed among the 23 students in the batch (Figure 17.5.). A structured questionnaire was shared among students and the following are the findings of the studio, as against a hypothetical academic project.

Improving Self-Confidence

The project addressed the public sector and students as beginners were having a foremost experience in interacting with the users. The surveys and interviews helped them frame the design requirements and start with the zoning. They got the opportunity to present their design thoughts and their findings in a public forum. The appreciation from the general public who attended the stage 7 (Table 17.2) meeting, made them believe that their ideas can have a positive influence on a common man. The survey amongst students reported that 74% of the students experienced a boost in self confidence due to those interactions.

Exposure to Architectural Professional Practice

The design-build program exposes students to a broader range of architectural practice and engages them in all aspects of the design and construction process; dealing with clients, codes, inspectors, contractors, product suppliers, engineers and neighborhood associations (Amer, 2015). Students were able to meet government officials, understand codes and produce drawings and specifications to be submitted for approval by Kerala PWD. All students showed their interest in exposure to real projects when compared to hypothetical academic projects.

Freedom of Expression and Communication of Ideas

48% of Students responded through the survey that there were limitations in expressing the ideas due to awareness regarding the budget for such projects. They were reluctant to explore new materials and technologies as well. The survey reports that 35% of students could find their freedom of design expression and communication.

Involving User Participation in Decision-Making

The Design-Build studio model allows for flexibility in design briefing by incorporating the opinions of the users, as against a hypothetical academic project. All the students believed that the survey helped them in finalizing the requirements and proceeding further in design.

Benefit of Teamwork

Design-build is a group project that benefits from diversity. It exposes students to working in teams, allowing them to grow in self-confidence in terms of working in teams and accepting that they don't have to be great at everything (Sokol, 2008). 87% of students agreed to this collaborative interaction, but 13% had the least or no preference as far as teamwork is concerned.

Convergence of Ideas

Even though each student had approached the project individually, the students were able to converge potential ideas into a single proposal for the final presentation by stage 7 (Table 17.2). The process happened during common presentations and healthy discussions in the classroom. 91% agreed to this idea, through the survey.

Knowledge Gain in Context Study

The realness of the design-build program gives a chance for students to be trained more responsively to a specific site and local conditions focusing on the climate and local culture. Moreover, the experience of the design-build program shows that the sources for design inspiration could be born from understanding culture and place (Dean, 2001). 61% of students could understand and respond with contextualized design solutions, but the rest couldn't. The reason could be their limited knowledge about climate and culture since they are in the first year.

Acquisition of Skills

The Design-Build studio model offers a learning opportunity for a wide range of skills like those of communication, decision-making, leadership, construction, experimenting etc. The majority of students in the class agreed to the fact that Design-Build Studio helped them improve their skills.

Exposure to Construction

The main motivation of the program is to introduce students to the art of building. Students' creativity is activated and enhanced through hands-on construction as they realize what is involved in taking architecture from a drawing to a building (Amer, 2015). The workshops on seating spaces, place-making and wall art helped students recognize the value of hands-on learning.

Knowledge on Practical Aspects than Scientific Knowledge

The exposure towards hands-on workshops helps students differentiate between theoretical and practical application of knowledge. Design-Build Studio helps students make appropriate decisions due to this practical know-how gained through the learning. The whole class preferred such kind of experiential learning for a better understanding of their design projects.

Observations and Conclusion

Design Education through real-life practical applications can enhance students' understanding of the building process on a deeper level. The demerit of Design-Build studio is its time consumption and budget sourcing. From the literature study and experiment, it concludes that the Design-Build studio method cannot be fully achieved in a semester duration, but may extend into the succeeding semester as well. Merits of Design-Build studio is in addressing four kinds of competencies that are academically allocated into skill, knowledge, technology and design streams, as mentioned in the interim report of National Education Policy 2020. Also, the study establishes the correlation between Experiential Learning Theory and Design-Build Studio method, to teach architecture design. The students' survey results also establish the achievement of targeted learning outcomes in a design studio. The Design-Build studio method extends the environment of learning beyond school and provides a new setting where students exchange knowledge, develop skills and form an alternative vision of professional practice. Author would like to acknowledge the support and cooperation from Thejus College of Architecture and the 2022 batch students in completing the academic work on time and the governing authority of Kadavallur Panchayath, for the constant follow up.

References

Amer, N. (2015). Appropriate technology and design-build program in architectural education. *Journal of Engineering and Applied Science*, 62(6):507–528. https://www.researchgate.net/publication/317029978.

Aydemir, A.Z., Sezgin, A., and İnceoğlu, A. (2020). A Design-Build Experience: Kilyos Boathouse. MEF University Faculty of Arts, Design and Architecture. http://fada.mef.edu.tr/en, https://www.researchgate.net/publication/339274837.

Balaram, S. (2005). Design pedagogy in India: aperspective. *Indian Design and Design Education*, Vol. 21, No. 4 pp. 11–22, https://www.jstor.org/stable/25224015.

Canizaro V., (2012). Design-Build in Architecture Education: Motivations, Practices, Challenges, Successes and Failures. https://www.researchgate.net/publication/287009615.

Charles., and Eames R. (1958). The Indian Report. *National Institute of Design, Ahmedabad, INDIA* http://echo.iat.sfu.ca/library/eames_58_india_report.pdf

Dean, A.O., (2001) Sam Mockbee: The hero of Hale County, Architectural Record, Vol. 189, No.2, (76-82). https://www.architecturalrecord.com/articles/12281

Hardy A., (2015) Theory and Practice of Temple Architecture in Medieval India: Bhoja's Samaranganasutradhara and The Bhojpur Line Drawings. Dev Publishers & Distributors

Salama, A. (2005). Skill-based/ knowledge based architectural pedagogies: an argument for creating humane environments. In 7th International Conference on Humane Habitat (ICHH-05), https://www.researchgate.net/publication/228672160.

Salama, A., and Wilkinson, N. (2007). Design Studio Pedagogy: Horizons for the Future. Design Studio, Design Pedagogy, Architectural Education. ISBN 1-872811-09-4.

Sokol, D., (2008).Teaching by Example: Design-build Educators Talk Pedagogy and Real Politick, Architecture Record, October p. 125. http://archrecord.construction.com/features/humanitariandesign/0810roundtable-1.asp

Stonorov T., (2018). The Design-Build Studio: Crafting Meaningful Work in Architectural Education. Routledge Newyork and London ISBN: 978-1-315-65074-6

Vriddhi, P. (2022). Investigating the contemporary architecture education challenges in India. *World Academy of Science, Engineering and Technology International Journal of Educational and Pedagogical Sciences*,10(3), 2016.https://www.researchgate.net/publication/360996593.

18 Review on the introduction of adventure education as catalyst for students' learning

Ankit Kumar[1,a] *and Narender Tanwar*[2,b]
[1]Prof., Vellore Institute of Technology, Vellore, India
[2]Commander, retired from Indian Navy, Surrey, Canada

Abstract

Adventure education and outdoor experiences play a pivotal role in shaping students' holistic development, fostering essential life skills, and providing a unique avenue for experiential learning. In the Indian context, where traditional rote learning has long dominated the educational Arena, even in architectural education sad reality of rote learning is clinging on to date, whereas in the name of experiential learning, more of holiday trips or travel learning is mostly professed.

Foster academic excellence, personal growth adventure education, and experiences hold the key to emphasizes that along with IQ measures cognitive abilities, EQ is equally vital in the complex landscape of human interaction and emotional understanding. Adventure education encourages students to recognize and manage their emotions, and collaborate effectively. By venturing beyond the classroom, students gain hands-on exposure to critical skills such as problem-solving, urban–rural fabrics context, feasibility assessment, and resilience. They can learn to adapt to the unpredictable challenges that nature and situations present, thus enhancing their ability to face real-world problems.

This review paper delves into the transformative potential of adventure education as an innovative pedagogical approach for architecture students. The aim is to highlight how adventure education and outdoor experiences can revolutionize the educational paradigm.

Indian cultural and geographical diversity provides a rich tapestry for adventure education. From the majestic Himalayas in the north to the pristine beaches in the south, India boasts a plethora of terrains, local sports, festivals, certain services as well as Territorial Army, and Citizens Emergency Response Teams. References from country approaches to education and services like Israel and Korea are also considered for case studies.

Keywords: Adventure learning, architecture education, emotional intelligence, traditional knowledge

Introduction

The educational environment in India is predominantly rote learning, an approach that emphasizes memorization and repetition. This approach, especially prevalent in architecture, tends to overlook the importance of experiential learning and critical thinking. This paper aims to explore the transformative potential of adventure education in Indian architectural education. By reviewing relevant literature and case studies, the paper seeks to highlight the benefits and challenges of implementing adventure education in the Indian context. Currently, architectural education in India faces the challenge of integrating practical, real-world experiences with theoretical knowledge. The current system often fails to prepare students for the dynamic nature of architectural design and problem-solving (Bongirwar and Das, 2022).

Approx. 25,000 architects graduate from all architectural schools each year in

[a]ankitkumar@vit.ac.in, [b]Aarch9347@gmail.com

DOI: 10.1201/9781003571018-18

India compared to 4 lakh civil engineers. Architecture, meanwhile, is going through a difficult time. More than ten new students are being accepted by several universities as a result of the sharp decline in admissions. Many people are being discouraged from pursuing this vocation by the average income for recent graduates, which ranges from 2.4 to 3.5 lacs annually. The number of architectural institutions has expanded from 137 in 2006 to 463 in 2020 according to the Council of Architecture, which has also standardized the course format to make it repetitious and rigorous, much like engineering courses. The importance of architectural education is further diminished to that of draughtsman ship training by the curriculum's absence of related courses like art, philosophy, music, etc. The reason for this is that the undergraduate education system and primary and secondary school design instruction are not aligned (Naralasetty and Ugrani, 2023).

Adventure Education and Its Potential Benefits

Adventure education presents an innovative alternative, focusing on experiential learning through outdoor experiences. This approach can significantly enhance students' cognitive abilities, emotional intelligence, and practical skills necessary for architectural practice. Adventure education has been widely recognized for its impact on both cognitive and emotional development in students. Studies have shown that engaging in challenging outdoor activities enhances critical thinking, problem-solving skills, and adaptability. Emotionally, adventure education contributes significantly to the development of self-esteem, resilience, and emotional intelligence. This is particularly relevant in architectural education, where students must often navigate complex, stressful situations (Priest and Gass, 2005).

Cognitive and Emotional Development through Adventure Education

The application of adventure education in architectural education is relatively novel. Literature suggests that incorporating experiential outdoor learning can bridge the gap between theoretical knowledge and real-world architectural challenges case studies in secondary education indicate that students who participate in adventure-based learning exhibit higher creativity, improved spatial awareness, and a better understanding of environmental and social contexts (Lee and Zhang, 2019).

Integration of Adventure Education in Curriculum

Various curricula can be integrated into adventure education as a method to enhance learning in architecture. This integration often involves field trips, interaction with diverse urban and rural environments, and engagement with local cultures and construction techniques. The literature highlights the importance of hands-on experience in understanding architectural design, sustainability, and the socio-cultural implications of architectural decisions.

Adventure education's impact extends beyond academic achievement, fostering personal growth, empathy, and a deeper connection with the built and natural environment as per according to Table 18.1.

Adventure Education in India: A Case Study Approach

Analysis of Adventure Education within Indian Cultural and Geographical Contexts as India's varied geography, from the Himalayas to coastal regions, provides unique opportunities for adventure education. These diverse settings offer practical learning experiences in understanding different architectural styles, materials, and environmental considerations. Case studies

Table 18.1 Comparative analysis of adventure education and traditional methods

Aspect	Adventure education	Traditional methods
Learning approach	Experiential, hands-on learning	Lecture-based, theoretical learning
Environment	Outdoor, dynamic settings	Classroom settings, structured environment
Curriculum focus	Practical skills, team building, personal development	Academic knowledge, standard curriculum
Student engagement	High engagement through activities and challenges	Variable, often passive learning
Assessment	Performance-based, often with real-world tasks	Standardized tests, written exams
Teacher's role	Facilitator, guide, coach	Instructor, source of knowledge
Skill development	Leadership, problem-solving, resilience, collaboration	Subject-specific knowledge, rote memorization
Risk and safety	Managed risks, safety training	Generally low-risk, controlled environment
Feedback	Immediate, based on tasks and challenges	Delayed, often through grades and formal assessments
Flexibility	Adaptable to different environments and situations	Structured, less adaptable to individual needs
Outcome focus	Personal growth, social skills, self-awareness	Academic proficiency, factual knowledge
Inclusivity	Accessible to diverse learners, adaptable to needs	Can be limited by rigid structures, less adaptable

Source: Research in developing a futuristic curriculum for architecture

of adventure sports from the Himalayan region highlight how students learn about mountain architecture, sustainability, and the challenges of construction in remote. Adventure education in India often incorporates local sports and festivals, which helps architecture students understand the cultural and social fabric of different regions. Such activities foster community engagement and provide insights into vernacular architecture. Programs involving participation in local festivals or collaboration with the Territorial Army and Citizens Emergency Response Teams offer real-life scenarios for students to apply their architectural knowledge in community service and disaster management.

The main challenges in implementing adventure education in architectural studies include logistical issues, safety concerns, and the need for specialized instructors. Additionally, there is often resistance from traditional educational institutions in adopting these unconventional learning methods. Despite these challenges, Future research could focus on developing frameworks for integrating adventure education seamlessly into architectural curricula.

There is also potential for cross-disciplinary collaboration, combining adventure education with technology, environmental studies, and cultural studies to enrich the architectural learning experience.

Nations Benefit from Adventure Education System Integration

With adventure education integrating with the normal curriculum for undergraduate

Table 18.2 Challenges and opportunities in implementing adventure education in architecture

Study focus	Author(s)	Key findings	Implications	Publication date
Self-confidence in mountaineering and skiing	Raghubanshi and Bhati (2023)	Found higher levels of self-confidence in male athletes in cross-country mountaineering and skiing compared to females	Highlights gender differences in adventure sports	2023-10-31
Mental toughness in adventure-sport curriculum	Moscati, et al. (2020)	Positive effect on undergraduate students' mental toughness; no significant difference between first and second assessment	Emphasizes the role of adventure sports in developing mental toughness	-
Anxiety levels in Kayaking and Canoeing	Raghubanshi and Bhati (2023)	No significant difference in anxiety levels among participants	Challenges the notion of heightened anxiety in adventure sports	2023-07-01
Information source preference in athletes	Hungenberg et al.	New athletes prefer social peer groups for information, while experienced athletes rely on memory	Shows the evolution of information reliance with experience	2019-02-01
Motivation in white water rafting	Gunawardhana, et al.	Seven motivational factors were identified, including thrill and nature connectedness	Highlights the diverse motivations in adventure sports	2018
Adventure sports coaching challenges	Berry, et al.	Covers various coaching challenges like skill acquisition and risk management	Provides insights into the complexities of coaching in adventure sports	2015-03-26

students learn and adapt to the challenges faced in society also have the enthusiasm and energy to serve the country which is useful to put the youth to a good cause, for example, the initiative of the government like *Agneepath*, *Agni Veer* can be utilized with the more volunteer serving the country *Agneepath* it is a system devised by the government as a tool to create interest among the young citizen of a country to serve the nation in terms of defense/military need. It's called a tour of duty which has four years of service and a normal student serving in the regular Army is called an Agni Veer. This is voluntary the student who undergoes adventure training has a better IQ patience, and overall personality which is essential for and healthy lifestyle and for serving the nation in difficult times here in the army through *Agneepath* system students/youth will be trained which puts them into a disciplined lifestyle after 4 years of serving the nation in a voluntary manner these students/youth are capable to start their journey in entrepreneurship as well as work in a professional job, best point is that they are trained in terms of adventure, physical fitness as well as national belief, servitude, sacrifice and terms of emergency they can be used as a source for a fourth line of Defense. Additionally, with the discipline philosophy in the military, they now have involvement in their nature which further enhances the sincerity discipline adventure health aspect of our country's youth. When they become parent they also follow this kind of philosophy and the coming generation will also have smarter emotional questions taking on adventure plus the aptitude to serve the nation.

International Perspectives

Israel's approach to education emphasizes practical skills and real-world applications. Adventure education in Israel often includes military training and survival skills, which instill discipline, resilience, and strategic thinking – attributes beneficial in architectural education.

Case studies from Israeli architectural schools show how military-style adventure training can enhance students' ability to work under pressure and develop efficient, practical solutions.

While Israel's model is more structured and discipline-oriented, the Indian approach is diverse and culturally rich. Both models provide valuable insights into how adventure education can be tailored to meet different educational and cultural needs.

South Korea's focus on technological integration in adventure education offers a different perspective. Their programs often combine outdoor activities with digital tools, enhancing students' technical skills alongside their physical and emotional development. This blend of technology and adventure is particularly relevant to modern architectural education, where digital tools play a significant role in design and construction (Rose, 2020).

Implications for Architectural Education

Adventure education offers a dynamic alternative to the static, rote-learning methods prevalent in traditional architectural education. It encourages active learning, critical thinking, and real-world problem-solving. By engaging students in diverse environments and challenges, adventure education enhances their understanding of context, sustainability, and the socio-cultural impact of architectural designs. Role of adventure education in fostering creativity and resilience exposure to varied and unpredictable outdoor settings fosters creativity and innovation in architectural design. Students learn to think outside the box and develop designs that are both functional and aesthetically pleasing. Adventure education also builds resilience and adaptability, skills essential for architects who often face challenging and changing project requirements.

Discussion

The review highlights the substantial benefits of integrating adventure education in architectural curricula. However, it also underscores the need for careful planning, resource allocation, and safety considerations. The success of adventure education in architecture depends on balancing practical field experiences with theoretical learning, ensuring a comprehensive educational approach. One of the primary challenges in implementing adventure education in India is the lack of infrastructure and trained educators who can safely and effectively lead these programs. There is also a need for educational policy reform to accommodate these unconventional teaching methods within the mainstream curriculum. Future research could focus on developing standardized frameworks and best practices for incorporating adventure education in

Table 18.3 Comparative analysis of the Israel model with the Indian model

Aspect	Indian model of teaching in architecture	Israeli model of teaching in architecture
Curriculum focus	Traditional, theory-heavy, focuses on architectural history and design principles	Innovative, often includes sustainability and resilience in design, practical problem-solving
Teaching methodology	Lecture-based, emphasis on theoretical knowledge, rote learning	Project-based learning, emphasis on real-world challenges, innovation and creativity
Survival skills Integration	Limited direct focus on survival skills; more emphasis on conventional architectural knowledge	Higher emphasis on integrating survival skills, especially in terms of designing for resilience and sustainability
Practical training	Less emphasis on hands-on training in survival scenarios	Greater emphasis on practical, hands-on training, including designing for extreme conditions
Research and innovation	Focuses on historical and cultural aspects of architecture	Focuses on new technologies, sustainable practices, and innovative design for survival in adverse conditions
Industry collaboration	Traditional industry collaborations	Strong links with tech-driven and sustainable design industries, often with a focus on survival and resilience
Global perspective	Often regionally focused, with some global perspectives	Strong global perspective, with a focus on international standards and practices in sustainable and resilient design
Student engagement	More passive, with emphasis on instructor-led teaching	More active, encouraging student-led initiatives and research in survival-focused design

Source: Author

architectural studies. Collaborations with international institutions could provide valuable insights and resources for developing more effective adventure education programs in India.

Recommendations

Development of specialized programs: Educational institutions should collaborate with adventure education experts to develop programs that are safe, educational, and relevant to architectural studies.

Training and certification: Invest in skilled training educators not only in architectural theory but also in leading outdoor experiential learning. Council of Architecture should play role as per expanding the establishment of COA-TRC and nata training module

Establish robust infrastructure and safety protocols to ensure that adventure education is conducted in a safe and controlled environment.

Collaborations with local communities: Engage with local communities to enrich adventure education programs with cultural and environmental contexts.

International exchange programs: Establish exchange programs with countries like Israel and Korea to gain insights into their adventure education models and implement best practices.

Adaptation to local needs: Policies should allow for flexibility in the implementation of adventure education, adapting to the diverse geographical and cultural landscape of India.

Optimal timing: Determining the optimal timing and frequency of adventure education modules is crucial. These should be strategically scheduled to align with relevant academic units and avoid overloading students. For instance, outdoor modules can be planned during semesters focusing on environmental architecture or rural planning.

Sustainable implementation: The frequency of adventure education activities should be sustainable, considering the academic calendar, students' workload, and available resources. Regular, but well-spaced, modules can provide ongoing experiential learning without disrupting other academic responsibilities.

Emotional and cognitive development: Evaluation of the impact on students' emotional intelligence, resilience, and adaptability.

Conclusion

This review paper has explored the transformative potential of adventure education in reshaping architectural education in India. "Why Adventure Sport Culture may Influence Information Search Behavior". Ouyang and Gray (2019)

The findings suggest that adventure education can effectively complement traditional learning methods, fostering cognitive, emotional, and practical skills essential for budding architects.

While there are challenges in implementation, the opportunities for innovation and improved educational outcomes are significant. Adventure education not only prepares students for the architectural profession but also contributes to their holistic personal development. Encourage educational authorities to recognize the value of non-traditional learning methods like adventure education in developing well-rounded architects.

Reference

Baratz, Lea. (2015). Israeli teacher trainees' perceptions of the term moral resilience. Journal for Multicultural Education. 9. 193-206. 10.1108/JME-12-2014-0041.

Berry, M., & Hodgson, C. (Eds.). (2015). *Adventure sports coaching*. Routledge.

Bongirwar, R. and Das, B. K. (2022). Evolving concerns in Indian architectural education - a way forward NEP 2020. *Journal of Northeastern University*, 25, 4926–4936. 10.6084/m9.figshare.21965642.

Council of Architecture (2023). https://www.nata.in/. https://www.nata.in/. Jun 03.

Gunawardhana, D. S. R. E. S., Ramanayaka, A. R. N. D., & Dompelage, P. T. (2018). Identifying Motivational Factors Affecting Adventure Sports with Reference to the "Kithulgala Adventure Base Camp, White Water Rafting. 19th Conference on Postgraduate Research, International Postgraduate Research Conference 2018, Faculty of Graduate Studies, University of Kelaniya, Sri Lanka.

Moscati, K., Boyea, J., Stamatis, A., and Maynard, S. (2020). The effect of short-term adventure-sport curriculum on the level of mental toughness of undergraduate students. *International Journal of Exercise Science: Conference Proceedings*, 2(12).

Nissim, Y.; Danial-Saad, A. The Resilient Teacher: Unveiling the Positive Impact of the Collaborative Practicum Model on Novice Teachers. Educ. Sci. 2023, 13, 1162. https://doi.org/10.3390/educsci13111162

Ouyang, L., and Gray, D. (2019). Measuring the information search behaviors of adventure sport tourists. *Event Management*, 23, 75–91.

Priest, S. and Gass, M. A. (2005). Effective Leadership in Adventure Programming. Human Kinetics.

Raghubanshi, A. and Bhati, C. (2023). Exploring gender-neutral self-confidence levels in adventure sports participants: a comparative study. *Journal of Integral Sciences,* 1–7. 10.37022/jis.v6i4.63.

Rose, P. (2020). Work-integrated learning in context: a South Korean perspective. https://files.eric.ed.gov/fulltext/EJ1250593.pdf.

(PDF) Rethinking Architectural Education in India - can the New Education Policy help us? (researchgate.net) Naralasetty and Ugrani, 2023.

https://eric.ed.gov/?id=EJ1216911 Lee and Zhang, 2019

https://www.researchgate.net/profile/Eric-Hungenberg/publication/329746034_Measuring_the_Information_Search_Behaviors_of_Adventure_Sport_Tourists/links/5ca4f4d0a6fdcc12ee911970/Measuring-the-Information-Search-Behaviors-of-Adventure-Sport-Tourists.pdf

19 Innovative initiatives: Architectural *Prayogshala* is an idea of a parallel institution bringing the mirage to reality

Poonam Solanki[1,a] *and Aakash Shah*[2,b]

[1]Bachelor of Architecture, CEPT School of Architecture, Ahmedabad, India

[2]Bachelor of Architecture, IPSA- Indubhai Parekh School of Architecture, Rajkot, India

Abstract

Knowledge of construction methodology and a sense of material properties are the undetachable components of architecture. It is a big concern that the architect making the construction decisions based on their design is aware of the building materials, their general and innovative construction techniques, and their properties. Even though it is a sensitive yet compulsory component of architectural education, most institutes have kept the practical construction experience aside as a part of theory only or a small practical session in the academic timetable. Students' complete understanding of the entire building design process, from sketch to actual, is incomplete. Sometimes, this problem grows even more significant as the unattended part of the building project in the architect's practice. There is no limit to learning the vast ocean of architecture. However, there is a lack of places where architectural learning can continue even after formal education. A platform has been created, allowing all types of exploring possibilities for those seeking the fraternity of architecture. The intention of this Architectural *Prayogshala* -a laboratory is to become the missing link in the architectural journey of the enquirer. It works as a junction where everyone can gain knowledge by using their hands on the building material and match their knowledge level through history to the contemporary. Through this paper, the idea of this initiative, which can be used at multiple locations, will be made open. Ultimately, the acceptance by the governing bodies can change the global scenario about architectural awareness.

Keywords: Architectural experiments, hands-on construction, hands-on in architecture education, traditional construction

Introduction

Importance of Architectural Education in India

We have come a long way in teaching and learning architecture since it came to our country as a drafting subject by the British. From only being a drafting degree to now a very intricate and insightful innovative design course. The course has, of course, evolved with time and generations for the better.

Studying architecture courses in the best schools in the nation to teach the course in different cities and states, technological advancement of the course was also observed. As a teacher, while being part of the syllabus-forming committee, subjects

[a]solankipoonam154@gmail.com, [b]aakash24385@gmail.com

DOI: 10.1201/9781003571018-19

evolving to sustain the testaments of time were observed. With the acceptance of technology and more branches of subject palate, only some time-tested subjects have taken a back seat. The course has become more choice-based with time. A few courses have become less popular; however, their importance can be understood once in the field. The struggle at the students' level to learn as many new and updated courses as possible and at the authority level to strike a balance between the needed courses like History of Architecture and Manual Technical Drawings, and the upgraded branches of courses like AI, intelligent facades, retractable architecture, etc., the task is not easy but not impossible as well.

Purpose of the Paper

Integration of Hands-on Construction Activity in Architectural Education

While conceptualizing a standard architectural practice, it became a complete idea only when construction experiments, tests, trials, and practical behavioral observations of building materials and elements became part of designing, on common grounds for the thinkers of the '*Prayogshala*'. It was ambitious that hands-on activities withstand the curiosity of students and professionals. Therefore, the deep exploration grew more and became a platform for explorations of all construction materials and techniques at an actual scale. The idea is to provide the scope for bringing the sketches and drawings to reality and boost confidence to create a better society. Classroom learning has limitations, and exploring on-site has exciting advantages.

Tools and Methodology

Finding Out the Topics

The topics are found as per the current market trends and requirements. All the issues selected for the hands-on construction workshop are associated with the academic architectural syllabus. The idea is clear: to cover up the modules from the syllabus with holistic skill development.

Building Materials

The most important topics are related to building materials and construction techniques. All the workshops are uniquely designed to develop innovative solutions and design processes, keeping the material properties in the center.

Construction Techniques

The construction techniques explored at the workshop amalgamatetime-tested or vernacular style and modern technology. The idea is to come up with solutions that are not only innovative but also sustainable. (tensile fabric structure with bamboo, space frame with bamboo)

Labors (Skilled, Unskilled)

Every structure is complete with adequate laborers. Finding skilled laborers for all the techniques is a time-consuming and lengthy process—often, they are not nearby. Therefore, it was initiated to create awareness and skill development in local laborers around our area. This is helpful to the designer, as well as the team, to share the skills further and faster. Exposed brickwork, bamboo work, earthworks.

Figure 19.1 Tensile fabric workshop Feb. 2023
Source: Author

Working Out Innovative Techniques, Spreading Awareness about Materials and Techniques

At the inception of this initiative, it was fundamental to stay connected to the vernacular architectural techniques and local and sustainable building materials. To survive and keep updated, efforts are made on many different things.

Following are the examples:

Prototyping

Experiments with modern technology: we have started prototyping space structure connectors for some furniture with a parametric technique.

Communication and Engagement with Industry Experts

As we accept that architecture and engineering go hand-in-hand in creating marvelous habitats, the knowledge sharing at the '*Prayogshala*' is done by industry experts. We believe communication is the key. Experts are invited for related topics that are required for experience sharing, and this has built a very healthy teaching-learning environment at the '*Prayogshala*'. This also bridges the gap between students and professionals in an informal way, which helps both sides bond stronger.

Architectural Katta

This is undoubtedly the most liked event at the '*Prayogshala*'. Architectural *kattas* are

Figure 19.2 Gabion wall workshop Nov. 2022
Source: Author

Figure 19.3 Discussion post katta activity
Source: Author

a platform for discussions, debates, team building activities, architectural quizzes, sketching, and doodling related to art and architecture, time problems, raising queries and solving together without any age bar, and so much more over CHAH and snacks. This is the rejuvenating weekly stress buster at the '*Prayogshala*'.

Stage-wise Courses (Beginner to Expert)

Considering the syllabus and the philosophy at the core, the courses have been curated to offer from beginners to expert level where any academic year students can join. Even many professionals have shown interest in learning the basics of earth architecture, bamboo, 3-D printing and more. The stages are designed to give participants enough exposure to the topic and build the confidence to prepare the participant industry that they can handle the related site flawlessly.

Exploring Vernacular to Modern Techniques

From earth architecture techniques to parametric architecture, 3-D printing and retractable and collapsible structures were explored at the '*Prayogshala*'. Most topics are covered under one umbrella so that participants can quickly get involved in the selected topics. One need not go to different places for different issues.

Figure 19.4 Tensile fabric structure with bamboo
Source: Author

A Platform for Experimentation before Execution for Architects and Students

There is always a fear of failure if someone innovates a new method. Difficulties arise when one has to stop or reject the idea only due to a lack of exposure towards experimentation. To increase the innovation and success of site implementation of the techniques, architects and students are invited to use the '*Prayogshala*' as a platform for experimentation, where we have a set-up with infrastructure, tools, and labor. This will help innovation in academic design studios and the architectural scenario in society, where most buildings look and are constructed similarly.

Public Participation

To spread awareness about architecture and the role of architects in creating more suitable livelihoods, participation in exhibitions was pivotal.

Conducting workshops for primary and higher secondary school students is another tool for spreading awareness about the course and skill development.

Figure 19.5 Rammed earth wall
Source: Author

Figure 19.6 Earth wall at colours innovation academy Sept. 2023
Source: Author

Giving Back to Society

Since the concept of '*Prayogshala*', the awareness of responsible and thoughtful architecture has been imbibed in all our decisions. As sensible architects, it is required to give back architectural marvels to the communities in need as our moral responsibility. The community programs have been initiated where identified small villages or areas where architectural involvement in terms of awareness or actual creation is required. As the idea is to create the architectural solutions at an accurate scale with participants' involvement, the creation gets installed at the place where it is necessary. Progressively, it goes with prior preparedness to design the solutions for the places and work on the site with the communities' involvement. This will bring the participants a sense of completion, acceptance and resilience. This is our way of making Architecture a festival to celebrate with people.

Specific Topics Covered in the Hands-on Construction Workshops

Since it explores traditional as well as modern and current construction techniques and covers topics from the architectural syllabus, the innovative various topics:

- Space structures: tensile, tensegrity and space frame
- Earthworks: all techniques
- Brick construction: arches, vaults, domes, jalies, flat slabs, walls with bonds etc.
- Stone construction: all techniques
- Retractable and collapsible structures
- 3D printing
- Bamboo construction and many more

Figure 19.7 Exposed brick work on-going
Source: Author

Benefits of Hands-on Construction Activity in Architectural Education

Through the extracts of academics as teaching faculties, it was realized the need for hands-on activities for the holistic skill development of architecture students. These hands-on activities bring in a lot of benefits to the students as well as a healthier work environment as they pass out. Involving body, mind and information together to create what they have only seen in pictures and drawings brings better understanding and clarity about the topics. Where all the architecture students seek to have tried their hands on one or more techniques, this is something which, if they get hold of one, will bring in solutions to many of the challenges at the execution level.

Promoting Teamwork and Collaboration

Working on any architectural execution is teamwork, positively changing participants' attitudes. One learns the importance of effective teamwork, cooperating, listening, collaborating, working towards a common goal, and becoming responsible for one's work. This platform offers students internal associations and opportunities to work under industry experts and learn from them. There are opportunities to make new and lifelong friends when the workshops

Figure 19.8 Collapsible building model in construction
Source: Author

Figure 19.9 National level workshop at DYPCOA Kolhapur Oct.2023
Source: Author

are organized at inter-city, state, or national levels.

Enhancing Communication and Interpersonal Skills

The hands-on activities create magic in bridging the gap between drawing-based information and experiential knowledge for participants. This is not only limited to architectural skill development; one can develop communication skills as there are a lot of discussions, group works, and presentations during the workshops. The boost in confidence was evident at the end of the workshops, a spark in participants' eyes speaks louder than words.

Figure 19.10 Activity and following discussion at katta Aug. 2023
Source: Author

Topics from Architectural Syllabus

At the '*Prayogshala*', all the topics are skillfully designed to cover one or more topics from the five-year architectural syllabus. Involving in the workshops gives students the practical application of the theoretical knowledge they have gained. This strengthens the participants' confidence and provides opportunities for innovative design ideas. Some modules are suitable for first- to fifth-year students depending upon the need of the syllabus and understanding. Pare offered the opportunities to continue the course and gain more in-depth knowledge about the topics. Since the workshops are currently participatory, a substantial group of participants is required for any workshops since we execute designs at an actual scale.

Improved Understanding of Construction Processes

All the workshops are designed to bring in rigorous involvement from participants; they become part of all processes like handling the tools and materials, solving the problems, and discussing with the team

Figure 19.11 Handling tools at Tensile workshop Nov. 2023
Source: Author

members and on-site laborers. They learn the on-site experiences and become part of the construction process. This helps develop technical knowledge about the topics and how the architectural concepts are shaped. Sometimes, at the time of execution, the design gets changed according to the requirement and site condition, allowing participants to see the design development and execution challenges. Participants also develop the ability to evaluate design decisions based on practical considerations when they complete required workshops.

Bridging the Gap between Theory and Practice

Exposure to various aspects of the construction industry opens up possibilities to think outside the box and create designs that are industry-ready and not only imaginary. As the outcome of many workshops, an enhancement of participants' critical thinking and problem-solving skills may be seen to grow. Participants achieve the application of theoretical knowledge in real-world projects as they understand the limitations and challenges in the construction process. They try to integrate practical experiences into future architectural projects. The goal is simple: Development of a well-rounded architectural professional. Along with the college students and fresh graduates, workshops are also conducted at primary and secondary school levels to bring awareness about the course and its value towards creating a better nation.

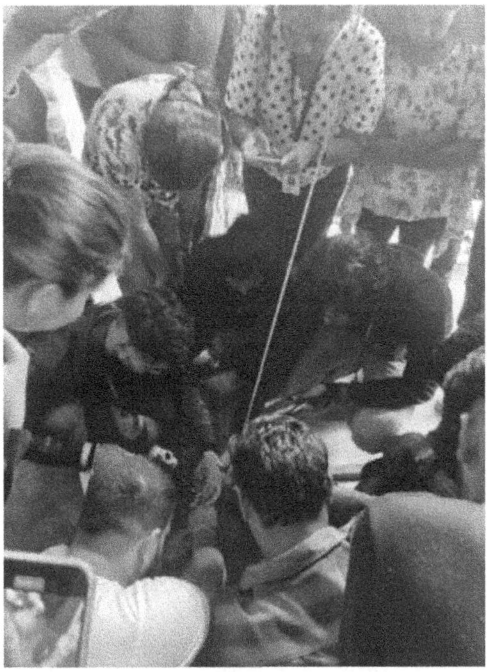

Figure 19.12 Handling tools at a national workshop Oct. 2023
Source: Author

Observations and Findings

Through all the workshops as well as activities like heritage walk, architectural *'katta'*, exhibition, interactive session, school program etc., few points are observed which are very relevant to improve the activities and performance if considered positively.

Following are our observations:

- Vertical modules are best as knowledge transfer happens with bonding.
- Girls need help to handle the tools and work on-site. We make sure to give them more challenges.
- Reluctance to stand and sit on-site or on the ground while working with hands.
- At least three or more days of workshops are needed to understand any topic.
- Out of college/at the workshop, participants enjoy and learn more due to location changes, which feels like working on the actual site.
- Combining workshops with other institutes increases teamwork, knowledge sharing, personality development, and new friends.
- Participating in workshops can be considered as academic credit.

- Integrated programs work best in different materials and styles, along with heritage walks, sketching trips, exploring local foods, traditionally celebrating special occasions, etc.
- Participants get involved quickly in hands-on and associate better with the outcome than the theory sessions.
- Participation in the innovation of techniques from vernacular styles and problem-solving is the most liked aspect as participants see the results evolving.
- Failing in some aspects encouraged the participants to develop better solutions, and they also registered the loopholes for the future. Developing problem-solving skills.
- Looking at the transformation of theoretical information to reality increases the confidence level of participants.

Conclusion

- In the paper, the benefits of hands-on construction activity in architectural education in India are discussed. Integrating practical experiences in architectural learning is the need of the hour now. We cannot remain with a one-semester internship/suitable experience course at a time when everyday innovations are done in the construction industry. This is when the '*Prayogshala*' seriously consider the practical implementation of the knowledge gained at the end of the course and before the degree award, just like medical courses.
- Now is the time for action for educational institutions and stakeholders to prioritize hands-on construction activities in architectural education.
- To conclude, a few points are proposed to be incorporated into the architectural syllabus of any institutions for initiatives like '*Prayogshala*' to achieve their goals of an architecturally sensible nation:
- Equal importance to hands-on and practical activities such as theory and technical drawings is given.
- Encouragement with credits in the academic syllabus will motivate students to participate more in such activities.
- Integration of hands-on activities within existing courses as part of elective or summer/winter school-like study programs.
- Introduction of specialized workshops and practical sessions as soon as the first semester to teach the ideology of holistic development.
- The institutional management and the National Council of Architecture can provide flexibility in the curriculum to accommodate construction activities.
- Along with the student's involvement in practical activities, there should be provision for professional development opportunities for faculty members.
- Collaboration with practicing architects and construction professionals for mentorship
- Making architectural festivals national events and promoting more student involvement, involving best practices and resources within the educational community, would be an added advantage.
- The Architectural '*Prayogshala*' has had an impact on the architectural community in India so far:
- Testimonials from past participants:
    - The tensile fabric workshop was very good, the hands-on practice was excellent: Fahad Khan 1st year AIKTC
    - It was boring during the theory session but interesting during the hands-on. I would love to participate more in this kind of workshops: Khan Lubna Abdul Kalam 3rd year AIKTC
    - Want to attend more workshops: Yash Bandal: 3rd Year ADT's MCOA

- More knowledge gained in hands-on than in theory: Ashika 4th Year, DYPCOA
- This is (hands-on workshop) required to be part of the syllabus: Prathmesh Kalate, a Fresh Graduate
- Collaborations with other organizations:
- MoU with National Association of Students of Architecture in the year 2023-24

References

http://www.hunnarshala.org
https://dev.earth-auroville.com

20 Integration of Indian knowledge system in architecture education – an innovative approach

AR. Sarika Joshi[1,a], *AR. Anuradha Bhute*[b], *Roopal Deshpande*[c] *and Lakshmi Rao*[d]

[1]"Jyotirmaya", SMMCA, 2nd Lane behind Coffee House, Gokulpeth, SMMCA, Nagpur

[2]Assistant Professor, Smt. Manoramabai Mundle College of Architecture, Nagpur University, Nagpur, India

[3]Principal, Smt Manoramabai Mundle College of Architecture, Nagpur University, Nagpur, India

[4]Professor, Smt Manoramabai Mundle College of Architecture, Nagpur University, Nagpur, India

Abstract

India boasts a rich heritage of the Indian Knowledge System (IKS), encompassing a diverse repository of treatises across various disciplines. However, since the post-independence era, Indian society has predominantly embraced contemporary architectural styles influenced by Western aesthetics. Consequently, architecture education curricula have predominantly focused on the knowledge base of the Western world. This paper argues for a paradigm shift towards integrating IKS-based design methods into undergraduate design studios, aiming to address the unsustainable nature of current development models that often conflict with the environment.

The study begins with a literature review, examining the syllabi of various universities to identify gaps in the integration of IKS within architecture education. Subsequently, through social surveys comprising structured interviews with students, faculty members, professionals in architecture education, historians, and conservationists, perspectives on the Indian Knowledge System are gathered and analyzed.

To practically explore the integration of IKS, students pursuing a Masters in Architecture Education (M. Arch) were tasked with designing assignments rooted in Indian knowledge systems for undergraduate design studios. The outcomes of these assignments are evaluated to understand the efficacy of incorporating IKS at the subject level, providing insights into the potential benefits and challenges.

The conclusions drawn from this study support the integration of IKS within architecture education to enhance students' understanding of foundational concepts. By analyzing the exploratory steps taken by the authors, this paper aims to stimulate discussions and prompt modifications in architecture education curricula toward a more holistic and culturally relevant approach.

Keywords: Architecture education, IKS, integration of IKS in architecture education

Introduction

Humans possess an inherent capacity to generate knowledge through personal experiences, experimentation, rational thinking, and reasoning. This accumulation of knowledge forms a repository that is passed down from one generation to the next, often through oral tradition or cultural practices. In India, this transfer of knowledge flourished under the '*Guru-Shishya Parampara*' until significant changes occurred in the education system around two centuries ago. The introduction of new rulers necessitated skilled individuals personalized to their job requirements, leading to a shift in education focus and a disruption in the continuous

[a]jsarika29@gmail.com, [b]anuradhaanu2@gmail.com, [c]roopal76@gmail.com, [d]venkylaks@gmail.com

DOI: 10.1201/9781003571018-20

transmission of knowledge. Consequently, much of this traditional knowledge now resides in manuscripts or the archives of libraries and institutes.

In the realm of architecture and architecture education, independent India gravitated towards contemporary buildings influenced by Western aesthetics. However, these models of development often prove unsustainable and conflict with nature. Recognizing these issues, governance and regulatory bodies are increasingly emphasizing the importance of education and awareness among the younger generation.

The implementation of the New Education Policy, 2020 by the Ministry of Human Resource Development marks a significant step towards revaluating the Indian knowledge system. The policy suggests guidelines for integrating Indian Knowledge System (IKS) into technical education curricula, including architecture.

Despite these developments, the current architecture education curriculum predominantly reflects the knowledge base of the Western world, overlooking India's ancient knowledge. There is a dire need to explore new paradigms for modifying architecture education curricula to incorporate IKS effectively.

Aim

This paper documents the process and analysis of an architecture studio conducted as part of the third-semester postgraduate Master's degree program in 'Architecture Education' at Shrimati Manoramabai Mundle College of Architecture (SMMCA), Nagpur. The primary objective of the P.G. M. Arch (Architecture Education) program is to provide professional training to architects, strengthening their expertise in design teaching, education, and practice.

The paper aims to explore methods of integrating IKS into architecture pedagogy and how these concepts can be incorporated in design exercises in undergraduate design studios. The authors mentored the postgraduate studio, where various exercises were formulated based on the principles of IKS. These exercises were then conducted for third-semester B. Arch students, aiming to reconnect students with historical architectural systems.

Objectives of the Study

- To stimulate interest in the value of the IKS among architecture teachers.
- To study various methods of transmitting Indian knowledge to B. Arch students.
- To revive and document IKS in architecture education through innovative approaches.
- To formulate a framework for the integration of IKS into architecture education.

Methodology

The studio stages were analyzed in the format of Kriya–Prakriya–Pratikriya (Prabhupada, 2019) (terms defined by the authors) which is as follows:

Kriya as 'understanding the concept' underlying as intent of each stage,

Prakriya as 'processing the learnings to arrive at the required observations'

Pratikriya as 'responses generated after deliberations to come to the conclusions' respectively.

Literature Review - Stage I

Kriya: In the initial stage, students were introduced to the vast educational heritage of India through selected readings from 'Revisiting the Educational Heritage of India' (see Figure 20.1). This book provided insights into ancient Indian systems of education, including the '*Guru Shishya Parampara*'. By analyzing chapters on topics such as women in education and the role of education in civilization, students gained

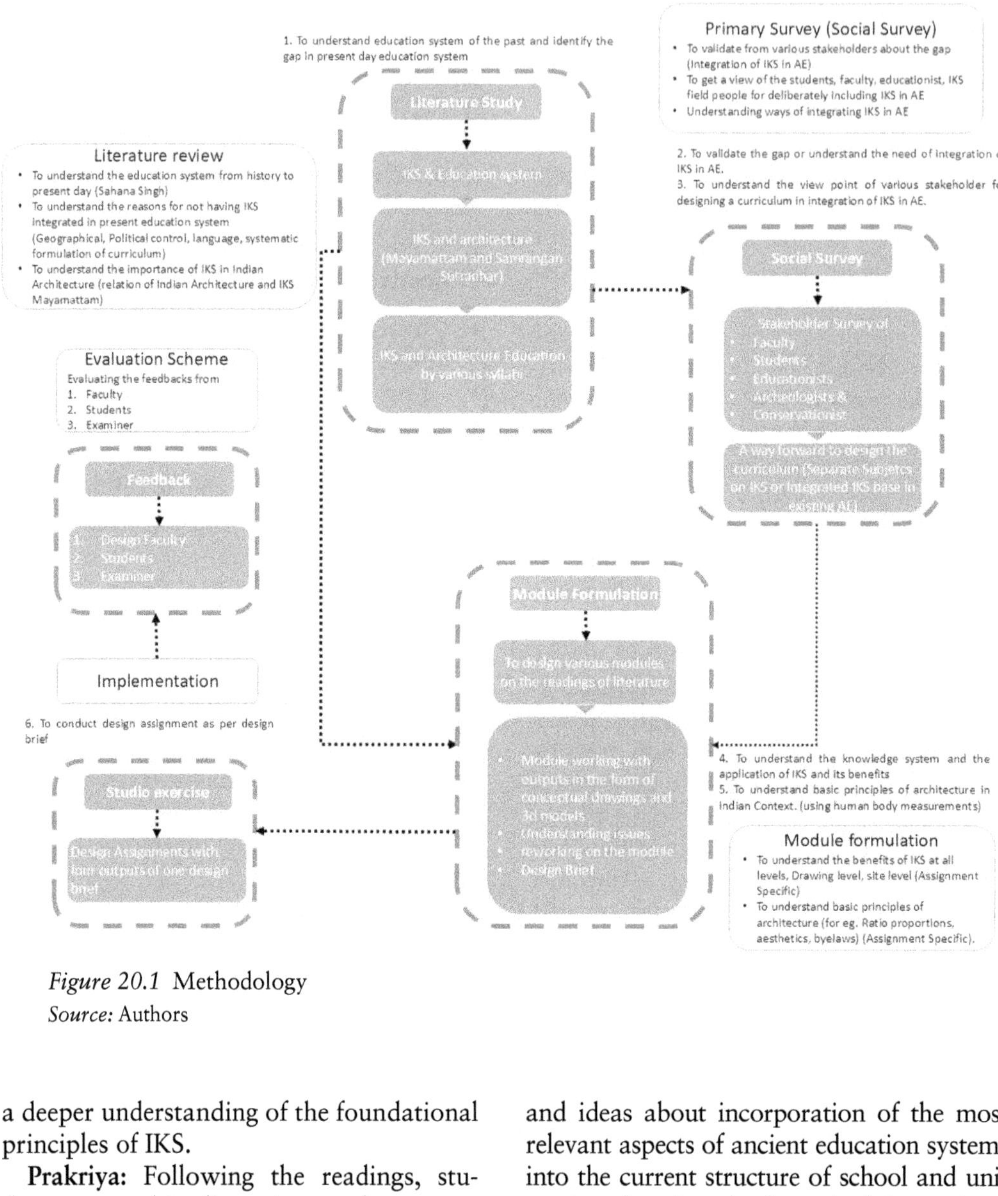

Figure 20.1 Methodology
Source: Authors

a deeper understanding of the foundational principles of IKS.

Prakriya: Following the readings, students engaged in discussions and presentations to investigate deeper into the concepts explored in the book. Post-graduate scholars selected chapters of interest to present their understanding, facilitating intense discussions with studio mentors (see Figure 20.2). This stage involved processing of the learnings from the readings and applying them to architectural education.

Pratikriya: Through the book review scholars could interchangeably discuss all the issues and ideas about incorporation of the most relevant aspects of ancient education systems into the current structure of school and university education. At the end of the sessions, there was a unanimous agreement about the potentials IKS has in revitalizing the education systems in India. The discussion focused particularly on architecture education and how it can be brought into main stream.

Literature Review - Stage II

Kriya: Scholars conducted an extensive analysis of B. Arch program syllabi across

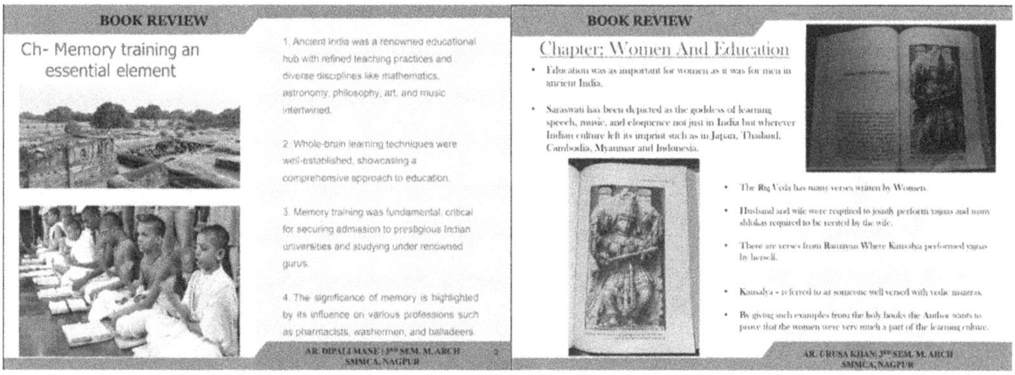

Figure 20.2 Book review

Source: Singh, 2022

diverse universities, including CEPT University, IITs, NITs, and standalone institutions (see Figure 20.1). This investigation was crucial for evaluating the integration of IKS in formal architecture education at the undergraduate level (Thapliyal, 2023). Findings indicated that while some institutions, such as IITs, engage students comprehensively through curricular, cocurricular, and extracurricular activities related to IKS, others offer only fragmented inclusion, primarily within subjects like History of Architecture and Design of Human Settlements.

Prakriya: Through the study, it was observed that CEPT University's teaching methodologies often align with the principles of IKS, despite not explicitly terming it as such. Similarly, IIT Roorkee had incorporated IKS as part of its course curriculum, introducing students to IKS in their second semester (see Figure 20.3). Notably, IIT Kharagpur had established a Centre of Excellence for Indian Knowledge Systems, facilitating interdisciplinary research on various branches of IKS. While some universities, like SPA Bhopal, integrated IKS through co-curricular activities, others, such as Mumbai University and RTM Nagpur University, had yet to introduce IKS into their curriculum.

Pratikriya: As a conclusion of the survey conducted, students suggested the hypothesis that there is need for separate course which would imbibe strong belongingness to traditional knowledge which would help in building a strong research base in the area.

Primary Survey - Stage III

Kriya: A social survey was conducted to gauge stakeholder awareness and perspectives on the integration of IKS in architecture education (see Figure 20.1). Stakeholders included various groups, including students, faculty members, (see Figure 20.4) archaeologists, conservationists, and educationists (see Figure 20.4). The survey aimed to strengthen hypotheses generated from curriculum reviews and discussions on IKS integration. These stakeholders involved a diverse range of perspectives, including students and faculty who actively engage with architectural knowledge, archaeologists and conservationists who hold deep insights into IKS, and educationists such as education officers, attorneys, directors, and principals who contribute significantly to the development of architectural education. By involving these stakeholders, the survey sought to gather comprehensive insights into the awareness levels and perceptions regarding IKS integration within architecture education. Through their diverse expertise and roles, these stakeholders provided invaluable contributions to the survey

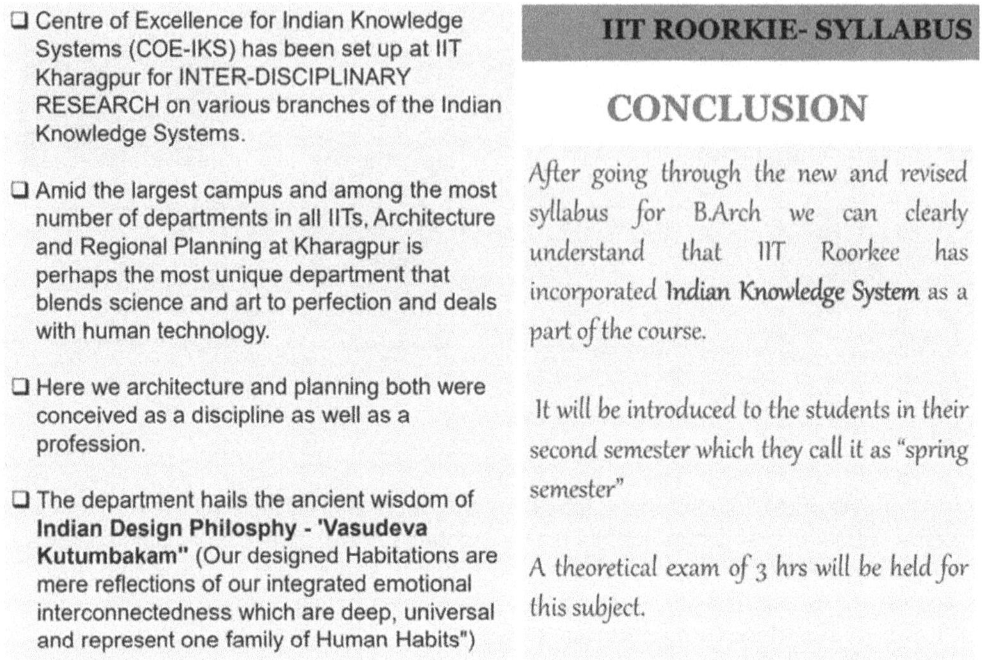

Figure 20.3 Book review
Source: Authors

Figure 20.4 Social survey
Source: Authors

process, elevating the understanding of IKS integration in architectural academics.

Prakriya: The main findings of the survey were:

- Students: Majority of the students felt that specific subjects like *Vaastushastra,* traditional architecture should be included in Architecture education as the late architect B.V. Doshi was the primary figure among Indian modern architects who effectively integrated traditional knowledge with contemporary design principles
- Among the educationists surveyed, the majority felt that a fusion of modern educational practices with IKS is essential for comprehending the essence of 'Indianness' on the international

stage. The majority of educationists advocated for a strong unification of ancient and modern educational approaches, proposing that it should be directed as a compulsory course for teachers and faculty members to enhance their understanding through research endeavors. Furthermore, 56% of respondents highlighted that the integration of IKS into architecture education could significantly enhance the architectural landscape of modern India. Additionally, some educationists suggested an unconventional approach, proposing the inclusion of IKS in career counselling sessions to benefit the Indian education system holistically.

- On the other hand, experts in Archaeology and Conservation stressed the importance of incorporating archaeologists, conservationists, and scholars into the Board of Studies of Universities to maintain the contextual identity in architectural practice. They highlighted that the relevance of IKS in conservation extends beyond its practical application, including contextual analysis, application techniques, and the future path of the conservation process. Additionally, they emphasized the inclusion of IKS in the National Education Policy to ensure cultural relevance and responsiveness to Indian society.

Pratikriya: After analyzing all the responses, the inferences were compiled under 5 heads:

1. Curriculum enrichment: Introduce dedicated courses or modules within architecture education that focus on Indian traditional architecture, construction techniques, cultural significance, and sustainability principles derived from IKS (AICTE).

2. Practical learning and field studies for collaboration and community engagement: Organize field trips to historical sites, traditional villages, and architectural landmarks, enabling students to observe and document traditional building methods, materials, and design elements rooted in IKS. Conduct workshops and encourage students to engage directly with artisans and local communities on real-life projects, encouraging hands-on learning of traditional construction techniques and craftsmanship (Mcpheeand Przedpelska, 2018).

3. Research and documentation: Support research projects focusing on IKS-related topics, encouraging students to explore, document, and analyze traditional architectural practices, materials, and sustainable design solutions. Promote the dissemination of research findings through publications, seminars, and conferences to contribute to the body of knowledge on IKS in architecture (Nimkulrat, 2007).

4. Faculty training and professional development: Arrange workshops and guest lectures by practitioners, traditional artisans, conservationists, and archaeologists specializing in IKS to provide faculty members with practical insights and real-world experiences (Guidelines for training of Faculty on Indian Knowledge Systems, 2022). Enhance faculty members' understanding and expertise in integrating IKS into architectural education.

5. Policy advocacy and institutional support: Establish dedicated centers within architectural institutions to promote research, education, and preservation of IKS in architecture. Advocate for the inclusion of IKS in national education policies to ensure cultural relevance and responsiveness to Indian society (AICTE, 2020).

Module Development - Stage III

Kriya: Approach adopted for assignment development (see Figure 20.1)

Following an analysis of all the responses from the social survey, it was inferred that there was a crucial need to orient the current generation towards the preservation and relevance of the Indian knowledge system. In response, Post Graduate scholars devised assignments rooted in the Indian knowledge system for undergraduate design studios. Each student was tasked with selecting one or all of the following books:

1. 'Mayamatam' the bilingual edition by Dr. Bruno Dagens
2. 'Theory and Practice of Temple Architecture in Medieval India: Bhoja's Samaranganasutradhara'
3. 'The Bhojpur Line Drawings' by Adam Hardy

Students supplemented their readings with relevant research papers aligned with their interests to inform the design of studio exercises. After comprehending the chapters from their chosen book(s), students engaged in a rigorous brainstorming session organized into four groups of two. The selected books and subsequent design exercises for each group were as follows:

a. **Group 1:** 'Mayamatam' - Focus on the chapter on Pavilions (Dagens, 2007a, b) for designing a related exercise (see Figure 20.5).
b. **Group 2:** 'Theory and Practice of Temple Architecture in Medieval India' and 'The Bhojpur Line Drawings' - Develop a design exercise centred on ratio and proportion in temple design, emphasizing visual balance in Vimanas (Hardy, 2015b) (see Figure 20.5).
c. **Group 3:** 'Mayamatam' and 'Samarangasutradhara' - Create an exercise exploring fractal geometry with reference to Temple Architecture (Hardy, 2015a) (see Figure 20.5).
d. **Group 4:** 'Mayamatam' - Explore the chapter on measurement to develop a design brief understanding the ancient measurement system through body parts (Dagens, 2007a,b). The brief entailed documenting building spaces, their components, and construction methods using this system (see Figure 20.5).

The analysis conducted by the authors revealed various embedded attributes of design in each exercise, including function, space, grid, aesthetics, structure, and principles of aesthetics. These exercises aimed

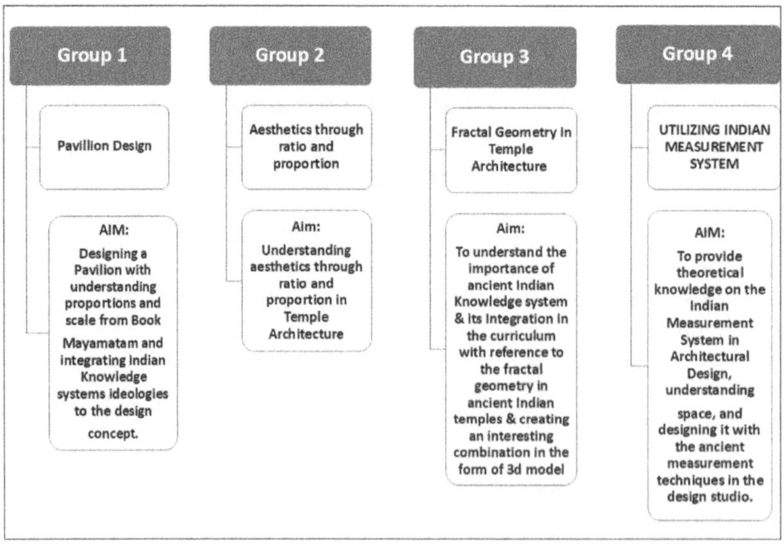

Figure 20.5 Aims of exercises

to deepen students' understanding of traditional Indian architectural principles while nurturing creativity and critical thinking in studio environments.

With this understanding, the detailed exercise to be conducted for the undergraduate students was formulated (see Figure 20.5).

Prakriya: All the assignments primarily centered on comprehending the practical application and relevance of Indian architectural texts. The focus was on:

1. **Understanding architectural elements:** Students researched into the intricate architectural elements described in the Indian texts, such as pavilions, temple plinths, Vimanas, and measurement systems. Through in-depth study and analysis, they gained insights into the structural components and design principles inherent in these elements.
2. **Analytical diagramming:** Students engaged in analytical diagramming of the architectural drawings, interpreting the details and proportions rooted within them. This process allowed for a deeper understanding of the spatial relationships and design details present in the Indian architectural tradition.
3. **Practical scope and relevance:** The assignments aimed to explore the practical scope and contemporary relevance of ancient architectural treatises. By applying the knowledge gained from the texts to real-world design exercises, students were able to distinguish the lasting relevance of traditional architectural principles in modern architectural practice.
4. **Scope for creativity:** Students were encouraged to explore their creativity within the framework of ancient architectural systems. Through design exercises and model making, they had the opportunity to reinterpret and innovate upon traditional design principles, both individually and group creativity.

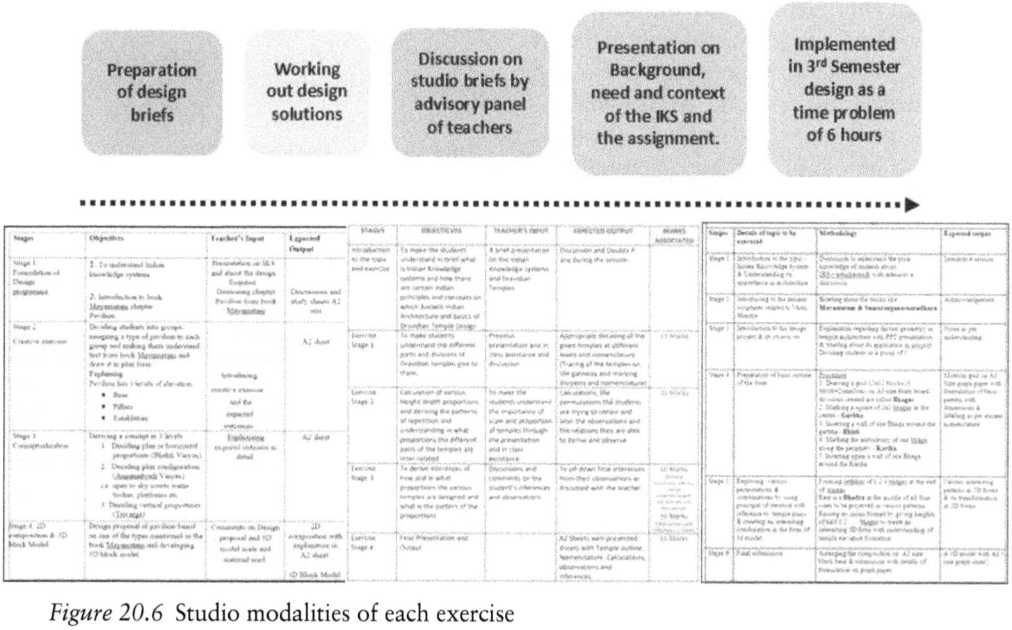

Figure 20.6 Studio modalities of each exercise
Source: Authors

Implementation Stage V

The Methodology

Pratikriya

Group 1: This group embarked on understanding the scale and proportion of ancient pavilions like Padmaka, Alankrata, Sub Husana types (Mandapas) based on the '*Mayamatam*.' They came across challenges in interpreting Sanskrit terms and decoding the plans from the text, which initially consumed a significant amount of time. However, they persevered and successfully decoded various types of pavilions based on parameters like the number of '*Stambha*' and '*Bhadra*' presence of inner courts, and ratios for length, width, and height. While the model-making process was time-consuming, students found the exercise both interesting and challenging, demonstrating patience and dedication throughout.

Group 2: In their exploration of temple plinths, students engaged in experimenting with various *Bhagas* to develop a comprehensive understanding of temple design principles. They delved into the terminology of Sanskrit related to different parts of the grid, enhancing their vocabulary and comprehension of architectural elements. Through their experimentation, students gained insight into fractal geometry in temple plans, expanding their awareness of *Vastushastra* terminology and its practical application. The culmination of their efforts materialized in a 3D model, showcasing their grasp of the subject matter.

Students understood the ratios and proportions used for *Bhitti* (walls) of Garbhagriha to Karika (ambulatory passage) to *Bahyabhitti* (outer walls). Understanding the ratios and proportions used in temple architecture, students successfully formulated elevations for various temple components, from the walls of the Garbhagriha to the corners of the plinth in Ngara style temples. However, formulating Rathikas in the Karna part presented a challenge, requiring students to navigate different proportions of recesses and projections within the given format. Despite this challenge, students demonstrated creativity by generating various plan variations, albeit limited to grid developments of 6, 10, and 12 *Bhagas*.

Group 3: Students were given inputs to understand the ratio and proportion principles governing the Vimanas of Dravidian Temples, as outlined in the '*Mayamatam*.' Their objective was to determine the pattern of visual balance within *Vimanas* by analyzing various height-to-width ratios. To facilitate their study, students were provided with analysis of proportions from different temples, each associated with specific ratios such as 10:7 for *Santika*, 9:6 for *Paushtika*, 8:5 for *Jayada*, and 6:3 for *Adbhuta* types.

Equipped with Dravidian-style temple elevations of varying storeys, students precisely traced these elevations on transparent sheets, annotating divisions and nomenclature for clarity. Subsequently, using graph paper, students calculated height-to-width proportions, translating patterns of repetition within Vimana designs. Through this exercise, students gleaned insights into the interconnectedness of different Vimana components, understanding them as integral parts of a cohesive whole. Interestingly, here the most important observation came from students' side was that- *Temple is designed as body of deity and not as abode of deity, so the proportions of body parts get reflected in the elevation of temples. So, there is a parallel which can be drawn and compared to anthropometric scale (The Modulor) of proportions devised by the master architect Le Corbusier* (Giri, 2015).

Group 4: The objective of the exercise was to acquaint students with the units of measurement system prevalent in traditional Indian architecture. Six students were

Figure 20.7 Measurement and discussions
Source: Authors

tasked with measuring and documenting the meeting room of the SMMCA office (see Figure 20.7). Employing a unique approach, each student utilized their body parts as units of measurement, such as *Hasta* (hand), *Angul* (palm width), and *Purusha* (height). With thorough attention to detail, students accurately measured and documented the meeting room, incorporating the principles of traditional Indian measurement systems into their analysis. They drew comprehensive plans and sections of the room, capturing the layout of furniture and other architectural elements.

By referencing body dimensions, students gained a practical understanding of how traditional Indian architectural measurements were derived, development of a deeper appreciation for the cultural and historical significance of these systems. This exercise provided students with a hands-on activity to engage with ancient measurement practices, bridging the gap between theory and practical application in architectural education.

Scientific Innovation and Relevance

The design studio for future B. Arch teachers, conducted by post-graduate students of M. Arch in Architecture Education, was centered on comprehending IKS and demonstrating to students how these systems facilitate easier learning. *Mayamatam*, a valued treatise among South Indian traditional architects, served as the primary source for formulating the design projects. This text holds particular significance in the contemporary era, as technical traditions from various domains are being reevaluated for potential modern applications.

Part 4 of the methodology (see Figure 20.1) outlined the development of studio exercises derived from the writings and ruchas (poetic verses) of Mayamatam. These exercises were distinctive, featuring nomenclature and labelling consistent with the canonical text.

The entire process was documented under three stages, which were described in Blooms taxonomy as Teaching-Learning Abilities (Chandio and Pandhiani, 2016). However, a major difference in these stages compared to Bloom's taxonomy is that this system is a cyclic process, where the output of one stage becomes the input or feeder of the upcoming stage (see Figure 20.9).

Kriya: Remembering and understanding the canonical text.

In this stage, students recalled and comprehended the foundational texts and concepts of Indian architecture. This formed the basis for further exploration and application in design exercises.

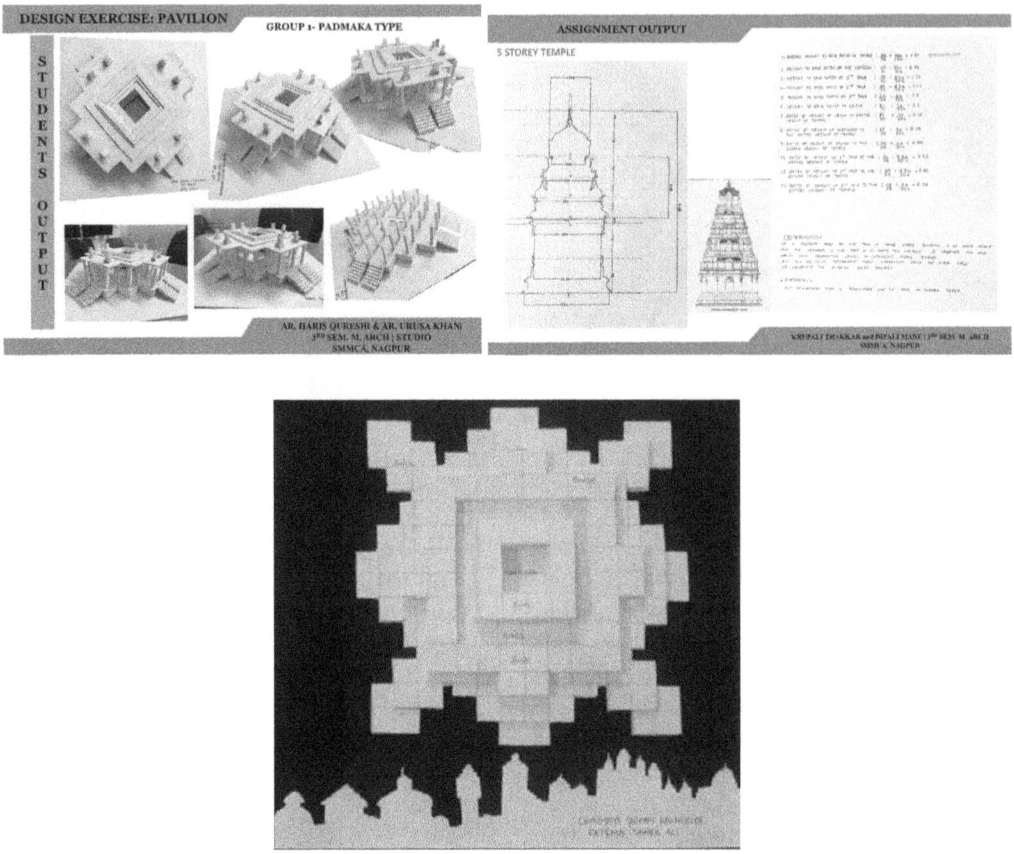

Figure 20.8 Studio modalities
Source: Authors

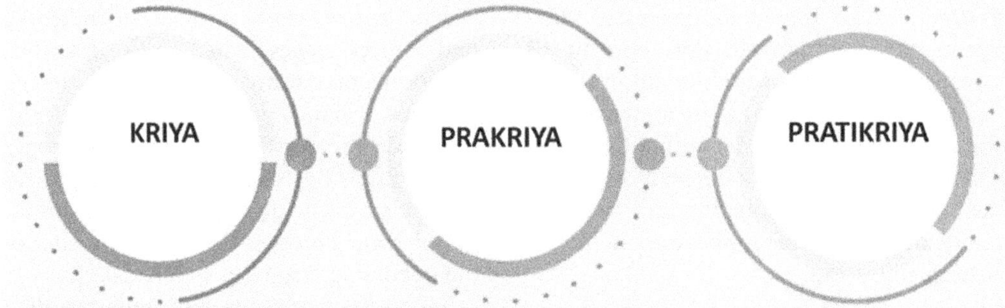

Figure 20.9 Conceptual process
Source: Author

Prakriya: Understanding, analyzing, and interpreting principles of architecture.

Here, students researched deeper into the principles and theories of architecture, analyzing the text and extracting key insights. They interpreted these principles within the context of architectural design, nurturing a deeper understanding.

Pratikriya: Applying and creating - Utilizing/reproducing concepts in new, relevant designs.

In the final stage, students applied their understanding by creatively integrating the learned concepts into new design projects. They produced innovative designs that show their understanding and application of the principles written in the canonical texts.

This cyclic process ensures that students continuously build upon their knowledge and skills, with each stage informing the next. The output of one stage serves as the input for the subsequent stage, development of a comprehensive approach to learning and design in architecture education.

The three design assignments clearly demonstrated a structured approach that aligned with the cyclic process described earlier. The design outcomes (see Figures 20.10, 20.11, 20.12 & 20.13) showcased a detailed understanding of various attributes of design and their significance in the overall architectural composition. These attributes formed the foundation of the design process, with each assignment rotating around these key points:

Attributes of Architectural Design

- **Function:** The purpose and intended use of the architectural space, ensuring practicality and efficiency in design solutions (Morris and Breidenthal, 2011).
- **Space:** The organization and manipulation of spatial elements to create meaningful and functional environments that cater to human needs (Vasudevanand McFarlane, 2008).
- **Structure:** The underlying framework and support system that ensures stability and integrity in architectural constructions, considering materiality and construction techniques (Yunitsyna, 2012).
- **Grid:** The underlying framework or structure that governs the organization and layout of architectural elements within a space (Yunitsyna, 2012).
- **Aesthetics:** The visual appeal and expression which is inherent in architectural design, encircling principles of beauty, harmony, and proportion.
- **Principles of Aesthetics:** Guiding principles and theories that inform the aesthetic decisions made in architectural design, including concepts such as ratio, proportion, balance, rhythm, and symmetry (Hekkert, 2006).

These attributes served as guiding principles throughout the design process, informing decision-making and shaping the final architectural compositions. By considering each attribute holistically and in relation to one another, the students were able to develop comprehensive and well-balanced design solutions that addressed functional, spatial, aesthetic, and structural considerations.

Results and Discussions

The output of the studio was in the form of graphical representation in 2 Dimensional and 3 Dimensional. The progression of importance of the attributes associated with any design is classified into Grid, structure, typology, function, space and principles of design. It can be analyzed by the design progression and the graph below (see Figure 20.13):

Hierarchy in Attributes of Architectural Design

Assignment 1

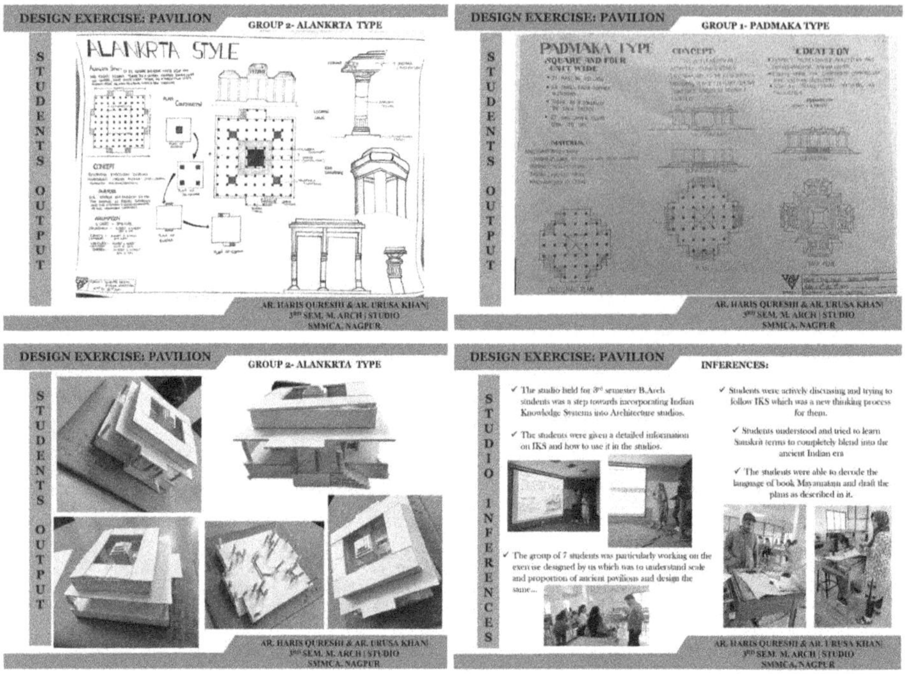

Figure 20.10 Output of group 1 (pavilion design)

Source: Authors (Dagens, Mayamatam, Treatise of Housing Architecture and Iconography, 2007)

Assignment 2

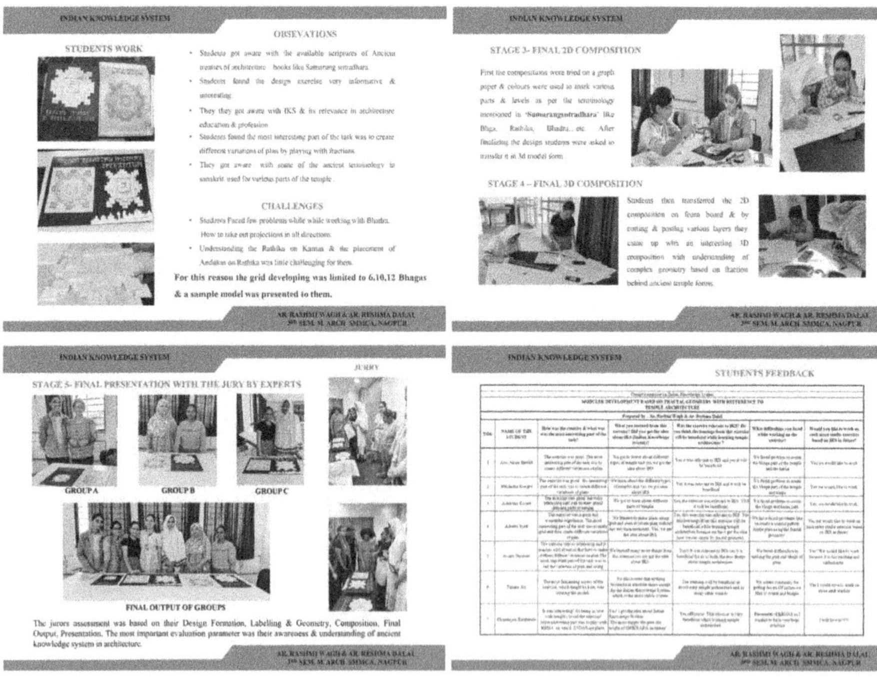

Figure 20.11 Output of group 2 (temple plan)

Source: Authors

Assignment 3

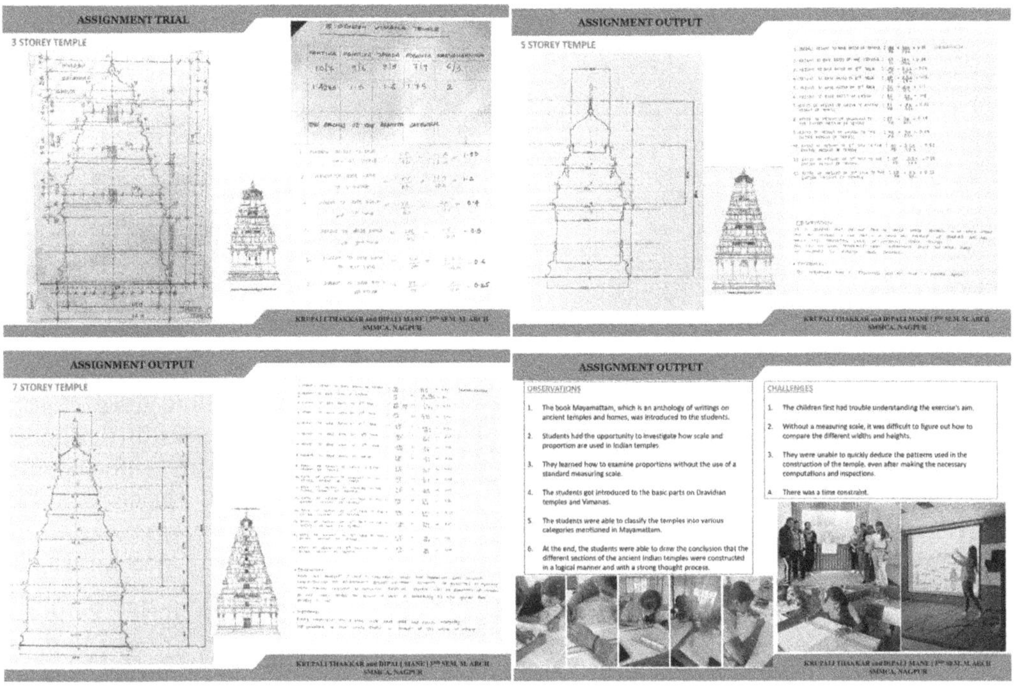

Figure 20.12 Output of group 3 (temple elevation)
Source: Authors (Dagens, Mayamatam (2 Vols.): Treatise of housing architecture and iconography, January 2007)

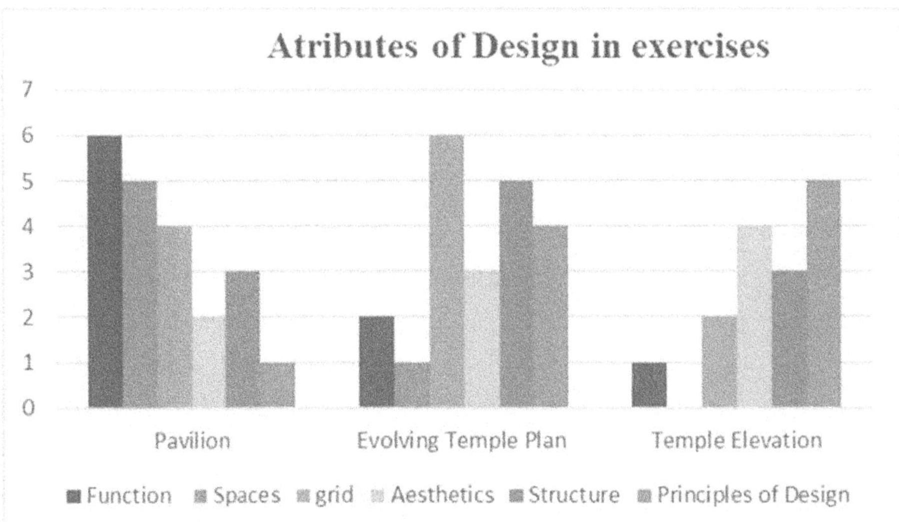

Figure 20.13 Attributes of design in various studio exercises
Source: Author

Table 20.1 Matrix to understand attributes of design in respective exercises

Attributes	Pavilion	Evolving Temple Plan	Temple Elevation
Function	1	4	5
Spaces	2	5	0
grid	3	1	4
Aesthetics	5	4	2
Structure	4	2	3
Principles of Design	6	3	1

0	
1	
2	
3	
4	
5	

Source: Authors

- Pavilions: Function (typology) – spaces – grid – structure – Aesthetics - Principles of Aesthetics
- Temple plan: Grid – structure - principles of aesthetics – aesthetics – function - spaces
- Temple elevation: Principles of aesthetics – aesthetics – function - grid – design

The undergraduate studio aimed to enhance an understanding of the design process and integrate applied subjects into design closely. The approach and outcomes of the studio demonstrated a seamless integration of various subjects. The method, which progressed through stages such as grid, structure, typology, function, space, and principles of design, (B. Mahadevan, 2023) provided a holistic view and motivated educators to delve deeper into the design projects.

The authors tried to understand the process and formulated a matrix of the assignments after seeing the progression and outputs of various groups (See Table 20.1).

The process of each assignment was mapped into a matrix and the matrix clearly explains the attributes of progression in various design assignments.

All the five attributes of design finally point toward the four words function, structure, aesthetics and space which draws parallels with the three anchors of design mentioned in the Vitruvian Triad (firmitas, utilitas and venustas) and also in Shriman Mahabharatam, sabhaparva-2 (Bhogadyam, Sukhadharshanam, Ramyam)

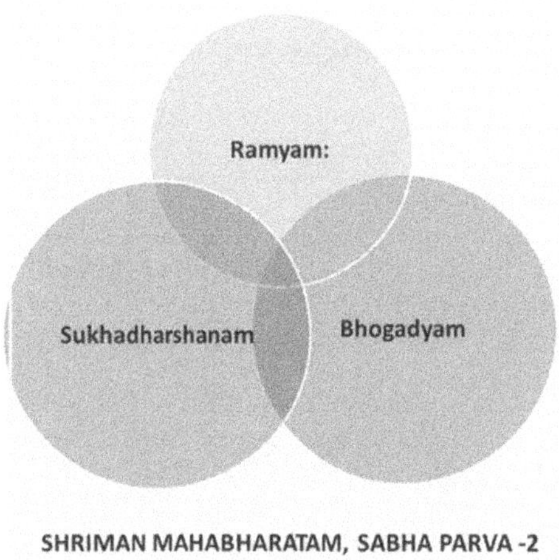

Figure 20.14 Bhogadyam, Sukhadharshanam, Ramyam
Source: Authors (Gupta, 2016)

Theory

Bhogadyam: Utility or usefulness, the user needs, climate, slope, topography, wind direction, temperature, needs of the person to live a comfortable life.

Sukhadharshanam: Aesthetics- visual aesthetics but it also encompasses all the other aesthetics that are created using color, texture, light and shade.

Ramyam: Spiritual satisfaction or inner delight.

The fourth attribute i.e. the space which is very well indicated by the third word mentioned in the Mahabharata 2a i.e. Ramyan which is the final aim of the Sthapati.

Conclusions

In conclusion, the design methodology integrating Indian Knowledge Systems (IKS) presents an innovative approach that has been developed through a systematic process of *Kriya*, *Prakriya*, and *Pratikriya*. While the outputs of students exhibit a detailed understanding, there is room for further strengthening their comprehension through a deeper metaphysical exploration of the subject matter. The design studios grounded in IKS have effectively included all necessary inputs within the design process itself.

The methodological framework highlights several key attributes of design, including grid, structure, typology, function, space, and principles of design. These attributes can form the basis of a structured framework for future integrated design studios, rooted in the knowledge of ancient India.

However, it's important to note that the design process is still in its nascent stages and requires further research and refinement before being fully adopted by design educators. Stakeholders, as identified through the social survey, have advocated for the proper integration of IKS into the regular curriculum, including credit-based courses as per the National Education Policy (NEP). This integration could be proposed for the architecture education design curriculum, with each subject initiated with an Indian knowledge base and reinforced with contemporary systems.

To adapt this concept three steps suggested by authors after taking in to account all the observations drawn at every stage of the studio are:

1. Curriculum development: Broad based IKS courses as electives, multidisciplinary courses, specialized courses to be designed for architecture students
2. Dissemination of IKS knowledge through teachers training and capacity building
3. IKS social outreach by collaborating with traditional scholars, artists and experts

The implementation strategy suggested by authors to adapt at all the levels of knowledge transfer which are as follows:

1. At institute level: Building institutional capacity and supporting IKS research centers, clubs, events, library reference books, workshops;
2. At faculty level: FDP to strengthen the faculty, Dedicated faculty to do research and teach the course;
3. At syllabus level: Program/course development with UGC so as to integrate IKS with CKS;

At subject level: Subject specific course development with university with practical content, infuse IKS into existing courses using relevant examples and concepts;

At student level: Encourage students to take up the course.

The incorporation of IKS into the curriculum not only improves students' understanding but also nurtures interdisciplinary research opportunities across various aspects of IKS. By incorporating these ancient wisdoms, the architecture education system can evolve to produce graduates who are not only skilled designers but also knowledgeable custodians of India's rich cultural heritage and architectural traditions.

Acknowledgements

The authors would like to place on record their deep appreciation to the Principal, SMMCA and the Post Graduate department of Architecture Education, SMMCA for encouragement in conducting the innovative studio.

The students of the post graduate program in Architecture Education have been strong building blocks in creating interesting studio briefs based on IKS for the third semester students of Allied Design.

The authors would also like to thank the Dean Design, Dean Academics SMMCA and all design teachers of third semester of SMMCA for their valuable guidance during the conduct of the studio.

References

All India Council for Technical Education (AICTE) Minor Degree in Indian Knowledge System.

Chandio, M. T. and Pandhiani, S. M. (2016). Bloom's taxonomy: improving assessment and teaching-learning process. *Journal of Education and Educational Development*, 3(2).

Dagens, B. (2007a). Mayamatam (2 Vols.): Treatise of Housing Architecture and Iconography.

Dagens, B. (2007b). Mayamatam (2 Vols.): Treatise of Housing Architecture and Iconography.

Giri, V. N., P. S. (2015). A Reformulation of the Indo-European System of Languages and a Recovery of the Deep Morphology of Varnamala. Cygnus Advertising (India) Pvt. Ltd. (Indian Institute of Technology Kharagpur).

Guidelines for training of Faculty on Indian Knowledge Systems (2022). New Delhi: University Grants Commission.

Hardy, A. (2015a). Dravida Temples in the Samarangansuthradhar.

Hardy, A. (2015b). Theory and Practice of Temple Architecture in Medieval India: Bhoja's Samarangana Sutradhara and the Bhojpur Line Drawings. Dev Publishers & Distributors.

Hekkert, P. (2006). Design aesthetics: principles of pleasure in design. Research Gate.

Mahadevan, B., and Bhat, V. R. (2023). Introduction to Indian Knowledge System Concepts and Applications. Delhi: PHI Learning Private Limited.

Mcphee, S. W., and Przedpelska, S. (2018). Experiential learning, community engagement, and student experience: undergraduate field school course in rural British Columbia: undergraduate field school course in rural BC. Research Gate.

Morris, A. T., and Breidenthal, J. C. (2011). The necessity of functional analysis for space exploration programs. In Conference: Digital Avionics Systems Conference (DASC), 2011 IEEE/AIAA 30th.

Nimkulrat, N. (2007). The role of documentation in practice-led research. Reserach Gate.

Policy, N. E. (2020). National Education Policy 2020. Ministry of Human Resource Development.

Prabhupada, A. B. (2019). Shrimad Bhagwat Geeta Yatharoop. Shree Paramhans Swami Adgadanandji Trust.

Singh, S. (2022). Revisiting the Educational Heritage of India. New Delhi: Vitasta Publishing Pvt. Ltd.

Thapliyal, D. P. (2023). Indian knowledge systems in the curriculum of higher education: a proposed model of a PG course in IKS. *International Journal of Research Publication and Reviews,* 4(7), 32963301.

Vasudevan, A., and McFarlane, C. (2008). Spaces of Enclosure. Elsevier.

Yunitsyna, A. (2012). Grid as an instrument of creating of the universal space in dwelling. In Conference: 1st International Conference on Architecture & Urban Design At: Tirana, Albani.

21 Exploring the ancient knowledge systems in contemporary architectural pedagogy as a means to true knowledge

Piyush Das[a]

Assistant Professor, School of Architecture, VIT University, Vellore, India

Abstract

The construction of architecture, regardless of typology, has been a longstanding practice among human societies for centuries, dating back to the early stages of settlement. Irrespective of the context, the act of designing and constructing is an inherently a skill. The concept under discussion pertains to knowledge. The information has been transmitted. Communication can occur through written or oral means. Throughout antiquity, the knowledge system gradually evolved to a level of sophistication wherein it assimilated various facets of disparate theories pertaining to the constructed environment.

In contemporary practices of architecture and pedagogy, there is a limited reference to ancient theologies, particularly within the context of Indian traditional knowledge of architecture.

An illustrative instance can be found in the *NyaaSutras*, an ancient Indian text, where the method of knowing is expounded upon. This pertains to the methodologies of cognition and perception, which are deemed pertinent and advocated for in the realm of architectural education.

The objective of this paper is to initiate a scholarly discussion regarding the importance of *Nyaya-Sutra*'s theories, particularly its examination of the techniques for attaining genuine knowledge, and its relevance in the present-day educational practices of architecture. This paper aims to analyze specific aspects of the topic at hand, leaving the remaining areas for further exploration by those who find them relevant.

Keywords: Architectural pedagogy, history of architecture, knowledge, new paradigm, nyaya sutras, traditional Indian knowledge systems

Introduction

What history of architecture? How does one engage with 'History' in the discipline of the built environment? A paradigm shift is required in the pedagogy of built environment, where history is understood from new lenses while questioning the not so contemporary relevant tools. It is also essential to thoughtfully unlearn the biases of non-Indian systems in interpreting Indian history in the context of the built environment. It is also necessary to understand what content goes and what tools are taught in the context of studying and teaching history of architecture in the Indian context.

One may find references of ancient Indian text and its content relevant to revisit the same with a lenses. *Nyaya Sutras* composed by Akṣapada Gautama dated probably between sixth century BC to second century was a comprehensive text of knowledge and logic.

Methodology

The paper uses primary translated texts as well as original Sanskrit texts (limited

[a]rupanpiyush@gmail.com

DOI: 10.1201/9781003571018-21

knowledge of Sanskrit that the author possess) to analyse the content and shares architectural relevant instances or usage of the same. This paper attempts to understand the book one of the *Nyaya sutras* and critically examine the sixteen categories of the knowledge of true nature in order to attain supreme felicity. Here attempt is made to use relevant conversations around different historic built systems in the Indian context in order to built a narrative of holistic understanding, hence helping with wright knowledge dissemination in the contemporary architectural pedagogy.

Analysis

Nyaya meaning 'from the root', also has many other similar ideas such as '*hetu-vidya*'-the science of causes, '*Anviksiki*'- the science of inquiry, '*pramanasastra*'-the science of correct knowledge, *'tarka-vidya'*-the science of reasoning. *Nyaya sutra* is considered to be the earliest work on the *Nyaya* philosophy (Basu, 1913).

The supreme felicity- knowledge of architecture can be attained by the knowledge about the true nature of sixteen categories.

1. Means of right knowledge (*Pramana*)
   1.1 Perception (*Pratyaksa*): It happens with the contact of the senses and the object.
   When, one studies a building (in this context of the paper, a historic building) the first perception happens when our senses contact the building. When our senses contact the surroundings or the context. That gives us three set of knowledge, that could be:
   a. Determinate: This distinguishes the perception from indeterminate knowledge. For example the built structures along with the ghats of Varanasi, when seen from the river Ganga with a distance, the knowledge acquired by the senses may be determinate but could be indeterminate due to various factors of time, season etc.
   b. Unnameable: This means the knowledge of a built or unbuilt entity, that is derived from the perception of the senses, with possible no connection of the reality how so ever.
   c Non-erratic: The JahazMahal a fifteenth century structure in Mehrauli, Delhi appears to be a ship in a water body, due to its reflection in the nearby water body named *Hauz-i-Samshi*. It is necessary to acknowledge two things here. One if the narrative of the floating ship is a later construct, then such perceptions to be eliminated in order to understand the architecture of the object of that time. Two, if evidences suggest that the narrative was intentional while the design of the same, then such perceptions probablybe considered.
   1.2 Inference (*Anumana*): *Anumana* is the knowledge that precedes perception. They are of three kinds:
   a. 'a priori': A priori is the knowledge that addresses the perception of the cause that effects the object or parts of the object. In order to explain the same one may take the example of some architectural systems that connects multiple built entities and to be seen as a whole. If one sees a gopuram, the that knowledge indicates there is an another architectural entity that is a temple. This is also subjective of the time of the observation and in particular the temple, its construction date. For example if one is talking about a

temple that was the first that had gopuram, then this process of knowledge access might not work (Das, 2017).

b. 'a posteriori': A posterioriis the knowledge of the cause that is derived by the perception of an effect. Similarly, using the previous example, if one sees the temple in the similar context, then one infers that there would be gopurams as well.

c. 'commonly seen': Commonly seenis the perception of the things that are together. This again as earlier discussed comes with the baggage of time.

1.3 Comparison: is the knowledge of a built systems that is previously known and understand the new systems from some of the achieved point of views. For example if one is studying temple architecture Bhubaneswar, and in particular the Parsurameswara Temple, then other built systems that were constructed before becomes the object for comparison. Comparing Parsurameswara temple, which was built around seventh century ad with Lingaraja temple, which was built in thirteen century AD does not make any sense, if the aim is to only study Parsurameswara temple comprehensively (Das, 2017). Otherwise if the aim is to study temple architecture of Bhubaneswar in general, that is a separate discussion.

1.4 Word (verbal testimony): Here one talks about the instructive assertion of a reliable person regarding the knowledge of the built system. Here one questions what references come under verbal testimony? The reliable person has to be or not from the same region? One must understand that the reliable person should not have any biases that might mislead the knowledge collection process. When one is discussing the idea of history of architecture, what are methods of procuring verbal testimony? For example if one is talking about Mehrauli, Delhi as a place and intents to study the historic water structure Hauz-i-Shamshi, then probably a reference of Iban Battuta, a fourteenth century AD traveller's writing may or may not be a suitable verbal testimony to demonstrate the architecture for reasons like, biases (he did talk about the water structure with. Bit of exaggeration of scale), not being local etc. or otherwise can be discussed, but the idea and the relevance of verbal testimony is certain.

2. Objects of right knowledge (*Prameya*): Objects of right knowledge or *prameya* can be of two types, that are the matter that can be seen and the matter that can be not seen. The tangible and intangible aspects of the built systems. It has to be understood a building system is a living being and hence the methods of studying them has to be in ways where, one identifies the idea of 'soul' in the built systems. The following are the objects of right knowledge:

2.1 Soul: Like a living being, the built system also has a soul. Similarly desire, aversion, volition, pleasure, pain an intelligence are the marks of the soul of any built systems.

2.2 Body: Both the body of the person intents to acquire the path of right knowledge and the object is the site of gesture, senses and sentiments.

2.3 Senses: Nose, tongue, eye, skin, and ear are the preceptors of elements such as earth, water, light, air and ether

2.4 Objects of senses: The above directing the senses of smell, taste, colour, touch and sound. Any built system has a body with senses and the perceives and displays the object of smells. It smells through a nose and made of earth and also make the observers smell. These philosophical construct of the built systems are relevant to analyse in order to have a holistic knowledge collection and dissemination.

2.5 Intellect (*budhhi*is one of the object of right knowledge. At any point of time a building possess certain intellect that is also the first thing that evolved from the matter (*prakriti*).

2.6 Mind: Perception doesn't just happen with the contact of the sense organs and the object. It needs the mind for the process. And hence it is not possible to perceive two things simultaneously. For example the mind of the perceiver makes a difference in the way one is assessing the built environment using his or her sense organs (Das, 2021).

2.7 Activity: When the voice, mind and body works together then activity happens. This is one of the crucial object of right knowledge in the context of assessing historic buildings. While referring the tangible built systems, it is noteworthy to observe activities of the original time. That allows to find new clues in the process of making of architecture. For example the in order to understand a temple precinct of Bhubaneswar with a water tank, probably one has to observe the activities that are dictated by the sacred systems and that leads to certain other ways of perception and activity becomes one of the object of right knowledge (Das, 2024). as per some rituals, one may have to first take a dip in the water, may be offer prayers, then move toward some particular directions etc. Activities are of three types, vocal, mental and bodily (Das, 2023a). Philosophically they are again divided into good activities and bad activities.

2.8 Faults (faults are the ones that cause activities). In this context architecturally while activities of varied types become the object of right knowledge, the causation of the activity or the fault is also an object of right knowledge. For example, a Choupal or a traditional building in rural setting, which was for gathering and other purpose etc, is a unique historic building typology. Now the activities, like gathering etc becomes an object of investigation architecturally. Now what caused the activity. Here interference of other disciplines such as sociology and anthropology may shed lights on aspects of the rural environment, social systems, architecturally an requirement that might reflect back to the kind of other typologies of architecture etc. that leads to new activities and again the causations. May be these processes of knowing leads to the true knowledge of the built systems.

2.9 Transmigration: It is the idea of death and birth and that happens in a loop. Birth happens when soul connects with body, sense-organs, mind, intellect and sentiments. Death happens when soul separates from all these.

2.10 Fruits is when activity and fault works together.

2.11 Pain is the hindrance, that tries to separate the soul from the body, sense-organs, mind, intellect and sentiments.

2.12 Release: is the absolute deliverance from pain.

3. Doubt is the non-similar or conflicting perception about an object. They are of various types.

3.1 Recognition of common properties: In architectural narratives often doubt arises and should be treated carefully regarding the commonality of architectural qualities that is recognised. In order to understand a specific built of a specific time it is necessary to identify and validate such commonalities. For example openness is an architectural quality, now is it appropriate to talk about the openness that one identifies in Quat-islam Mosque (one of the early masque in Delhi, 13th century AD) and the openness one identifies in the Jama mosque in Delhi of Shahjahan times. Here though the typology is the same but the time context and many other parameters may not be appropriate to recognize such common properties. At the same time may be some common properties of two different typology of a particular time could be a subject of discussions, for example if one tries to identify the idea of openness in some region and explore the same in the vernacular architecture and traditional architecture.

3.2 Recognition of properties which are not common: Some cases doubt arises when some architectural properties which are not common but recognised in one context. An example of that would be while understanding a historic garden in Indian context where probably there would have been some metaphysical planning principles based on sacred systems, which might be perceived to be "informal', 'not planned' or 'organic' etc and at the same time trying to understand with rather out of context lenses of grid etc becomes properties that are not common and should be understood as doubt (Das, 2019). As metaphysical planning principles are quite evolved and sophisticated methods to be understood and a built system should be understood from the valid lenses.

3.3. Conflicting testimony: Does knowledge system of architecture in the Indian context has strong testimonies, like in some other fields? When one talks about historic buildings in the Indian context, were there many disciplines top offer in the knowledge systems or it was so holistic that it was a system of living in that everything was articulated including the systems of the built. However in the context of conflicting testimony, one may observe the case of Zafar Mahal, where the last Mughal King, bahadur Shah Zafar II lived, in Delhi, that was a summer retreat for the king probably build during the late Mughal time. In this case, it is evident that the built systems of Zafar Mahal had many layers of time. From the Slave dynasty till the British time, there one would find built interventions in the building. All these layers of different time, different dynasties and probably a blurred line differences in so called testimonies somehow partly evident and one explains the architecture partly. What about build systems of earlier times, where such conflicting testimonies are not evident and one narrates the architecture? How about an biased notation of a testimony has continued and one finds conflicts and doubts in order to explain the architecture of a certation building of a certain time. Its noteworthy to identify such doubts and critically respond.

3.4 Irregularity of perception: In some instances doubt arises when there is an irregularity in perceptions. A philosophical example of this would be one may perceive water in a tank that exists but at the same time one may perceive the same in the case of a mirage. In this context architecturally it could be any components of any scale.

3.5 Irregularity of non-perception is the opposite of the stated doubt in the section 3.4.

4. Purpose (*Prayojana*): Identification of the purpose, with an eye that one proceeds to gain the true knowledge. In this context one may take the example the UNESCO world heritage site of Bhimbetka rock shelter. Here there could be multiple eyes of investigations and identifications of the same is probably the one of the early steps. It is noteworthy to understand, to decode the idea of a pre historic site, the 'eyes' has to be in harmony of the time of the same. The causation of the rock art? The knowledge systems of that time? The requirement of knowledge retention? The adaptation to the nature and techniques? And many more.

5. A familiar instance (*Drastanta*): An observation made by an ordinary man and an observation made by an expert, if it is the same, then that's the familiar instance. A path to true knowledge of the architecture of a particular entity may address and recognize such familiar instances. Here probably one may try to understand the simplicities of complex architectural entities that a common man and an expert both agrees upon. It is also about how architecture is explained! Does it has to be with architectural jargons or one may find ways to explain the same with simplicity that everyone understands?

6. An established tenet (*Siddhanta*)is a set of principles or a dogma based on certain school of thought, hypothesis or implications.

6.1 A dogma of all schools: The diverse range of historic architectural resources (all natural, cultural, tangible and intangible) in the Indian complex context, do we have universal tenet to explain the same? And from the contemporary relevance point of view does the discipline requires one?

6.2 A dogma of some schools but rejected by some: A best example of these would be the built systems in the complex sacred cultural landscape sites in India, irrespective of any religion. The making of the built did relay on sacred systems, beliefs and rituals. Probably the understanding of the same should refer the same school of thoughts. As in the current times there is this still ongoing conversations regarding the same (Das, 2022).

6.3 A hypothetical dogma: in continuation with the previous idea, a hypothetical dogma is a principle or believe, where if a system is accepted, then that leads to the acceptance of many other principles that precedes it. For example if a temple is accepted as a residential typology, where a god lives, the many other principles of tenet has to be accepted. Such as his daily and weekly calendar that effects the built systems of the temple. His movements pattern that effects the larger landscape, for example the case of sacred landscape of Bhubaneswar.

6.4 An implied dogma: Is one of the relevant idea in the discipline of history of architecture, as it implies on case specific principles of investigations.

7. Members (*avayava*): In the quest of true knowledge of architecture, along with all the sixteen measures, one may also inquires the *avayava*, which are the proposition, reason, example, application and conclusion. For example a proposition (what is to be established?), does early temple architecture of Bhubaneswar did borrow some references from the architecture of the Buddhist period? Reason (means of establishing what is to be established) being in some early temples, there one would find Buddhist iconographies. Example (two types, 1. heterogeneous or negative, 2. homogeneous or affirmative) could be such scenarios have been noticed in a different context. Application is finding more evidences to understand the reason holistically. Hence the conclusion that yes or no.
   In this context the process may seem simpler but when all parameters of a built system are addressed, that demands other parts of *avayava* that are, Inquiry: questing the proposition (the need?), question the reason (*samsya*), relevance and capacity of the example, purpose of drawing conclusions etc.
8. Confutation (*tarka*) is carried out to identify certain not known features of the historic built environment.
9. Ascertainment (*nirnaya*) is the removal of any doubts and determination of the real question by hearing two contrary principles.
10. Discussion (*vada*), 11. Wrangling (*jalpa*), 12. Cavil (*vitanda*), 13. Fallacy (*hetvabhasa*), 14. Quibble (*chala*), 15. Futility (*jati*), *16.* Occasion for rebuke (*nigrahashana*) are the other ways of true knowledge to attain a holistic understanding of history of architecture for both teaching and learning process.

New Knowledge and Conclusions

Interpreting the relevance of such ancient texts that does thoroughly talks about the paths to attaint true knowledge in the context of history of architecture is a new attempt and helps with new knowledge and methodology for better understanding as well as better dissemination. In the paradigm of architectural pedagogy, one may find the process relevant in order to revisit the content that to be taught in history of architecture. More importantly a paradigm shift is needed where the tools and skills expected by the students in the context of history of architecture, where it becomes relevant and holistic.

References

Basu, B. D. (1913). The Nyaya Sutras of Gotama (Vol VIII). Alladabad: The Panini Office.

Das, P. (2017). Memory of a Stream. Copal Publishing.

Das, P. (2019). Metaphiscial planning principles: case of puri. *Verandah*. Vol 1, Issue 1, Published by SAA, Sushant Univerity.

Das, P. (2021). Built from the Puranas. Copal Publishing. New Delhi, 978-9383419999

Das, P. (2022). The sacred-secular-divide: myth or reality? case of ekamrakshetra: Bhubaneswar. *The Quarterly Journal of Mythic Society*, 113(2).00478555, pp 80-88.

Das, P. (2023a). Revealing architecture: case of the sacred cultural landscape of ekamrakshetra: Bhubaneswar. *Malaysia Architectural Journal*. 27166139, Vol 5, issue 1, pp 17–29.

Das, P. (2024). Sacred geography and belief systems: unveiling the asta ayatanas of ekamra kshetra. *The Quarterly Journal of Mythic Society*. 00478555. Vol 114, issue 4, pp 149–154.

22 *Educating the Educators*: Studies in *a*rchitecture-*s*ynergy ... *e*volving *v*isionaries toward *p*rofessionalism

AR. Anil Bhingarde[a]

Prof. AR. Anil V. Bhingarde, Graduated from Rachana Sansad, Mumbai and Development Planning at CEPT, Ahmedabad.

Worked at National Institute of Design, Ahmedabad.

Abstract

Here is an attempt to define and initiate healthy academic progression for architecture studies- 'Educating the Educators'- by all, the Teaching Fraternity- Student community - the Trustees of institutions; to develop Synergy for evolving Visionaries!

The revelations in the Monograph- the observations-generalizations- concepts are described for the follow up actions by discerning stakeholders And this is to envision well-structured durable environs! Unknowingly or otherwise, we adhered to international trends in ideologies and typologies in our architectural ensembles; and consequently, legacy to the members of Gen Next! Let us strive to evolve a *'Successful business model'* at the respective college that also contributes to building *'Healthy Growth'* of communities and settlements. There is immense scope for Innovations at all levels. The upcoming institutes must put in sustained efforts to shape Skylines of urban and rural environs in the Indian sub-continent within the realm of Ethical practice. It is possible to cultivate innate design sensibilities by exploring various issues explored here!

Keywords: Architectonics, architectural significance, data-layering process, design dynamics, ethical norms of architecture, indicators of creativity, knowledge systems, natural academic progressions

Introduction

Intents

The progression is conceptualized as a giant step forward to Induce: Energize- Radicalize- Formalize the processes of 'Educating the Educators' in Architectural field. This is to reorient practice of environmental designing and also, fuse scientific-technological skills adequately as professed by various institutions scattered across sub-regions of Indian subcontinent- that is our Bharat.

Strategizing various subsystems of architectural practice and academics is extremely necessary wherein all the stakeholders must strive for a fairly realistic and sensible attitude and approach towards an array of training courses. It is to inculcate 'Indicators of Creativity' with reasonable balance of tangibles and intangibles of design grammar in our learning processes.

To a great extent- all these aspects contribute to individual's intuitive power in resolving and evolving successive layering in designing process imbued with profound sense of confidence and regenerative spirits lasting to eternity. Intuition is the ability to perceive the truth about everything. Can we nurture basic instincts of curiosity and inquisitiveness to decipher relative merits of design equations for typical built form?

[a]bhingardeanil@gmail.com

DOI: 10.1201/9781003571018-22

The Preamble

The Action Plan (7.1... Addendum) illustrates the Overview and follow up actions to reorient our knowledge systems in Syllabus Designing; the environmental equations thereby, making rational choices about constants and variables for interactions among teachers and learners with profound sense of certitude and exactitude; a type of clinical approach with absolute convictions! It is a long-term exercise.... DIY- do it yourself-for any individual or compact and wellnit team members; or even a whole class as a think-Tank.

The revelations in the Monograph- the Observations and concepts are described for the follow up actions by discerning stakeholders, and this is to envision well-structured durable environs! Unknowingly or otherwise, we tend to adhere international trends of ideologies and typologies in our architectural ensembles; and consequently, legacy to the members of Gen Next.

Value Premise

Intense research and self-introspection is necessary that focuses on the discovery and identification of meaningful patterns of 'Architectural Significance'. The following terms signify shades and hues of *'significance'*. the relative merits of variables and constants we dwell to accrue and balance of tangible and intangibles of any environmental equation for typical built form.

> *The merits of successive stages*
> *in thinking process*
> *(for evolving solutions)*
> *Enable one to carryout*
> *designing activities to various*
> *stages of fruition*
> *With a critically appropriate prudence*
> *to present designing progression;*
> *... Of due Reference with Reverence*
> *and Relevance-*
> *... w.r.t. Ambience wit Exuberance*
> *and Excellence-*

> *... Diligence with Omniscience*
> *and Omnipotence-*
> *... Opulence with Sustenance*
> *and Subsistence-Substance with*
> *Pursuance and Preservance-*
> *... Assurance with Semblance*
> *and Resemblance –(typologies*
> *and precedence ...?)-*
> *Practice to master with great*
> *Patience and Surveillance*
> *for 'Self-Realization'!*

How to enhance and exfoliate Knowledge systems and contents over a period of time? Should all take initiative in scripting the grass-root ideas and concepts of learning material in local languages for Architecture as 'Noble Profession'-being social art in the socio-cultural context?

As practitioners, all should contemplate over the Ethical norms of architecture. The TRC-COA, our Institute of Excellence, and array of Institutes of Education Research scattered all over the country have great role to play to fuse intelligently and forcefully with NEP-National Education Policy.

Review of Literature

In General

Few landmark books stirred the imagination towards resolving various issues of environment and ecology across built-ensembles, the warp and weft of systems for sculpting fenestration systems or outer skin-for system of resilient envelopes.

A comprehensive matrix can be devised to list *'Development and design concepts'*- at micro and macro level- to specify range of basic reading material essential to cultivate comprehensive outlook towards socio-cultural-technical undercurrents.. Possibly thereby build the set of interrelationships between varied aspects of tangible and intangible elements of constants and variables in systemic environmental design equations. The contents may be reviewed periodically. If working in a team, the data may be compiled as research and development activity. especially by academics and serious student groups.

Methodology

The Thought Processes

The happenings in profession enliven and invigorate as one grows in stature, to experience and recreate all the ground realities about emerging Urbanscapes-Suburbanscapes–Rural scapes across human settlements, with inherent technical and Design inputs, while deriving resilient built ensembles of appropriate design dynamics.

Accrued through the study of relevant sstreams in humanities, all tend to imbibe and master 'The Social art of Architecture' by intense creativity, adopt socially conscious attitude which lasts through our career as thinkers and innovators. Infuse our rich local inherent vernacular and contextual inputs and interventions using appropriate tangible and intangible elements with sensitivity and subtlety into our ensembles while evolving habitation patterns to reflect contemporariness in creative ensembles and urban-rural skyline in general.

Professional Landmarks

The overview of projects and deliberations at various symposia in the decades of sixties and seventies may be recollected. This is how any sensible individual designer moulds oneself.

In contrast- the ghastly happenings- the trends of implanting 'pseudo classical-modernistic the stylized facades to residential and public buildings vitiated the academic progressions.

This was in contravention of thinking modes associated with natural progressions of academics and tend to demoralize budding designers and aspiring academics in search of our roots. The matter needs a great soul searching to evaluate and substantiate relations between successive events.

Awards

In appreciation for creative teachings or projects, every year awards are also conferred to academia as 'The Best Teacher' as well as the Best Student project-thesis etc.' at graduate and postgraduate levels. The professional market is flooded with awards for iconic designers for innovative displays. The array of professional publications-magazines, Journals-Print media etc. largely contribute to the process.

Contextual Inputs

All have to infuse our rich local inherent vernacular and regional constants and variables with sensitivity and subtlety into our ensembles while evolving habitation patterns. The faculties in leading architecture colleges and students at National Association of Students of Architecture (NASA), India, in their annual assemblies have generated gargantuan of research work.

After pioneering efforts in initial phases by stalwarts, we see hectic activities toward professional excellence in recent times, we ought to spare time to chronicle recent events and logical progressions; the transitory phases tend to nudge professional's approach toward project conceptualization.

The Noble Profession and Public Service

Do we all build a successful business model (at individual institute) that also contributes in building the harmonious growth of our settlements and communities? Certainly, there is tremendous scope for refinement and innovation to enhance livability patterns.

Introspection

Entry Level aptitude and agility-it is a great struggle to initiate-inculcate and cultivate aptitude, once you opt for architecture; the schooling and nurturing for Science Technology Engineering Arts Management

(STEAM) disciplines is lot to be desired wrt inherent patterns about sensitivity and sensibility. English being the medium of instructions, mostly as a way of fashion and rarely as passion, it is imperative that grammar of design (arts etc.) and appreciation skills of beauty, aesthetics, and excellence imbibed invariably are of rudimentary nature.

To a great extent home management experiences reveal various sensitive and sensible issues in 'Man – household resources synergy' with subtlety within the domain of Domestic engineering - Home making - Family abode. The synergy between Householder and Office goer irrespective of individual sex is necessary which we learned in recent pandemic; also imbibed skillfully by elder younger members of any family unit.

Interpreting Environmental Functions and Systems

It is necessary to revitalize learning trajectory. A focus on exploring and monitoring information systems with series of topics to broaden the perspective of 'place making' i.e. Built ensembles - Community clusters- Habitat planning is ideally desired to present a comprehensive treatise about appropriate 'Learning and Teaching' model for replicating by new colleges. Certainly this process will enhance thinking capabilities of budding faculty members from our institutes of excellence.

Should not COA-TRL with its inherent power and capabilities strive to form *a 'Repertoire of themes'* interacting with places and communities? In the turbulent professional climate it will be safer to glide through if we evolve thinking processes.

Cultivate Instincts

Intuitive Power

Innovations in teaching processes with comprehensive contents enable to derive Design progressions at graduate and post graduate level studies. Here normally architecture design studios extend up to three to four months each semester with six to twelve hours for formal interactions every week. Yet, studio interactions ought to be supplemented by Contemplation: observing and thinking processes all-round the day for deriving interrelationships with appropriate sets of constants and variables of relative merits to complement the environmental designing activities and in turn our environs, i.e. experiential learning.

Site and Space Planning

The matrix explains the Ambience and Spirit of the place - (7.2-Addendum) - to derive Comparative analogies of Landscape activities: Space- Form- Volume and profile of Outdoor landscape elements.

Organizing and updating the project-features periodically for future reference as one grows in stature- Internship & takeoff for future practice in the form Log book?

All should plan semester studio programme to define work culture; series of activities to imbibe agility- multi-tasking- time management! Can we insist and manage to keep them engaged, enthused, enthralled every day or perennially for invigorating events and experiences?

Is architectural literacy possible at the end of five years needed for graduation with reasonable qualitative attributes? Or one can afford to struggle for a further two years for the same in post graduate deliberations? Information systems for design development diagrams with data- layering process as studio activity is illustrated in Process-Data Layering Systems (7.3 Addendum).

Layering

Can all responsible design teachers attempt to learn and present design scenarios inclusive of successive layers for design interpretation as anticipated – from students during a period of one semester on a preconceived Format: brief-plans- and sections as because, are precursor of subsequent multi-fold layers of details for execution (7.3-Addendum).

Discussion

R & D Activities

Each discerning learner should prepare his own data base of concepts and vocabulary at macro and micro level of current ideas and thinking applicable in the field of environmental designing and planning activities as illustrated earlier in matrix (7.1-Addendum).

Current Thinking

How do we derive environmental equations through use of constants and variables from array of aspects within Design grammar- Humanities- Natural sciences and Technologies, etc., albeit with Sustainability guidelines, a new 'Mantra' or Buzzword in recent times to evolve Spirit of inquiry cutting across disciplines?

Is layering of concepts as regards various inputs possible; preferably circular approach as compared to conventional...the linear type. The Prerequisite for offering any 'Didactic Dialectics' and precise good constructive 'crit' by any learned teacher is clean and neat diagrams from prospective learners- as exchange platform to reflect-on in AD studio. Thus, it is imperative that constructive measures are necessary towards demystifying scope of preaching architecture.

Innovation and Embellishment

As regards resultant built-ensembles, mostly many of us all limit to skeleton, structure etc. There are many more steps to contemplate over!

It is all about sculpting interiors and exteriors and also outdoors with appropriate environmental parameters. Should not it be integral part of studio activities?

Do we ever derive Outer skin or Fenestration system?

- Do we ever contemplate over Green Products-Systems-Devices?
- Ever attempted seriously fusing basic concepts for layering process...i.e. ergonomics and anthropometric (data) checks-
- Concept of outdoor recreation spaces and design grammar-Typology of public places-the Relative scale of spaces for environmental functions-
- Landscape activities, elements and plant material-
- Landscape Design Grammar to enable composing landscapes-
- Common sensory stimuli- received by the senses and elicit response from a person! ....

  To name a few!

  All collectively contribute to evolve appropriate Architectonics for built forms!

CONCLUDE

Conclusions and Recommendations

Visualisation

In fact, thorough discussion is necessary to explain comprehensive contents and relative scope and processes- methods- system of deriving or visualizing organically 'Plan-Section- Elevation' and later 'Built Ensemble' building envelope, with its multifarious Products- Systems-Devices to match preset brief and site context for a project proposal. Does enlightened fraternity contemplate and introduce appropriate measures to the prevailing ways and means?

The Transition via Evoke

A course can be designed by universities or University Grant Commission for nurturing faculty all over India and can be initiated with guidance from COA- India for interpreting syllabus with necessary preamble and postscript for every subject – to assure a fair amount of comprehensiveness to suit regional context.

It is possible to adopt Circular process than Linear one as a rule, an healthy practice while evolving built forms and designs, by checking and rechecking often with creative instincts- all along documenting the progress for further reference and self-evaluation.

No doubt, all have imbibed thinking processes in this direction, but in varying degrees- to reach finishing lines! Please take this note in right spirit and move ahead as per individual leanings and awakenings! Enjoy Designing activities! It is a pleasant experience discovering and honing our communication skills while inventing creative talents! Is it possible to inculcate artistic, literary and creative instincts?

Think Tank

It is a great challenge for 'Architecture faculty! Ceratainly there are hopes of resurgence in near future- within a period of Semester...Year...Decade... or so. even century as is the case of AIA and RIBA in far off developed countries;if all care to understand gravity of situation and work together diligently.

The paper can give different perceptions to serious and conscious readers. A systemic approach may help in imbibing multifarious information system to derive a set of layering in designing activities for overbearing experience; joy of creating designed systems for our masses. It is a process of turning architectural engineering into Architectural designing.

Legacy of civil engineering practices and "Bible" scripted in PWD handbooks governed the past vast pool of Built Profiles by design professionals!

Healthy practice is precursor of 'Wealthy practice' as measure of enhancement and success.

Addendum

Three matrices-7.1, 7.2 and 7.3 are illustrated here.

Numerous matrices can be evolved to start thinking process at different levels of academics and professional practice; ... a sort of spadework for future R & D activities in established practices.

To name a few specimens-

- Interpreting Environmental functions and systems
- Development and a design concepts and vocabulary
- Typology of Public Spaces
- Landscape Design Grammar
- Outdoor Landscape Elements

Action Plan

The steps enable to evolve correlations; ease of monitoring any project design-development and comprehensive approach to syllabus designing tasks.

<table>
<tr><td colspan="4">7.1. Preamble
'Educating The Educators'- Synergy for evolving visionaries –
The Action Plan</td></tr>
<tr><td colspan="2">Part-I:
OVERVIEW

MONOGRAPH
Academic Scenario- Learning And Teaching Synergy! Define And Initiate- Healthy Academic Progression For Architecture
'Educating The Educators'-
By Honourable Teaching & Student Community...And Importantly- Trustees / Promoters Of Institutions - The Modern Age Maharishis'!!!....
Synergy For Evolving Visionaries!!</td><td colspan="2">Part-II:
FOLLOW UP ACTION

DECODING ARCHITECTURE!
An Approach To Basic Research
Enriching Pedagogy! Through...Information Systems -
Evolving syllabus Topics and classroom lessons-
As Checklist for Concepts and Vocabulary-</td></tr>
<tr><td>INSTILLING SPIRIT OF ARCHITECTURE IN ACADEMICS</td><td>Environmental Planning And Designing Progressions:

Mentoring and Monitoring wart. information systems in teaching tasks- Aided By Case Studies / Research / DIY- Experiential Learning-</td><td>EE RESEARCH DATA- INFORMATION SYSTEMS</td><td>Index-Array Of Notes

Design Grammar – Making Concept Designing and Detailing – Design Theory
Landscape- Environmental Planning – Urban Planning</td></tr>
<tr><td>LEVELS OF STUDIES AND PHASES</td><td>Evaluation By Academic Teams From Individual College?

S. No. 1 & 2 - 1st To 4th Yr. Ad Studio Project - Sem. 1 To 4 / 5 T0 8th Sem.
S. No. 3 & 4 - My Own Academic (Learning) & Public Places
S. No. 5 & 6 - Academic Subjects & Teachers- 1 To 4 & 5 To 10 Sem.
S. No. 7- Extracurricular Activities-Sem. 1 To 10 Sem.
S. No. 8 - Dd1 & Dd2 Activities-Sem. 9 & 10
S. No. 9 - Post Graduate Studies
S. No. 10 - Ph.D. and Post-doctoral Studies</td><td>CRITICAL THINKING – OBSERVE- IMAGINE-LEARN: 1

CRITICAL THINKING- OBSERVE- IMAGINE-LEARN: 2</td><td>Teach Thought – Verbs – (Bloom's Taxonomy)

Knowledge – Understand – Apply Analyze – Evaluate – Create....

Tools for Taking Charge Of Your Learning & Your Life - Ref: Repeal & Lender

Clarity – Accuracy – Precision – Relevance – Depth – Breadth Logic – Fairness – Significance</td></tr>
</table>

7.1. Preamble (*contd.*)

INTERPRETING ENVIRONMENTAL FUNCTIONS AND SYSTEMS	**Understanding Comprehensive Measures** For Monitoring Communities And Settlements – Information Systems And Processes For Environmental Planning And Designing Activities –	**EDUCATION: HOW WE ARE SET TO LEARN ONLY POINTERS FOR OWN INTUTIVE RESPONSE**	**Brain Storming- Intuitive Response** Common sense approach- to begin with – Try to decipher meanings from the Pointers enumerated by the Author –
ARCHITECTURE AND BEYOND	**Statutes-Standards- Legal Sanctions- Professional Bodies Etc.** Consultancy- Execution And Management Miscue.- Research- Feedback From Institutions Implications (Synergy With) -Other Quasi Govt. Bodies- Statutes- Consultancy- Management- Execution- Monitoring: A Charter Thro' Dialogue On The Same Wavelength By Bare Minimum Code Of Conduct (E.G. As MOE - GOI- Interpreting Contents)?	**EDUCATION: TRACKING** **CHANGES IN HOW WE ARE SET TO LEARN**	in the light of NEP- New Education Policy **The Author's Observations** – Logical explanation for individual pointer Media Article –
COMMUNICATION TO PROFESSIONAL BODY- IIA	**Response To The Communication From Our President-** **IIA:** Indian Institute of Architects- No acknowledge as yet..! Similarly from COA..!! To Individual Centres/ Chapters/Members Etc.- On The Occasion Of Anniversary Message- May 12, 2018 –	**INDIA OF FIVE NATIONS**	**The Five Indian Worlds** do relate to each other and interact to varying degrees but they are also strangers to one another- How can politicians representing these five worlds AGREE ON ANYTHING? And How can they GOVERN? **EDUCATE?????**

7.1. Preamble (*contd.*)

DEVELOPMENT-CONCEPTS AT MACRO & MICRO LEVEL:	**Compendium Of Articles & News Snippets:** 1.Natural Processes wart Life Sciences- 2.environment & Ecology wrt nature Cycles &Transition-3. Growth & Development wrt economics and Technology-4. Learning Guidelines wrt Establishing Co-relationships- 5.Professional Practice wart. Building Vision-6-Media Snippets	**GOVERNANCE AND IT'S RELEVANCE TO SUSTENANCE**	**Ecological Balance** – Economic Performance- Institutional Capacity- - Viable Governance... the key dimensions of Sustainability- At All Levels and In All Contexts???

7.2-Site and Space Planning									
Ambience & Spirit of The Place: Comparative Analogies of Landscape Activities-Space-Form-Volume and Profile Outdoor Landscape Elements**									
Planning / Designing Aspects-S.NO. & Type-Project Groups	Site context:	Circulation:	Recreational zoning:	Intensity of development:	Skyline: Long. Section & Cross section	Activities & landscape elements:	Services & Installations:	Overlays For site analysis:	Remarks general:
1- **Studio Projects: A.D. Landscape Master Planning Town Planning**	1.1								1.9
2- **Own Residential Environs**				?					
3- **Own Public Spaces-Squares-Plazas**									
4- **Own Study-Learning Environments**				?					
5- **Public Parks and Playfields**				? Updating & Innovating Knowledge Systems					

7.2-Site and Space Planning									
Ambience & Spirit of The Place: Comparative Analogies of Landscape Activities-Space-Form-Volume and Profile Outdoor Landscape Elements**									
6- Residence to College-Streetscape				?					
7- Classical Period Landscapes				?					
8- Notesd-Class Lectures and Studio Interaction				?					

** Organising and updating the project features for future reference as one grows in stature-internship & takeoff for furure practice-as Log Book

7.3- Layering of Data Systems

Process Chart-I

Built Forms - Embellishment
With Design Grammar And Environmental ethics

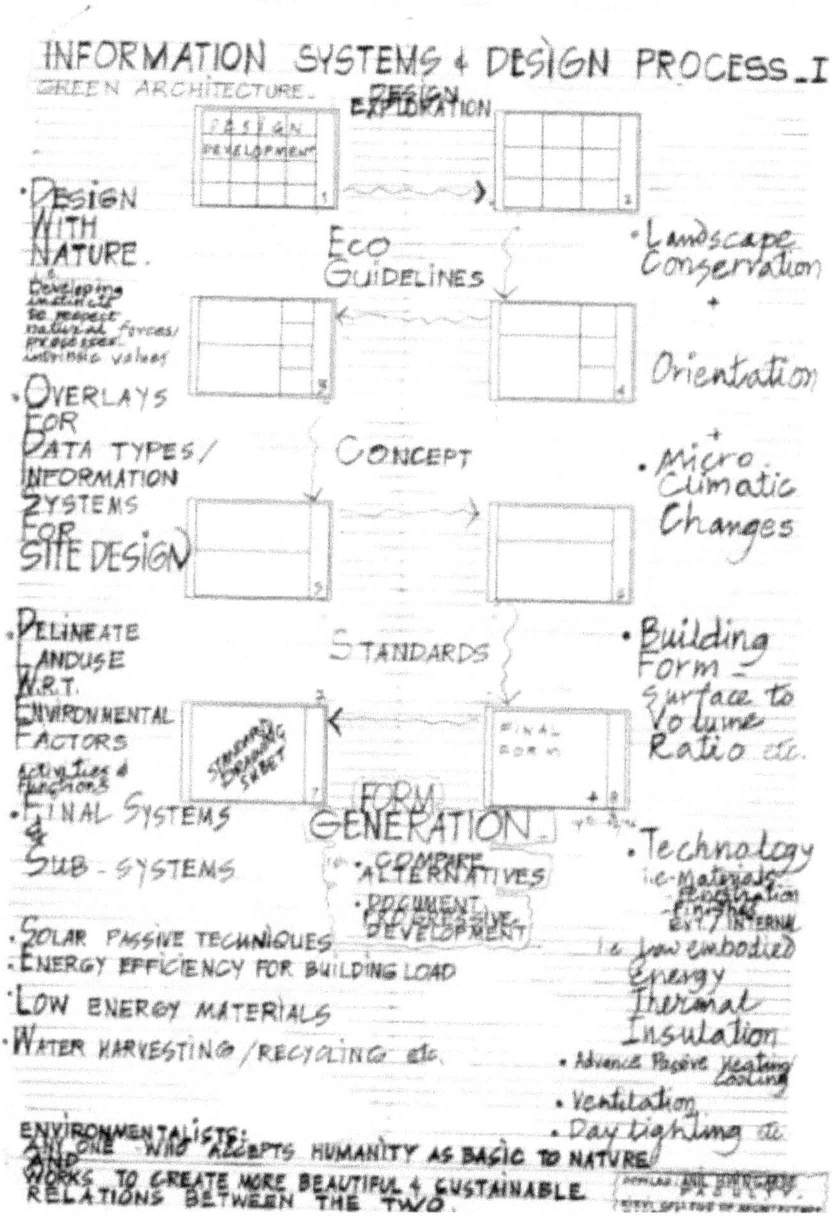

Process Chart-II

Design Exploration -
Layering of Ideas With Ecological outlook

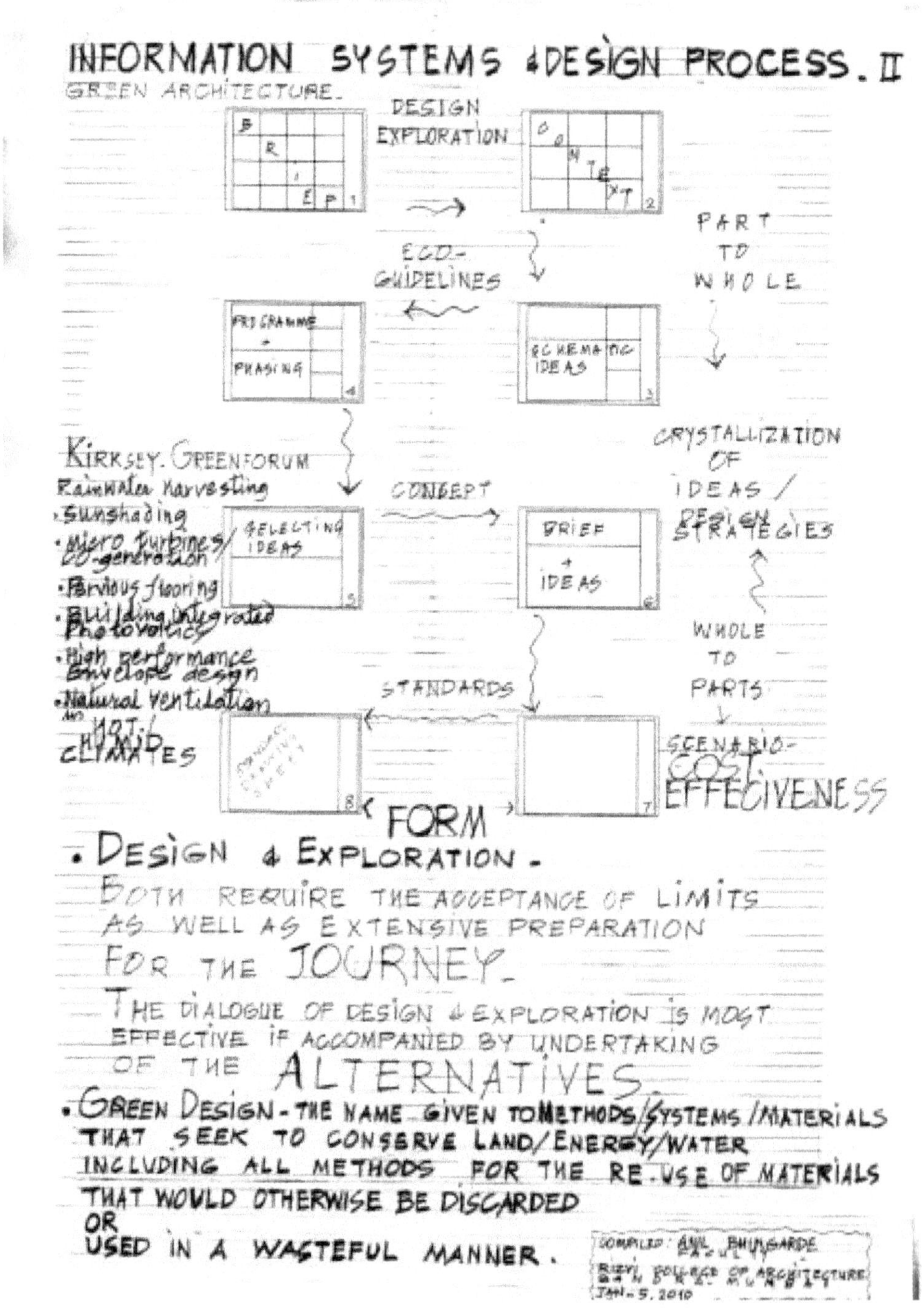

Process Chart-III
Information Systems -

Evolving green framework In Office studios as R & D activities-
How shall our communities judge us As
Responsible professionals?

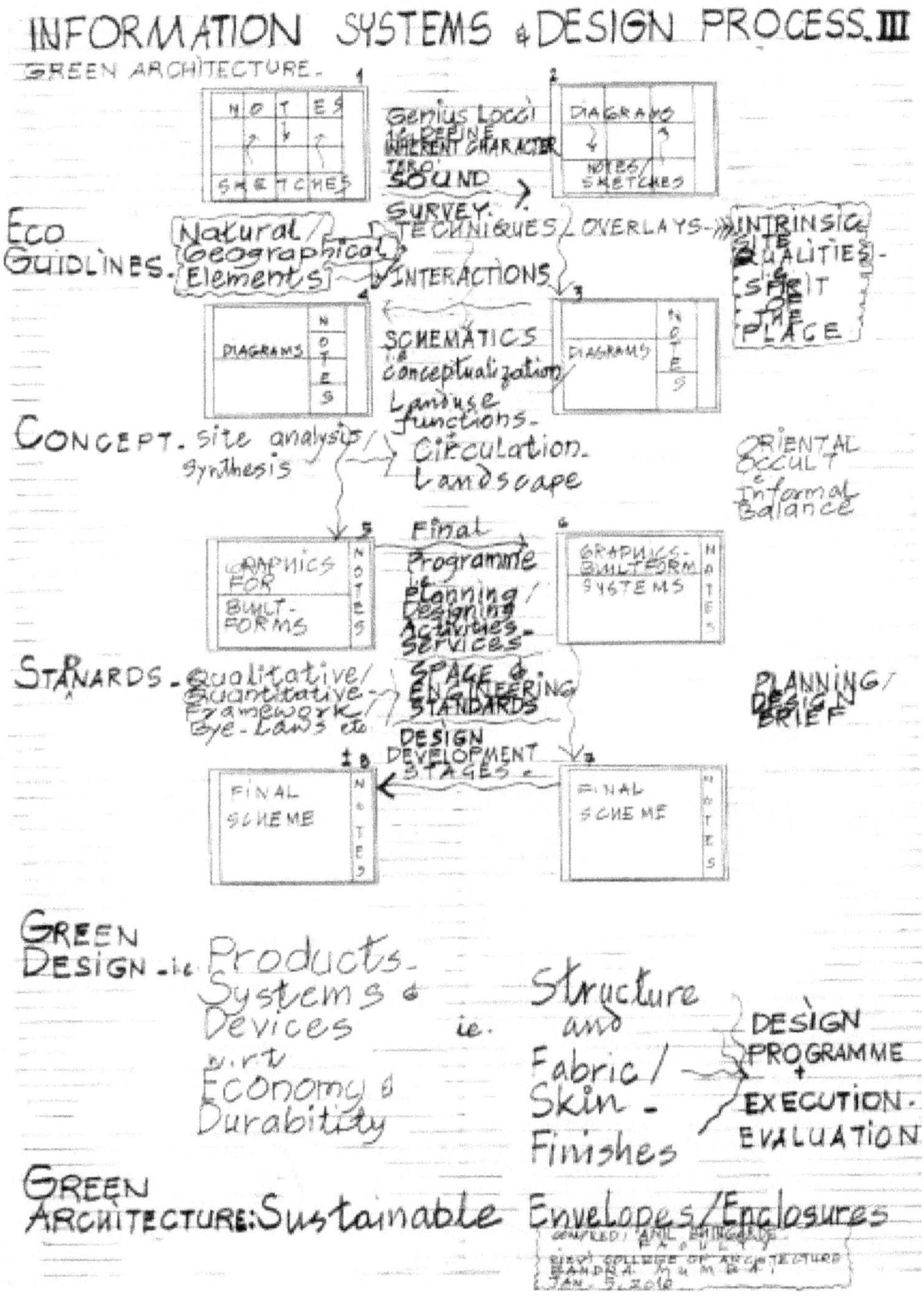

23 Generating the design program: Collaborative, connective, and creative perspective in design of architectural projects

Apurva Deshpande-Dhuri[a]

Assistant Professor, Full-time Faculty, School of Architecture, Anjuman-I-Islam's Kalsekar Technical Campus, New Panvel, Mumbai, India

Abstract

The design program is a crucial stage in architectural design, defining not only physical and functional aspects but also the qualitative essence of the resulting structure. It serves as a foundational blueprint guiding architects in translating abstract concepts into tangible spaces. This research paper explores the significance of the design program, examining its role in decision-making, fostering creativity, and delivering spaces that resonate with users and their environments.

In an architectural design project, the resultant product comes with the physicality of built form, along with inbuilt experiences of comprising immeasurable qualities. The tangible and intangible qualities of the architectural product form the overall experience within and around the structure. In order to achieve a multi layered, enriching, and progressive experience, the design and architectural program play a significant role.

This present paper shall analyse the design program for its spatial construct and the architectural program for its experiential construct, for any architectural project. The appropriate use of referencing, analysis of values from social, cultural, and behavioural examples taken from the architecture of the past, and their application within the core values of the design project through built structure, spatial arrangement, and operational mode, are achieved through constructive thinking and testing strategies. The core design function with its supporting functional spaces forms the primary program of the project. The expansive and profound nature of this program must be designed with an elaborate, meaningful method.

In the design of a project brief, an approach that is collaborative in nature, generates this extended set of uses, that expand the main and supportive programs, thus increasing the outreach of the main function of the project. The spatial program is further developed keeping in mind, these categories of uses, i.e., connecting spatially, proximity and scale of the main, supportive, and collaborative uses. The further part brings forward a creative expression for designing experiences through transition, to create a sequential order of experiencing the project for users.

The strategic working of the design process with academic and practice cases shall be explored in this paper. Keeping the core method of the design process, the paper shall explore ways to generate a wholesome design program for architectural projects.

Keywords: Collaborative architecture, design methodology, design value, experiential architecture, strategic process

[a]aapurva.deshpande@aiktc.ac.in

DOI: 10.1201/9781003571018-23

Introduction

People feel spaces: spaces that are imagined, drawn, and then built. The most basic elements of building, i.e., walls, roof, floor with the added quality of light and ventilation, lead to experiences which may be unidimensional, or experiences of heightened sensorial expressions. The memories of a generic, single-dimensional experience are many-a-times lost, but the memories that triggered a response from an individual remain with them for a lifetime. This is the power that architecture holds. The design of an architectural project that focuses on creating a deep experience must consider tangible and intangible elements along with a methodological approach to root the building in its context.

This research paper focuses on the framework of a design program conducted in architectural design studios, trying to identify its potentials, assess its lacunae and examine the deliberations to make it intensive and expansive. While a generic design program consisting of site data collection, site study, case study, and design drawings, can create a design project, it requires certain more parameters to turn it into a sensible and sensitive design project. This paper shall discuss and deliberate on the parameters that shall help achieve a better architectural outcome, and its value for the architectural and design program.

Understanding the Design Program

For some time in the 1980s and 1990s, the stage of creating a design program in the process of designing was replaced by post-modern and deconstructivism concepts of form-making. The design of radical forms that triggered emotional responses took over the priority given to users' needs. In present day academic studios, the design process allows multiple approaches of interpreting the design project brief. Students may choose to begin with form-based studies, operational case studies, site and context-driven designs, technology and material-based approach or even hypothetical concepts, justified for the project.

Background

The Whole Building Design Guide (WBDG) defines architectural programming as a design discipline. As described by authors Edith Cherry and John Petronis, on the research page of the WBDG website, architectural programming is a research and decision-making process that identifies the scope of the work to be designed in the architectural project (Edith Cherry, 2016). The WBDG states the process of any design project in the six stages as shown in Figure 23.1.

The Design Program in Use

The process of designing a built form has both convergent and divergent stages in its methodology. The documentation stage collects evidence of physicality of the site, and analysis is carried out for mitigating these site aspects. The stage of case studies looks at the similarities in the program and selected projects are studied for their overall aspects, to derive tools for the design. Next, the process jumps to the concept stage as an expression from the designer. The concept may be individually driven, or it converts the design into a problem- solving equation. The programs in the project become functions and the users merely exist within these programs as actors. In academics, the area statement in the brief decides the width span and sizes of the box that the program shall fit into. A large amount of time is spent on the planning process, arranging the sequence of the circulation and programs that cohesively fit together. When left to the student, the vertical explorations shown as sections are drawn at the evaluation stage, and elevations are an outcome derived from 3D explorations of the project.

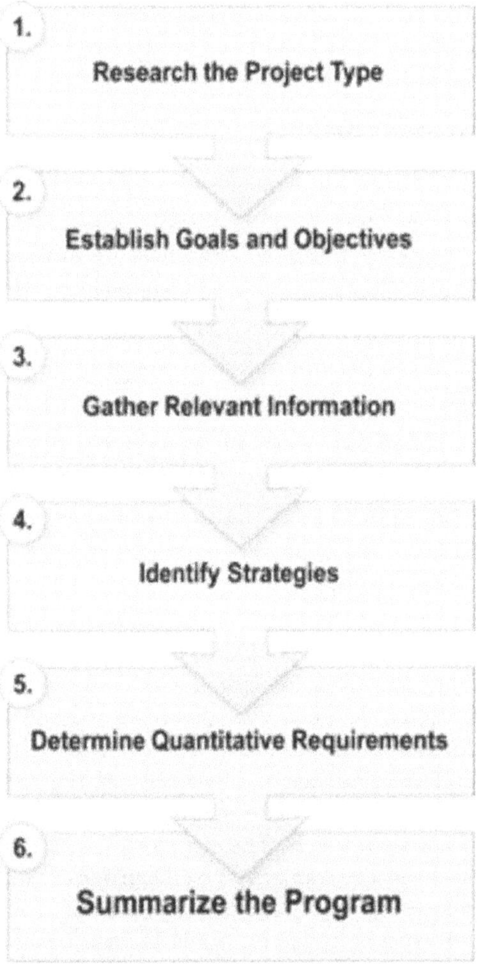

Figure 23.1 The six-step process for programming

Source: Whole building design guide article on architectural programming

Let us assume an academic situation of designing a community centre with workshop as the main program. The project is situated in an urban area, in the older part of the city that attracts tourism. The first user assumed is the artist/craftsman who create the product of the workshop. The secondary user is the tourist, the visitor in the role of audience for the activity of workshop. The area statement quantifies the workshop area as 600 square meter space. The first response of the student/ designer is to divide the area into a sizable box, mostly a rectangle of 15 meters on one side and 20 meters on the other. This 15 × 20 m box remains just that, an empty box, for most of the time in the semester. At the end of the design drawing stage, certain furniture is fitted inside with a few other empty rooms for storage and toilets. Barring a few exceptional cases, this is a common practice by students in the design studio.

Programming a Project Sequencing

Envisioning Learning Outcomes

The first impression of an architectural product, the building, is seen through its form. The form of a built mass has a direct impact on the viewer, generating a preliminary response from them regarding the approachability of the building. The architect, in the role of an artist, many- a times, initiates the design process by responding to the existing context through the design of the form. At other times, the design erupts through the planning of the spatial scheme, the diagram of arranging functions. This process of planning involves the trailing of circulation of people or vehicles or both in an effective manner to 'use' the spaces well. In majority cases, the design process starts and ends with the generation of the 'plan', the ultimate goal of the studio process, with the sections and elevations followed up on through three-dimensional models of the plan.

Why then, does architect B. V. Doshi, the late pioneer of Indian modernism, believe architecture to be synonymous with life? Where do the architectural marvels of master architect Le Corbusier stop being a building and transform into timelessness? Does this change happen on the drawing board during the site analysis or do the 3D explorations infuse life into the masses generated on them?

Louis Kahn's structures bring together space, form, light, and ventilation to create ephemeral and tranquil experiences. They make the 'dweller' pause, slow down and indulge in their silence. Doshi's architecture is known to be a vehicle for constant celebration, a medium for heightened experiences. The simplicity in his forms allow for easy and enjoyable moments for its people, making buildings into vessels for action and interactions, creating memories. His buildings tell stories to the viewer, the narratives of the occupants or the history of the place.

In his book, 'Spaces for Engagement' (Burte, 2008), author and architect Himanshu Burte, talks about place-making as an outcome of the architectural process. The architectural product cannot survive in isolation in the form of a building. A parcel of land has certain protocols that make it a potential site for building of a program. The context is full of rituals and practices that form the everyday for the place. The 'designed' building should knit itself into these existing protocols and practices, using these support programs to empower the primary program. Burte understands place, as a weave of physical space, the matrix of social structures and procedures and a tumult of actual actions related to either. This theory leads to the confirmation that one of the major reasons for buildings that fail to survive their programs stand in isolation or bear a huge disconnect with their physical and intangible contexts.

The Architectural Design Studio

The design process in an architectural design studio, is a balance of convergent and divergent procedure. It defines areas of exploration that allow the designer to think and experiment radical ideas, thoughts, and processes, and then converges the content by defining limitations to bring the project closest to reality. The production of design brief is as significant here as the conception of the project, both envision parts of the project that shape the future of that place.

The design brief contains, in its barest form, the aim of the project, the selected site with its study aspects, the main programs, the auxiliary and support programs, their respective built-up areas and the schedules for submissions and assessment. They sometimes include a list of projects that work as references or case studies for the similar programs. The built-up areas for various programs in the project are specified as calculated divisions of the overall built-up area of the project or sometimes from the National Building Codes (NBC), as specified standards for every program. The NBC states the areas for certain capacities of persons, based on individual anthropology studies and functioning of each space as separate individual programs. The NBC specifies actions, not the engagement of the people with the space. The occupancy standards from the NBC may, however, be the first parameter for understanding the minimal areas for certain functions. It does not account for the transition of one program to another, the multiplicity of the same space offering different experiences every time, and time spent by people in engaging with these spaces.

As an academic example, in the design of institutional buildings such as schools and colleges, programs standardize the areas for classrooms and lecture halls, as the provision of lecture delivery spaces is numerically specified in design briefs. In addition to this, when institutions are envisioned as spaces for learning, every space beyond the lecture rooms, transforms into a self-learning space, and spaces for group learning and explorative learning. Scaling these spaces as individuals, small groups and larger groups, creates opportunities for the design of pockets of spaces that may be corridors and passageways in the program, but can be utilised for creating informal libraries or niches of silent reading spaces for individuals. The cafeteria becomes space for larger group events of an informal nature, bringing together students and teachers on an informal learning platform. In experimental labs, the teacher may be helping slow learners with understanding of concepts in hands-on or practical mediums, while the same lab can help a teacher learn about the latest apps and software's on digital media from the student. These equations that create spaces as containers of memories and bring life to the institution, are not accounted for in a design program brief that specifies lecture halls as 80 square meters and corridors as 2.5 meters wide.

This brings us to question whether the assertion of numerical values to area programs in the brief, is empowering the student to generate the areas for their own projects in their design dissertation and thesis. In which stage are students exploring their hypothetical outcomes against the areas that would be appropriate for bringing out ephemeral qualities or layered experiences in their design?

In most studios, the site and case studies are carried out as mere steps or stages of the process. The lack of post study analysis carried forward to the next step in the process, breaks the continuity in exploration as well as disconnects the process from the end product. The post- study analysis in a comparative matrix form leads to the selection and connection of the collaborative-connective- creative approach to generate a design program and statement, that defines the building in all aspects.

Architecture is a form of storytelling. Sothen the question arises, why do students, in the architectural viva or jury, explain the sequencing of spaces to explain their design? Why is the student unable to create a unique story for their design process? Why do majority of the stories begin from site study, case study, go on to concept, zoning, architectural plan, section, elevations, and end with some form of rendered or raw 3D views?

In practicing architecture as a professional, the designer is responsible for every inch of space that is encompassed into the built form. Building design in practice is no less than a revenue generating model for the client/owner. This poses a challenge for designers as the creation of uni-dimensional spaces affects the feasibility of that area of space. It is not uncommon to provide for waiting areas in commercial or corporate setups. If one observes a waiting area in majority of the corporate/IT offices in Mumbai itself, the spaces hold many more functions than just the act of waiting and sitting. The new age waiting areas act as open cafes with free Wi-Fi facilities, where people wait, but avail the internet services with an optional cup of coffee. Apart from ensuring that work does not halt, this makes the user/person in- waiting engage with the surroundings, and more importantly, locks the space in the memory of the phone as well as the mind. The next time the same person passes through the same space, their phone automatically registers their presence through logged in Wi-Fi and the mind registers the positive experience of the space, probing another visit. The space becomes a pause space in the everyday of the user.

Design solutions vs Learning Outcomes

From Building to Architecture

Problem- solving is a significant part of the design process. When the designed building goes beyond its core functions and creates opportunities to give back to the place or the city it is within, the experience of the merged programs defines architecture. When people are not brought to the building for the key reason, instead they choose to engage with the building in various configurations, the influx of life that the building gets, is architecture.

This is the definition of the vast body of work done by renowned architect Bjarke Ingels. The design programming and analysis for many projects from Ingels' practice, result into outcomes which merge the new and old built fabric, and create liberating experiences for the people. One such project from his native city Copenhagen, is the Amager Bakke Waste- to Energy power plant with its radically designed iconic ski slope rooftop. Described by Ingels as 'a manmade mountain of activity' Energy Plant (*Copenhill: The Story of Big's Iconic Waste-to-energy Plant*, 2019), it fulfills the city's and its people's aspirations to indulge in the activity of skiing, which is an unimaginable activity in the short duration of snow-covered flat terrain of Denmark. Another one of Ingels' projects, the ((*Maritime Youth House*, n.d.)), also brings together two contrasting programs of Sailing Club who needed to moor their boats and a youth centre which needed outdoor spaces to play. This was alongside a challenge of very polluted layer of topsoil. So, instead of investing expenses in removing the topsoil, the architect chose to raise the built form on a wooden deck that became the outdoor play, whereas the sail club could dock their boats under the deck in a sheltered area.

These projects show that in order to develop the relationships of the people with the 'created' spaces, the program must create associations which the people can relate with. The program must collect the desires of the people, whom Burte describes as the 'dwellers' instead of users.

The Collaborative – Connective – Creative Perspective (Figfure 23.2)

When documentation conducted on site takes into consideration the tangible protocols of the site along with equal importance to the unseen, intangible aspects, such as ritualistic practices of the people and place, the dimension of time becomes a significant aspect of analysis. Annual, bi-annual, day-to- day and special events add an influx of life to the physical and fixed aspects of the site. This analysis adds the layer of potential programs to the list of main programs proposed for the project. These collaborative programs include defining the edge

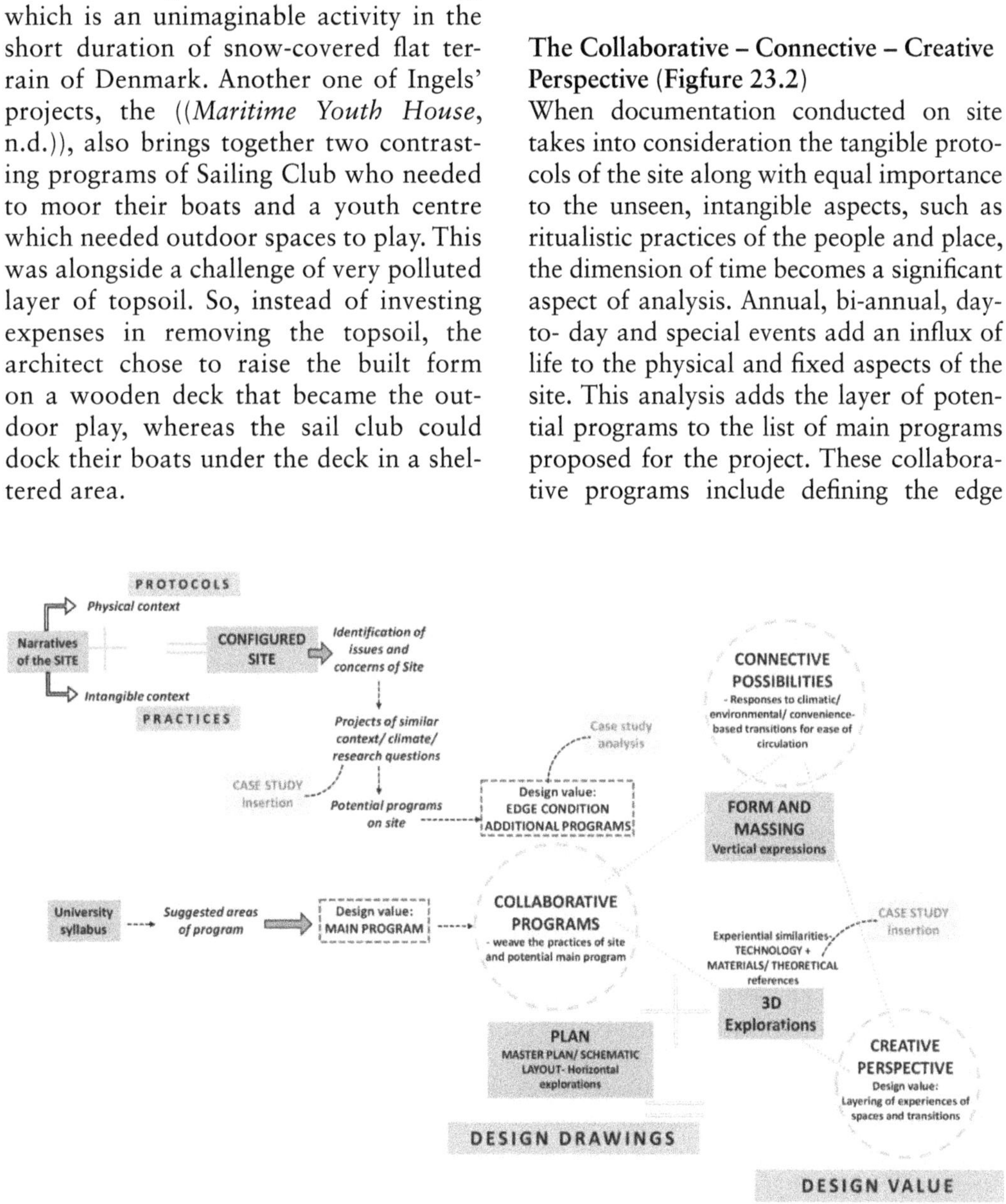

Figure 23.2 The Collaborative – Connective – Creative perspective diagram
Source: Author

conditions of the site and using them to weave the new intervention into the city's fabric. This act of giving back to the city, not only creates fluctuating edges for the designed building, but also avoids it from disconnecting with the context with empty side open spaces.

The design value of the main program and the added value from collaborative programs brings a transformation in the way the building functions, with activities and energies visible at different spaces at different times of the day and night. The connective possibilities of arranging the syntax of cohesive functions and contrasting functions creates an impact not just on the building alone, but also in the making of the place. The design of the program should allow cohesive spaces in close proximity, to create smooth transitions of experiences. It should also allow for the possibility of conflicting programs and activities to co-exist together, in order to create new, liberal experiences for the dweller. While understanding that the user is a human being and an individual, the creative perspective of generating the design program must add value and growth to the overall development of the individual, in isolation and as part of a community or group.

Challenges in Academic Realm

Time is the fourth dimension in architectural design. Time shapes the variety of experiences that an architectural design product offers. Time, however, is also a constraint in academical project processes, with co-curriculars and extra-curriculars balancing the bare minimum requirement of studio hours for design and technology studios. How then, would the students with slower learning capabilities be able to provide the resultant project requirement for sessional evaluation?

There are multiple ways to look at the time factor in the sessional scheme. One approach is to divide and collaborate with subjects of elective, college project to use a certain ratio of hours for conducting analytical studies. The architectural theory subject requires a paper writing module for enhancing the students design vocabulary and reading prowess. The college project subject may create a short module for analytical diagramming of the project undertaken, where the collected data of site and case study could be formulated into a matrix of analysis with appropriate conclusion. Pre- design studios may be used for creating design values through a design matrix that defines the building in its entirety.

Another approach that would help in working with slow learner student groups or limited design studio hours would be, where the faculty provides the information in design matrix form. In case of hypothetical sites for projects, the design team could create a pre-study matrix for design values and provide it for students to analyse individually. The student's role will then be to choose the scheme that resonates with their aspirations for the project and work it through application strategies.

Conclusion

The methodology and learnings above have been applied to the following two cases of students' design projects in Semester 6 architectural design. The students worked with the framework to achieve the outcome discussed below.

Case 1: Design of workshop area in a centre for an occupation- based community.

While the primary user of the workshop area would be the artist/ craftsman, it will be inhabited by many other users at different times of the day. The spatial syntax must include the artist's work area along with a small group, but also scale it to their individual expression space, their studio. There could be a hypothesis of the artist having skill-workers for specific detailing, administrative staff for their logistical affairs, maybe even interns or learners of the art/craft. Work area would then be scaled to individual (private studio), small group (learning spaces for interns/skill workers) and large group (gathering spaces for the entire team). The secondary user, the visitor, is a temporary, yet major stakeholder of the program, as a client or potential buyer of the artwork.

Spaces for the visitor would be formal and informal meeting spaces with the artist, where physical connection would be made with the artwork, but for a tourist, the workspaces could have only visual connections at specific hours, to maintain the privacy of the workers/learners. Storage would go beyond the storeroom into specified raw material storage and finished product storage, and toilets would bring in the hierarchy of private and public. Spaces would be carved into the same area, flexibly shifting between learning spaces, demonstrating spaces, individual niches for the musings of the workers, breakout spaces for informal activities, event spaces for special access to the visitor to engage with the art hands-on, and large gathering spaces for events for the public.

Case 2: Form based approach for the same project – Centre for occupation-based community.

The above framework (Figure 23.2) when applied to the design, the architectural form shall transform from a box with openings to spaces interspersed with activities depending on the intensity of engagement, with the display space weaving through the spatial arrangement, instead of being a separate box. Conscious decision and action would be taken to involve the local community with jobs to learn the art or contribute to parts of food preparation, maintenance or security, thus creating collaborations with the social context.

The following outcomes (Table 23.1) were observed in the design studio by multiple faculty/design mentors, upon using the expanded program framework as design process. The outcomes were collected at various stages of the design studio duration.

In conclusion, from applying the framework to design studios, it is observed that, the design program of any project is not just the spatial requirements of the main function, or the needs of the client or the wants of the user and stakeholder. It is not just what the architect or designer wishes

Table 23.1 Outcome observed in design studio

Design approach	Student could create multiple possibilities to approach the project, through theoretical references, architecture styles, technological innovations and form-based designs.
Research stage	It was observed that students spent more duration on research and related studies to increase the intensity of spatial design. This was reflected in the final panel as direct connections could be made from research sheet to design drawings sheet, justifying the design queries better.
Case- study	Case study was not limited to one stage in design, as student could refer and connect to multiple reference points for their design stages. Analytical framework for case study to understand all aspects of multiple buildings with same parameters helped achieve a comparative study.
Planning stage	Students took time to come up with a plan/design drawing, yet the outcome was justified and logical than a plan derived from generic design stages.
Jury stage	The jury discussed the research and analytical work in depth, trying to understand the project's value and students could confidently put across reasonings for multiple decisions taken in the design process. Use of appropriate vocabulary through referenced cases allowed the student to clearly put across their thoughts about the experience of the project.

Source: Author

to experiment with, neither is it a direct result of the site analysis and existing issues derived from the SWOT diagram. The design program is not just a table format of spaces- areas – user capacity that the student uses as an information manual and starts drawing.

The design program is something from all of the above, but it is much more than just the above. The design program is the identification of the value, that the project may be able to create. It is the knowledge of the terms design value, and Program statement, that will lead to an Architectural value of the project. It is the knowledge of selecting primary functions, supporting them with secondary functions, based on the need of the stakeholders and then further on, identifying activities that shall allow the aspirations of the upcoming generations. The design program must take into consideration the possibility, that the built form may have certain innovative values to survive beyond time, and the significance of heritage it would stand for, in the future.

The design program in an architectural design process should be expanded with an approach to create layers of experiences. This approach should be:

1) To introduce collaborative programs in architectural space seeking to catalyse engagement and bring in newer experiences.
2) To introduce connective approach to arrange these experiences in a manner that shall generate new responses.
3) To introduce a creative approach in order to recreate variations of experiential transit through multiplicity of spatial schemes.

References

Burte, H. (2008). Spaces for Engagement. Mumbai: Seagull books.

Edith Cherry, J. P. (2016). Architectural Programming. Washington DC: Whole Building Design Guide.

Eric Baldwin. “CopenHill: The Story of BIG’s Iconic Waste-to-Energy Plant” 07 Oct 2019. ArchDaily. Accessed 25 Jun 2024. <https://www.archdaily.com/925966/copenhill-the-story-of-bigs-iconic-waste-to-energy-plant> ISSN 0719-8884

Maritime Youth House. (n.d.). *Maritime Youth House*. Website of Architect Bjarke Ingels, BIG Architects. Retrieved January 25, 2024, from https://big.dk/projects/maritime-youth-house-2123

24 Enhancing graduate-empowerment in architectural education: A comprehensive analysis

Pashmeena Ghom[1,a] *and Abraham George*[2]

[1]Professor, School of Architecture, Urban Development and Planning, Symbiosis Skills and Professional University, Pune, Maharashtra, India

[2]Professor, Department of Architecture and Regional Planning, Indian Institute of Technology Kharagpur, West Bengal, India

Abstract

This in-depth research examines the transformation of architectural education from the 1940s to 2022, shedding light on pivotal changes and the hurdles in meeting modern requirements. The study thoroughly analyses how historical, economic, scientific, and cultural elements have shaped and sometimes hindered the progress of architectural teaching methods, revealing a concerning downturn in educational standards. It charts the journey from traditional, mentor-led apprenticeship systems to more formalized institutional education, scrutinizing the move from a curriculum driven by passion to one focused on pragmatic results. The impact of various regulatory organizations and a push toward a uniform educational approach in architecture are meticulously examined, pinpointing challenges such as the dilution of mentorship, the homogenization of course content, and the struggle to merge age-old educational philosophies with current trends. Research highlights the urgent necessity to blend intrinsic passion and creativity in architecture with the demands of modern education and professional practices. The study advocates incorporating interpersonal skills and cutting-edge technology into architectural education to align with changing industry benchmarks. The study suggests a novel curriculum that will effectively prepare graduates and emphasizes developing flexible, imaginative architects suitable for the rapidly changing architectural landscape. This study offers vital insights and actionable strategies for rejuvenating architectural education and practice for future generations.

Keywords: Institutional learning, job market demands, mentor-guided apprenticeships, soft skills development, standardization in education

Introduction

The evolution of architectural education has undergone significant shifts from its earliest forms to modern era. Originally, architecture was more of a craft and trade, with skills passed down through apprenticeship models rather than formal education. This approach was prevalent globally, with various cultural adaptations (Davis, 2017; Woods, 1999).

In India, the *Gurukul* system exemplified this apprenticeship model. Here, the educational framework was deeply rooted in a relationship between the *Guru* (master craftsman) and the *Shishya* (disciple). This model emphasized hands-on learning and direct transmission of knowledge and skills from the master to the apprentice (Prabhu and Achyuthan, 1996). In contrast, the Western world adopted a similar apprenticeship approach, particularly through

[a]pashmeenaghom@gmail.com

DOI: 10.1201/9781003571018-24

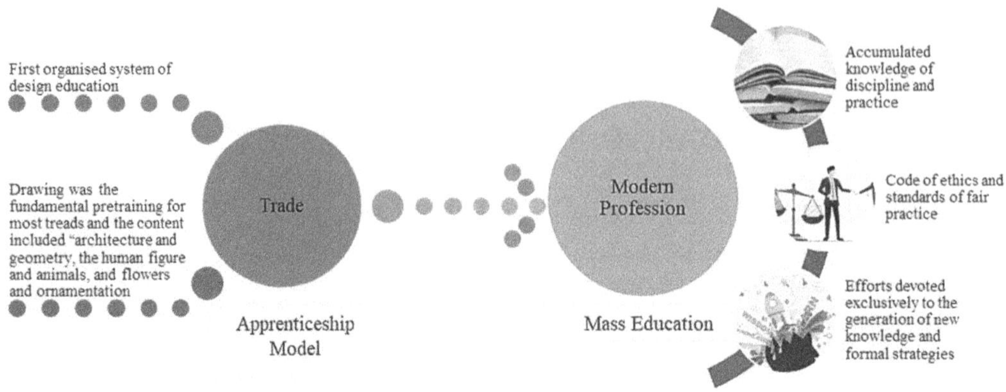

Figure 24.1 Apprenticeship system to mass education
Source: Authors

guilds. These guilds were exclusive, often limited to young, unmarried males who lived with their master's family during their training. This period of apprenticeship was a prerequisite before one could attain full membership in the craft (Meyer and Ramirez, 1992; Ramirez and Boli, 1987).

The concept of mass education, which later became a driving force in national development worldwide, first emerged in Europe. This institutionalized form of education marked a departure from the traditional apprenticeship models, emphasizing structured, formal learning environments over the personalized, hands-on approach of apprenticeships (Figure 24.1). These historical approaches to architectural education reflect the diverse ways in which societies have transferred knowledge and skills across generations, shaping the architectural landscape through both formal institutions and traditional craftsmanship.

The Historical Trajectory of Architectural Education

The historical trajectory of architectural education in Europe and its global influence is rich and multifaceted, drawing from diverse cultural, religious, and scientific sources. In Europe, architectural training and its curriculum have roots in three major influences: Greek culture, Jewish religion and ethics, and the principles of industrialism emerging from modern science. The core knowledge of architectural education in ancient Greece, Rome, France, and Italy was significantly shaped by these influences. For example, in Greece and Rome, architecture was heavily influenced by their respective cultures and philosophies. In France and Italy, the Renaissance period brought a renewed focus on classical forms and principles (Figure 24.2) (Cunningham, 1979).

In contrast, Indian architectural education was deeply intertwined with the cultural and religious fabric of the society. The concept of *Vaastupurusha Mandala*, a cosmological model based in Hindu religion and mythology, played a central role in shaping architectural practices. This approach emphasized strict geometric rules, alignment with natural forces, cultural considerations, and the leadership of a master craftsman (Prabhu and Achyuthan, 1996). Notable figures in European architectural history, such as John Shute and Sir Henry Wotton, contributed significantly to the body of architectural knowledge. Shute's work, influenced by Vitruvius and Serlio, and Wotton's comprehensive 'The Elements of Architecture' are examples of how architectural thought evolved in Europe, blending practical knowledge with philosophical underpinnings (Gerbino, 2010).

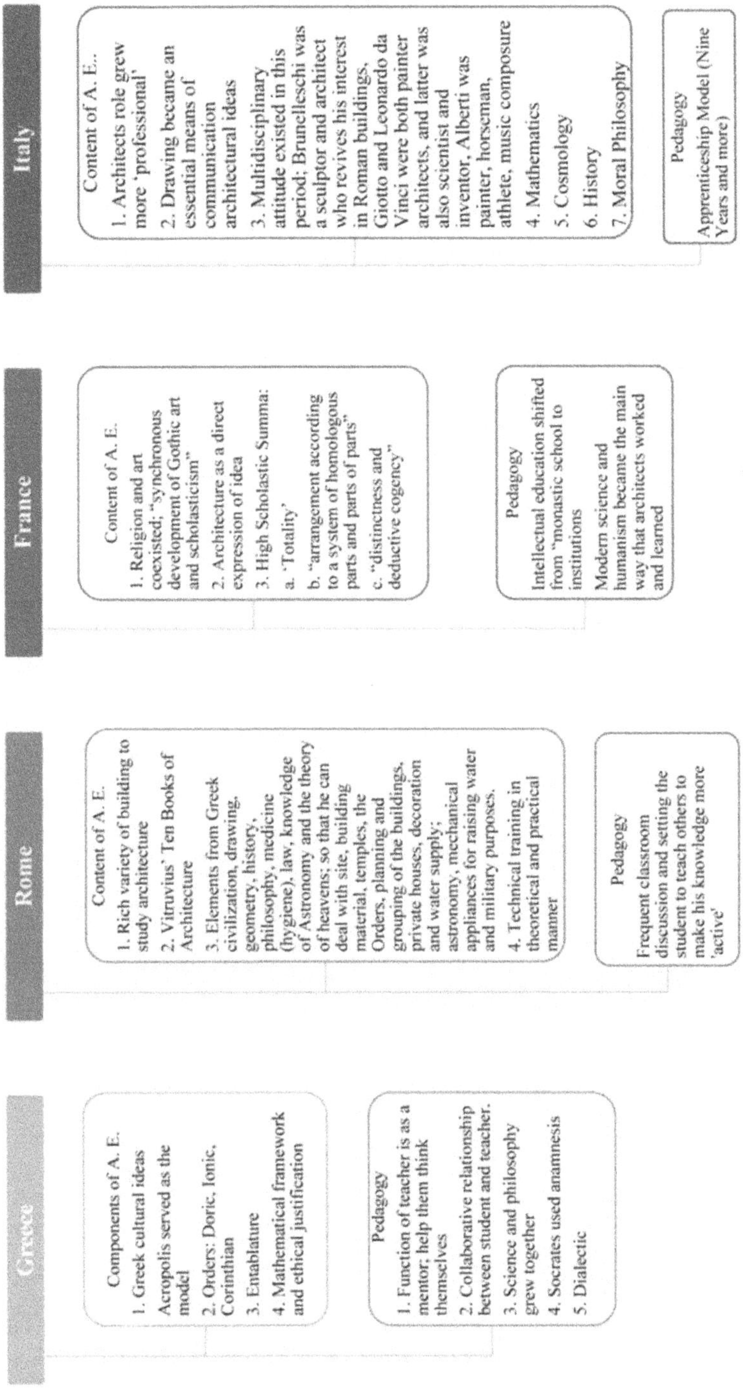

Figure 24.2 Architecture education with reference to content and pedagogy
Source: Authors

The Industrial Revolution in the 17th century marked a significant shift, leading to a rise in specialists and a move toward more institutionalized forms of architectural education. This era saw an increase in the number and variety of buildings, reflecting the growing needs of an industrialized society and shifting tastes among the affluent classes. The apprenticeship model, which traditionally required a lengthy period of training, gradually gave way to a more formalized educational system. This shift is exemplified by the establishment of the five-year architecture program in London, which became a standard in architectural education. By the 19th century, a systematic professional education for architects was developed. The École des Beaux-Arts in France played a pivotal role in elevating architectural study from a craft to a higher level of philosophical and aesthetic discourse, heavily influenced by ancient Roman and Italian Renaissance architecture. The curriculum emphasized drawing, mathematics, geometry, art and architectural history, local construction methods, and regional cultural traditions, based on the teachings of classical authors (Bhattacharjee and Bose, 2015; Cunningham, 1979; Salama, 2016).

The influence of the École des Beaux-Arts extended globally, including to American universities, and its principles continue to be practiced in various forms around the world. This evolution from apprenticeship to specialized education reflects the broadening scope and complexity of architectural practice, mirroring broader societal and cultural shifts (Chafee, 1977; Cret, 1941).

The development of architectural institutions and education from the 17th century through the mid-20th century reflects a dynamic evolution in teaching methodologies, social inclusion, and design philosophies (Figure 24.3). In the mid-1980s, a significant shift occurred with the emphasis on computer-assisted design, marking a transition towards more technical

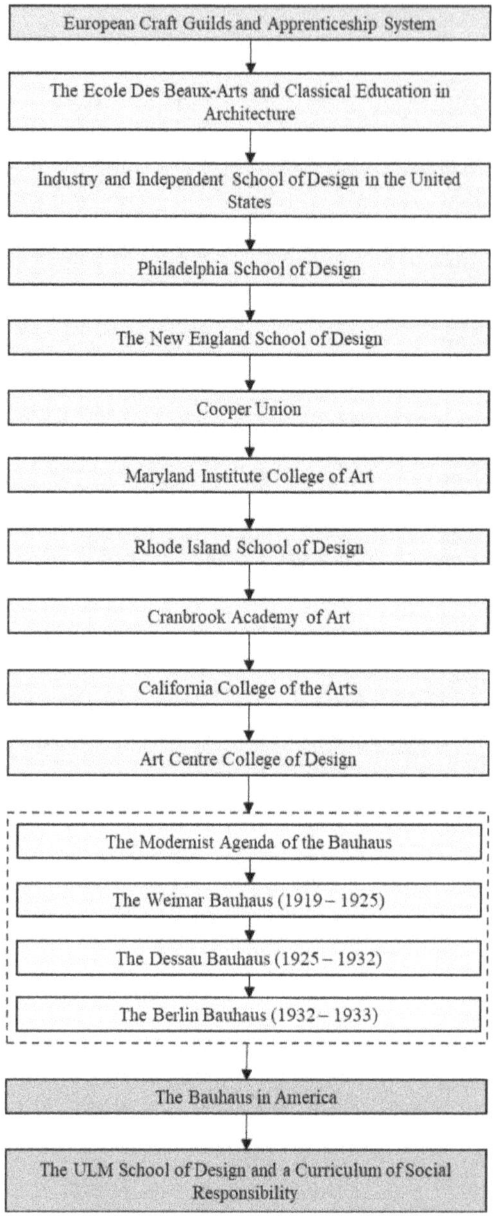

Figure 24.3 Design schools in the United States

Source: Compiled by the author from Davis, 2017

education in architecture. This era saw the integration of emerging digital technologies into architectural design, altering the way architects work and think about design. During this period, institutions like the

Philadelphia School of Design made notable contributions to social inclusivity in architectural education. Despite its restrictive admissions criteria, which included native birth, churchgoing Protestantism, and the absence of a male breadwinner, the school provided unique opportunities for women in architecture. Similarly, the Cooper Union School in New York demonstrated progressive social views by accepting students of color and engaging with the community through initiatives like a public reading room for children. The Art Deco style, emerging in the 1930s, was another influential movement in architectural education. This style, rooted in the art and craft tradition, brought a unique aesthetic to architectural design. The Cranbrook Academy of Art, led by Eero Saarinen, played a pivotal role in this era, influencing American architecture and product design in line with the Arts and Crafts tradition. Notable figures associated with Cranbrook, such as Charles and Ray Eames, Florence Knoll, and Harry Bertoia, contributed significantly to the development of the modern furniture industry in Michigan. The Art Center College of Design further exemplified the evolving educational landscape by preparing designers for leadership roles in various fields like advertising, publishing, and industrial design. The college established a tradition of having a faculty composed of design professionals, a practice that continues in many institutions today (Davis, 2017; De Angeli Walls, 1994).

The Bauhaus movement, although often associated with the 20th century, had its roots in the 19th century, as a response to the Industrial Revolution. The movement sought to reconcile the art and craft tradition with the demands of industrialization, which the earlier 'Art and Craft' movement had struggled to meet (Findeli, 1995, 2001). Founded by Walter Gropius, the Weimar Bauhaus (1919–1955) revolutionized architectural education. It emphasized artistic expression in design and restructured education around modern ideas of art school, with handicraft as its foundation.

The Bauhaus employed renowned artists and thinkers like Lyonel Feininger, Johannes Itten, Wassily Kandinsky, Paul Klee, Oskar Schlemmer, and Laszlo Moholy-Nagy, who

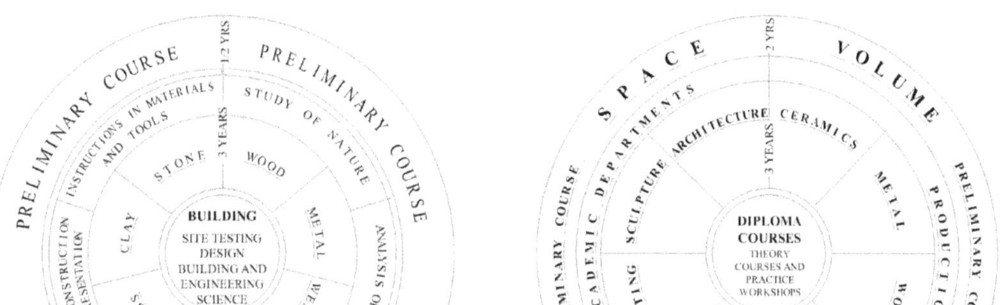

BAUHAUS
Teaching program diagram, Gropius W., 1922

VKhUTEMAS
Teaching program diagram, 1923

Figure 24.4 Preliminary and diploma *Bauhaus curriculum* (Bokov, 2017)
Source: Bokov, 2017

brought innovative teaching methods to the primary course (Vorkurs) (Figure 24.4). Unlike the École des Beaux-Arts, which focused on copying exercises, the Bauhaus emphasized understanding fundamental concepts through analytical drawing, encouraging students to discover through exploration (Aynsley, 2009; Droste, 2022). Gropius also believed in the importance of the interconnection between the academic world and the commercial industry. He saw commercial commissions as a way for institutions to reduce their dependency on public subsidies, thereby fostering a more practical and industry-aligned curriculum (Droste, 2022). This overview of architectural education's development reflects a trajectory of increasing technical sophistication, social inclusivity, and a blend of artistic and practical approaches, shaping the field as we understand it today.

The evolution of architectural education throughout the 20th century, as seen in the curricula of the Dessau Bauhaus, Berlin Bauhaus, *VKhutemas*, and ULM School of Design, reflects a shift towards integrating practical skills, social responsibility, and scientific understanding into design disciplines. At the Dessau Bauhaus (1925–1932) under Walter Gropius, the curriculum included tracks in 'building and interior decoration', focusing on function, technology, and economy. This period emphasized workshops in metals, textiles, joinery, and mural painting, and introduced systematic exploration of materials and theories of form and color. Under Hannes Mayer's leadership, the Bauhaus incorporated scientific and social criteria into design, advocating for designs that meet every day human needs and serve the social good. Mayer's approach transformed the curriculum to prioritize practical and rational design, though the school was ultimately closed by the Nazi Party. The Berlin Bauhaus (1932–1933), led by Ludwig Mies van der Rohe, introduced a curriculum emphasizing practical construction experience. This curriculum was structured in three stages: technical fundamentals, apartment and town planning, and a focus on drawing as an essential intellectual and artistic tool in architecture. Mies van der Rohe's philosophy of architecture as a confrontation with space, proportion, and material differed significantly from Mayer's social and practical discipline. However, the school closed due to the Nazi regime's opposition to its principles (Davis, 2017; Wingler, 1978).

The influence of the Bauhaus spread to the United States post-World War II, but not without controversy. The American Bauhaus, influenced by the ideas of John Dewey, saw disagreements between Joseph Hudnut and Walter Gropius over the balance between liberal education and modernist design. Critics of the Bauhaus education pointed out its lack of emphasis on broader intellectual stimulation and understanding of economic, political, and social sciences (Aynsley, 2009; Davis, 2017; Wingler, 1978). In Russia, the *VKhutemas* school combined craft tradition with modern technology, offering courses in aesthetic principles, color theory, industrial design, and architecture. Its curriculum, a pioneering approach to mass design education, emphasized objective methods and a unified approach to disciplines like graphics, color, volume, and space (Figure 24.5) (Bokov, 2017; BXYTEMAC, 2020).

Post-World War II Germany saw the emergence of the ULM School of Design, which emphasized social responsibility in design, focusing on practical, affordable, and honest architectural productions. The ULM curriculum integrated science and analytical studies, emphasizing the designer's social responsibility and a rational design process. The teaching faculty at ULM viewed design as a civilizing culture, counter-art, and a work of civilization. These developments in architectural education reflect a growing emphasis on practical skills, social and ethical considerations, and

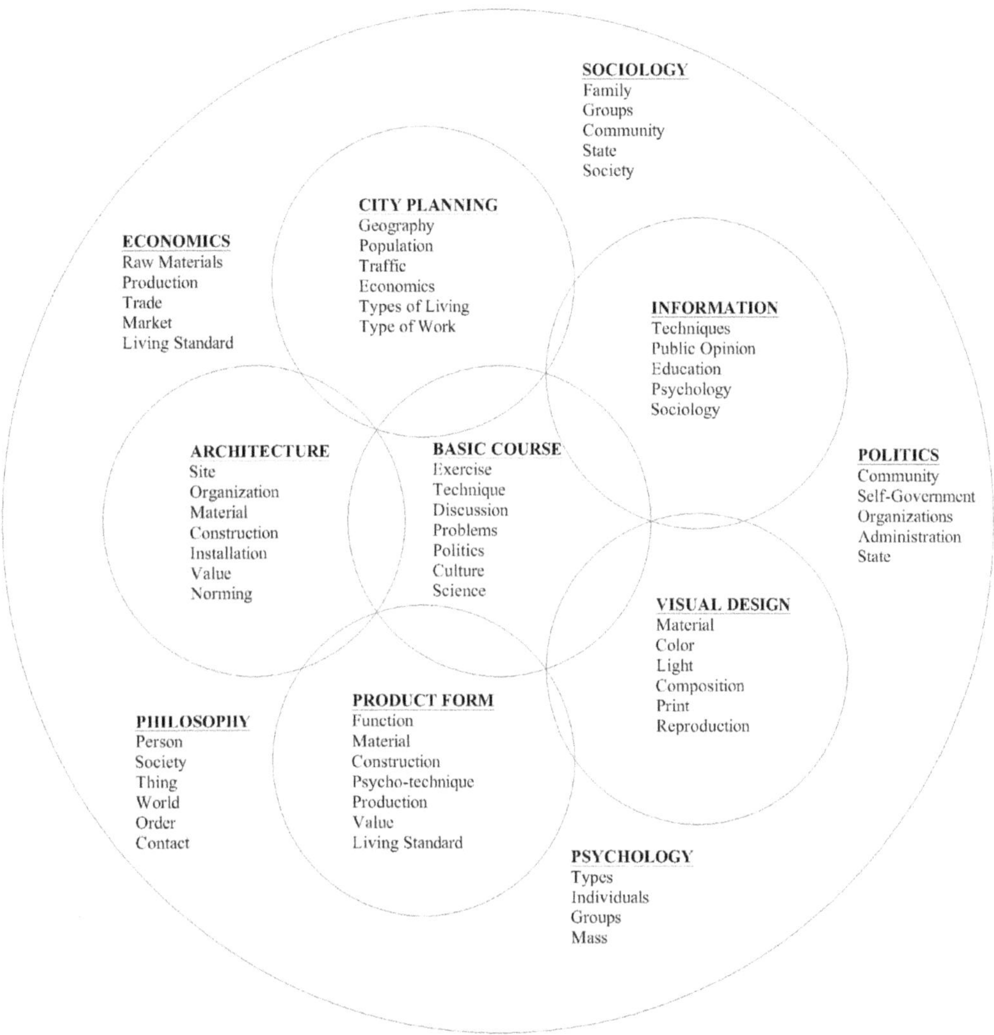

Figure 24.5 Curriculum of the ULM school of design

Source: Oswald, 2022

a scientific approach to design. These shifts in educational philosophy demonstrate the evolving nature of architectural thought, balancing the need for functional, socially responsible design with artistic expression and technical expertise (Cunningham, 1979; Findeli, 1995, 2001; Spitz, 2002).

Architectural Education in India

The history of architectural education in India illustrates a significant transition from traditional, informal methods to formalized, structured systems, influenced by both indigenous practices and colonial impacts. Traditionally, architecture in India was taught through an apprenticeship model, where a master craftsman (a polymath embodying various roles such as architect, engineer, sculptor, artist, and philosopher) imparted knowledge to his disciples in a Gurukul setting. This informal system of education, known as Parampara, emphasized the transfer of knowledge from one generation to the next, deeply rooted in vernacular wisdom and regional cultural identity (Chakraborty, 2015; Ghom and George, 2023).

Contrasting this, the British colonial period in India saw the establishment of formal art and architecture schools, adopting a mass-education model. Institutions like the Madras School of Industrial Arts (1850), the Government College of Arts in Calcutta (1854), and the Sir J. J. College of Fine Arts in Bombay (1857) were pivotal in this shift. The Sir J. J. College of Architecture, established in 1913, was particularly significant as it introduced a structured, European-style curriculum in architectural education in India. This period also saw rapid advancements in technology and construction methods, influenced by European developments (Chakraborty, 2015; Dua and Chahal, 2014).

After India's independence, there was a further evolution in architectural education. The establishment of the Indian Institute of Technology Kharagpur, in collaboration with the Massachusetts Institute of Technology, marked a shift towards a scientific and technological approach in architectural programs. This trend was mirrored in other premier institutes across the country. From having only four architecture institutes in 1947, India now has a significantly larger number of institutions offering architectural education (Dua and Chahal, 2014).

The regulation and oversight of architectural education and practice in India are managed by several statutory and regulatory bodies:

1. Ministry of Education (MoE): Established in 1985, it oversees higher education programs, formulates national education policy, provides financial assistance to disadvantaged groups, and encourages national and international cooperation in education (Ministry of Education, 2021).
2. University Grants Commission (UGC): Constituted in 1953, it coordinates, determines, and maintains standards of university education (UGC, 2020).
3. All India Council for Technical Education (AICTE): Responsible for planning, formulating, and maintaining norms and standards in technical education, along with funding, accreditation, and monitoring technical education courses (AICTE, 2021, 2022; Borkar, 2021).
4. Council of Architecture (CoA): Established under the Architects Act of 1972, it sets minimum standards for architectural education, accredits schools of architecture, maintains a registry of architects, and regulates professional conduct (CoA, 2022a).

Additionally, architectural bodies like the International Union of Architects (UIA) and others like IIA, NASA, ARCASIA work towards the betterment of the architectural profession and education globally. They emphasize multidisciplinary education and the development of employability skills, essential for graduates to successfully complete the five-year training period in architecture (Fintikakis, 2022).

This evolution reflects a blend of traditional knowledge and modern educational practices, aiming to prepare architects who are not only skilled in design and technology but also sensitive to cultural and environmental contexts.

Market Demands and Economic Factors

The discourse surrounding the National Education Policy (NEP) 2020 in India reveals an evolving educational paradigm that seeks to harmonize with the dynamic requisites of the 21st century workforce and economy. Within this framework, there is a discernible shift towards emphasizing quantifiable accomplishments and skills that are directly aligned with labor market expectations and economic imperatives. This shift is manifest in the policy's advocacy for a multidisciplinary approach in higher education, underscored by the integration of industry-relevant fields such as Artificial Intelligence and Sustainable Technologies into the curriculum. The policy's design, with its flexible entry and exit points in the four-year undergraduate program, further reflects a pragmatic orientation towards enhancing employability and adapting to the fluidity of the modern employment landscape (Bhattacharjee and Bose, 2015; Ghom et al., 2023; Ghom and George, 2021). However, this pivot toward market-oriented competencies raises critical reflections on the place of artistic and creative disciplines within the educational spectrum. The emphasis on tangible outcomes and vocational skills, while addressing immediate economic and employment needs, potentially marginalizes the intrinsic value and transformative potential of creative endeavors. Artistic and creative fields, which thrive on innovation, critical thinking, and cultural expression, risk being undervalued in an educational ethos that prioritizes economic utility and measurable success. This trend suggests a nuanced tension between the demands of a rapidly evolving global economy and the enduring importance of nurturing creativity, imagination, and holistic human development (AICTE, 2021; CoA, 2022b; National Council of Architectural Registration Boards, 2012; NCARB, 2010; RIBA, 2016, 2020; UNESCO-UIA, 2017).

In essence, while NEP 2020's strategic focus on employability and technological proficiency is aligned with contemporary global trends, it is imperative to foster an educational environment that equally cherishes and cultivates the arts and creative disciplines. These fields are not only vital for cultural and personal enrichment but are also essential for fostering a well-rounded, innovative, and empathetic citizenry capable of contributing to a diverse and sustainable global community.

Empowerment through Educational Synergy

Empowerment in education is not a one-dimensional attribute but rather the culmination of a synergistic relationship between core knowledge and employability skills. The diagram we consider illustrates this relationship as an equation where the sum of core knowledge and employability skills equates to empowerment. To achieve true empowerment, educational systems must transcend the binary of vocational versus academic and embrace a holistic model that fosters intellectual growth and practical acumen.

This conceptual model serves as a guiding framework for developing a mitigation strategy that addresses the concern of educational reforms prioritizing market-driven skills over artistic and creative pursuits (Figure 24.6). A strategy for empowerment based on this conceptual model will be:

Core Knowledge as the Bedrock of Education

Core knowledge is the foundational element that underpins the entire learning experience. It comprises the essential theories, principles, and facts across various disciplines. In our mitigation strategy, we advocate for the infusion of creative disciplines into this bedrock, ensuring that the arts are not sidelined but interwoven with scientific and technological studies. The diagram's left-hand circle, representing core knowledge, must thus be envisioned as a diverse palette that colors the canvas of a student's mind with a blend of humanities, sciences, and arts.

Employability Skills Beyond the Conventional

Moving to the right of the diagram, employability skills traditionally emphasize

Figure 24.6 Empowerment in architectural education
Source: Author

technical prowess and job-specific competencies. However, these skills must also encapsulate creativity, communication, critical thinking, and adaptability. This paradigm shift necessitates that curricula nurture innovation as much as they do information technology, treating both as indispensable skills for the 21st century labor market.

The Fusion point: Holistic Approach to Assessment

The plus sign in the diagram is not merely an operator; it represents the fusion of core knowledge and employability skills. This fusion must be mirrored in assessment methods that evaluate students' ability to integrate and apply their diverse learning in complex, real-world situations. Such holistic assessments validate the multidimensional capabilities of students, thereby encouraging educational practices that promote analytical acumen and creative expression.

Creative Projects: the Nexus Between Theory and Practice

As we look toward the equal sign in the diagram, we see the realization of empowerment. In fields like architecture, this translates into curricular structures that provide creative projects, requiring students to apply their multidisciplinary learning practically. This nexus between theory and practice ensures that students are job-ready and equipped to push the boundaries of conventional design and problem-solving.

Evolving Curricula for Dynamic Futures

The educational ecosystem is ever-changing, and so the curriculum must evolve to stay relevant. By incorporating the latest in creative and artistic fields into the curriculum, educational institutions ensure that core knowledge and employability skills reflect current and future landscapes.

Cross-Disciplinary Collaboration as a Catalyst for Innovation

The diagram, while seemingly linear, implies a dynamic and interactive process of learning. As part of this process, cross-disciplinary modules can create a melting pot of ideas and skills, encouraging students from diverse fields to collaborate and innovate. This approach enriches the educational experience and mirrors the collaborative nature of the modern workplace.

Guidance Toward Diverse Career Pathways

Finally, the journey towards empowerment is marked by informed choices. Career counselling must present a broad spectrum of career paths highlighting the value of technical skills and creative prowess. Understanding the diverse opportunities available can inspire students to pursue paths they might otherwise overlook, especially in the creative sectors.

Conclusion

This article encapsulates the evolving dynamics and inherent challenges within architectural education, scrutinizing their influence on the architectural realm's academic and practical facets today. It reflects upon the historical progression of architecture from a craft to a recognized profession, examining the critical roles played by various statutory and regulatory bodies in India. These institutions are pivotal in shaping the educational landscape and professional standards of architecture.

The research's central claim is that the empowerment of the architect is essential to the viability and success of an architectural career. Such empowerment is attainable through a solid foundation in core knowledge and relevant employability skills. The urgency of this matter calls for a clear delineation of what constitutes core knowledge and vital employability skills in the context of the twenty-first century.

Presented as a visual equation, 'Core Knowledge + Employability Skills = Empowerment,' the article advances a blueprint for reforming architectural education. This formula champions an educational ethos that equally appreciates quantifiable and creative skills, merging the tangible with the visionary and the actionable with the innovative. Implementing this educational strategy promises to nurture professionally capable individuals and impassioned creators and innovators, thus shaping a future where architects are adept in their field and pioneers of change and progress.

To further this endeavor, the research identifies a significant gap that must be addressed: the need for further investigation to define the parameters of core knowledge and employability skills. An effective empowerment framework hinges on a clear understanding and precise identification of these elements. Therefore, ongoing research is imperative to establish detailed indicators to serve as benchmarks for educational curricula and professional development. This ensures that empowerment through architectural education is not only an aspirational goal but a tangible reality.

References

All India Council for Technical Education (AICTE). (2021). *Approval process handbook 2021-22*. New Delhi, India: All India Council for Technical Education. Retrieved from https://www.aicte-india.org/

AICTE (2022). AICTE History. All India Council for Technical Education. https://www.aicte-india.org/about-us/history.

Aynsley, J. (2009). Designing Modern Germany. Reaktion.

Bhattacharjee, S. and Bose, S. (2015). Comparative analysis of architectural education standards across the world. *Agricultural Research Communication Centre*, 579–589. https://www.brikbase.org/content/comparative-analysis-architectural-education-standards-across-world.

Bokov, A. (2017). *Vkhutemas and the Bauhaus: On common origins and 'creation with fire.'* Walker Art Center. Retrieved from https://walkerart.org/magazine/institutionalizing-the-avant-garde-vkhutemas-1920-1930

Borkar, P. (2021). What is AICTE: role of All India Council For Technical Education (AICTE). MasterSoft. https://www.iitms.co.in/blog/role-of-all-india-council-for-technical-education-aicte.html.

BXYTEMAC (2020). The History. Vkhutemas. https://www.vkhutemas.ru/en/the-history/.

Chafee, R. (1977). The teaching of architecture at the ecole des beaux-arts. In The Architecture of the École Des Beaux-Arts. The Museum of Modern Art, pp. 61–109.

Chakraborty, M. (2015). Designing Better Architecture Education. Copal Publishing Group.

Council of Architecture (CoA). (2022). Approval Process Handbook (2022-23). New Delhi: Council of Architecture. Retrieved from https://www.coa.gov.in/showfile.php?lang=1&level=1&sublinkid=1070&lid=942.

Council of Architecture (COA) (2022b). https://www.coa.gov.in/index1.php?lang=1&level=0&linkid=3&lid=4.

Cret, P. (1941). The Ecole des Beaux-Arts and Architectural Education. *JSTOR*, 1(2), 3–15. https://www.jstor.org/stable/901128.

Cunningham, A. (1979). The genesis of architectural education. *Studies in Higher Education*, 4(2), 131–142. https://doi.org/10.1080/03075077912331376917.

Davis, M. (2017). *Teaching design: A guide to curriculum and pedagogy for college design faculty and teachers who use design in their classrooms*. New York, NY: Allworth Press.

De Angeli Walls, N. (1994). Educating women for art and commerce: the Philadelphia school of design, 1848-1932. *History of Education Quarterly*, 34(3), 329. https://doi.org/10.2307/369956.

Droste, M. (2022). *Bauhaus 1919-1933: Updated edition*. Cologne, Germany: Taschen.

Dua, S. and Chahal, K. S. (2014). Scenario of architectural education in India. *Journal of the Institution of Engineers (India): Series A*, 95(3), 185–194. https://doi.org/10.1007/s40030-014-0087-6.

Findeli, A. (1995). Design history and design studies: methodological, epistemological and pedagogical inquiry. *Design Issues*, 11(1), 43–65. https://doi.org/10.1162/07479360152103796.

Findeli, A. (2001). Rethinking design education for the 21st century: theoretical, methodological, and ethical discussion. *Design Issues*, 17(1), 5–17. https://doi.org/10.1162/07479360152103796.

Fintikakis, N. (2022). International Union of Architects (UIA). E-FIADE. http://www.efiade.org/uia/#:~:text=Built on the basis of level. Promote their creative%2C.

Gerbino, A. (2010). François Blondel: Architecture, Erudition, and the Scientific Revolution(1stedn.). Routledge.

Ghom, P. V. and George, A. (2021). Dynamics of performing aesthetics in architecture: acritical study. *Vitruvio-International Journal of Architecture Technology and Sustainability*, 6(2), 82–101. https://doi.org/https://doi.org/10.4995/vitruvio-ijats.2021.16424.

Ghom, P. V. and George, A. (2023). Decolonizing south Asian architecture: sustainable and community-oriented social housing in India. *Vitruvio*, 8(1), 46–57. https://doi.org/10.4995/vitruvio-ijats.2023.19499.

Ghom, P. V., George, A., and Bharule, S. (2023). Socio-economic aspects affecting architectural education and profession: strategies and tactics. *New Design Ideas*, 7(1), 152–170.

Meyer, J. W. and Ramirez, F. (1992). World expansion of mass education, 1870-1980. *Sociology of Education*, 65(2), 128–149.

Ministry of Education (2021). About MoE. Government of India. https://www.education.gov.in/en/about-moe.

National Council of Architectural Registration Boards (NCARB). (2012). *2012 NCARB practice analysis of architecture*. Washington, DC: National Council of Architectural Registration Boards.

National Council of Architectural Registration Boards (NCARB). (2010). *NCARB and how it relates to architectural professional associations* (Issue June). Washington, DC: National Council of Architectural Registration Boards.

Oswald, D. (2022). Cybernetics, operations research and information theory at the Ulm School of Design and its influence on Latin America. AI and Society, 37(3), 1045–1057. https://doi.org/10.1007/s00146-021-01339-1

Prabhu, B. T. S. and Achyuthan, A. (1996). A Text Book of Vastuvidya. Vastuvidyapratisthanam.

Ramirez, F. and Boli, J. (1987). The political construction of mass schooling: european origins and worldwide institutionalization. *Sociology of Education*, 60(1), 2–17.

Royal Institute of British Architects (RIBA). (2016). *What clients think of architects: Client survey*. London, England: Royal Institute of British Architects. Retrieved from https://www.architecture.com

Royal Institute of British Architects (RIBA). (2020). *The way ahead*. London, England: Royal Institute of British Architects.

Salama, A. M. (2016). *Spatial design education: New directions for pedagogy in architecture and beyond* (1st ed.). London: Routledge. https://doi.org/10.4324/9781315610276

Spitz, R. (2002). *Hfg Ulm: The view behind the foreground: The political history of the Ulm School of Design, 1953-1968*. Ulm: Edition Axel Menges. Retrieved from https://books.google.com/books?id=t41eqR5r1rEC&pgis=1

UGC (2020). Genesis. Union Grant Commission. https://www.ugc.ac.in/page/Genesis.aspx.

UNESCO-UIA. (2017). *Charter UNESCO/ UIA for architectural education*. Paris, France: UNESCO-UIA.

Wingler, H. M. (1978). Bauhaus: Weimar, Dessau, Berlin, Chicago. The MIT Press.

Woods, M. N. (1999). *From craft to profession*. Berkeley, CA: University of California Press.

25 International collaboration in the digital age: Transforming architectural education for a dynamic future

Abd Muluk Bin Abd Manan[a]

Universiti Tunku Abdul Rahman, Bandar Sungai Long, 43000 Kajang, Selangor, Malaysia

Abstract

This paper explores the transformative impact of international collaboration in the digital age on architectural education, aligning with the theme of "Architecture Education in the Changing World - Reviewing and Rethinking Forward." In an era defined by technological advancements and global connectivity, the dynamics of architectural education are evolving, necessitating a critical examination of collaborative practices. It delves into the pivotal role of digital tools, inter university collaboration, global perspectives, and ethical considerations, shedding light on how these elements collectively contribute to shaping the future of architectural education. Globalization and technological advancement have revolutionized the dissemination, exchange, and application of architectural knowledge. Central to this paradigm shift is the concept of cross-border collaboration, where institutions and individuals from diverse geographical locations converge to redefine the boundaries of architectural education. This paper explores the transformative potential of international collaboration in the digital age, elucidating its impact on architectural pedagogy, practice, and the future of the profession. It transcends traditional constraints, fostering a dynamic exchange of ideas, perspectives, and methodologies. Through digital platforms and virtual studios, students and educators from different corners of the globe collaborate on design projects, research initiatives, and interdisciplinary endeavours. This collaborative approach enriches the educational experience, exposing students to diverse cultural influences, alternative design philosophies, and innovative solutions to global challenges. Virtual classroom, augmented reality (AR), cloud-based design platforms, and collaborative software and apps empower real-time collaboration, transcending physical distance and fostering a culture of innovation and experimentation within architectural education. Cross-border collaboration extends beyond the classroom, bridging the gap between academia and practice. Collaborative research initiatives, industry partnerships, and professional exchanges enable students to gain first-hand experience in real-world architectural projects, fostering lifelong learning and professional development. In conclusion, cross-border collaboration in the digital age represents a transformative force reshaping architectural education. By leveraging digital tools and embracing global networks, institutions prepare architects for dynamic challenges, driving innovation, diversity, and excellence in the profession. It forms a cornerstone for a sustainable future of architecture education, addressing complex challenges through collaboration and innovation technique.

Keywords: Architecture education, digital age, collaboration, curriculum integration

Introduction

In an increasingly interconnected world driven by digital technologies, international collaboration has become essential for transforming architectural education to meet the demands of a dynamic future. Furthermore, a transformative, student-centered design pedagogy in architectural education, emphasizing the importance of design studios for creative exploration and interactive

[a]abdmuluk@utar.edu.my

DOI: 10.1201/9781003571018-25

learning, while promoting inquiry-based, active, and experiential teaching practices to address contemporary social and environmental challenges through democratic and inclusive education (Salama (2015). This paper addresses the critical role of international collaboration in reshaping architectural pedagogy, with a focus on the opportunities and challenges presented by the digital age. By conducting a comprehensive literature review and employing a mixed-methods research approach, this paper aims to fill the gap in current research by exploring the theoretical underpinnings, methodologies, and outcomes of international collaboration in architectural education. The findings of this study shed light on the transformative potential of digital globalization in fostering creativity, innovation, and global citizenship within the architectural profession.

As the world becomes more interconnected through digital technologies, the field of architectural education is undergoing a profound transformation. International collaboration has emerged as a key driver of innovation and excellence in architectural pedagogy, enabling students and educators to engage with diverse perspectives, cultures, and ideas from around the globe. This paper explores the role of international collaboration in shaping the future of architectural education in the digital age, highlighting its potential to foster creativity, expand horizons, and address global challenges. By examining the theoretical frameworks, methodologies, and outcomes of international collaboration in architectural education, this paper seeks to provide insights into its transformative impact on the profession and the built environment.

Research Metodology

A mixed-methods research approach was employed for this study, combining qualitative and quantitative data collection methods. Qualitative data was gathered through a comprehensive review of relevant literature on international collaboration in architectural education, encompassing scholarly articles, books, reports, and case studies from diverse sources. Quantitative data was collected through surveys and interviews with students, educators, and professionals involved in international collaborative projects in architecture. The data collected were analyzed thematically to identify key insights, trends, and patterns related to the theory, practice, and outcomes of international collaboration in architectural education.

Related Research/Literature Review

The literature on international collaboration in architectural education underscores its significance in promoting cross-cultural understanding, knowledge exchange, and innovation (Duff, 2018). Scholars have highlighted the role of digital technologies in facilitating virtual collaboration, enabling students and educators to collaborate seamlessly across geographical boundaries (Abdelmohsen et al., 2019). Moreover, international collaboration has been shown to enhance students' intercultural competence, critical thinking skills, and global perspective (Groat and Wang, 2013). However, challenges such as language barriers, technological limitations, and cultural differences may hinder effective collaboration and communication (Kim et al., 2017). Universiti Tunku Abdul Rahman (Figure 25.1) own experience in International collaboration is analyzed. The aims is to explore the theoretical frameworks and methodologies that underpin successful international collaboration in architectural education, as well as the potential benefits and challenges associated with this approach.

This paper is grounded in the theoretical framework of digital globalization and transnational education. Ansell and Gash's model of collaborative governance,

Figure 25.1 UTAR Kampar Campus
Source: Universiti Tunku Abdul Rahman (website)

emphasizing trust, shared understanding, and mutual learning among stakeholders, can be applied to architectural education by fostering collaboration between students, educators, and professionals to address complex design challenges and integrate diverse perspectives into the design process. (Ansel and Gash 2008). Digital globalization refers to the process of integrating digital technologies into global networks of communication, commerce, and collaboration (Kurbalija, 2019). Transnational education, on the other hand, encompasses educational programs and initiatives that transcend national boundaries, enabling students and educators to engage in cross-cultural learning experiences (Watters et al., 2016). By leveraging these theoretical perspectives, this paper seeks to examine how international collaboration in architectural education is mediated and facilitated by digital technologies, as well as the implications of this phenomenon for the future of the profession.

Results and Discussion

The analysis of the literature and empirical data revealed several key findings regarding the role and impact of international collaboration in architectural education. Firstly, digital technologies such as virtual reality, online platforms, and collaborative software have facilitated cross-border collaboration, enabling students and educators to work together on design projects in real-time (Kolarevic, 2020). Secondly, international collaboration has enriched architectural pedagogy by exposing students to diverse cultural perspectives, design traditions, and urban contexts (Jabi et al., 2018). However, challenges such as language barriers, time zone differences, and technical constraints may pose obstacles to effective collaboration and communication (Carmona and Garcia-Esparza, 2021). There are strategies for overcoming these challenges and maximizing the potential benefits of international collaboration in architectural education, including the use of multilingual communication tools, cultural sensitivity training, and project management techniques.

Transforming architectural education using digital technology involves leveraging digital tools, platforms, and methodologies to enhance teaching and learning experiences, foster collaboration, and prepare students for the evolving demands of the profession. Here are some key strategies to achieve this transformation:

1. **Integrate Digital Design Tools:** Incorporate digital design software such as computer-aided design (CAD), building information modelling (BIM), and parametric design tools into the curriculum. Provide hands-on training and projects that allow students to develop proficiency in using these tools to create complex architectural designs and simulations.
2. **Virtual Reality (VR) and Augmented Reality (AR):** Utilize VR and AR technologies to create immersive learning environments where students can explore architectural spaces, visualize design concepts, and experience real-world scenarios. VR can also be used for virtual site visits, allowing students to study existing buildings and urban environments remotely.
3. **Online Learning Platforms:** Implement online learning platforms and digital repositories to deliver course materials, lectures, and resources to students anytime, anywhere. These platforms can facilitate asynchronous learning, collaborative projects, and peer-to-peer interactions, catering to diverse learning styles and schedules.
4. **Digital Fabrication and 3D Printing:** Introduce digital fabrication technologies such as 3D printing, laser cutting, and CNC milling into the curriculum to enable students to prototype and fabricate architectural models and components. Encourage experimentation with materials and fabrication techniques to explore innovative design possibilities.
5. **Collaborative Design Studios:** Foster interdisciplinary collaboration and teamwork by organizing virtual design studios where students from different disciplines and geographical locations can collaborate on design projects in real-time. Use collaborative software and project management tools to facilitate communication, coordination, and feedback exchange among team members.
6. **Digital Portfolio Development:** Guide students in creating digital portfolios showcasing their design projects, research work, and professional achievements. Emphasize the importance of digital presentation skills, visual communication, and storytelling to effectively communicate design concepts and narratives.
7. **Digital Research and Visualization:** Encourage students to conduct digital research using computational tools and data visualization techniques to analyse architectural precedents, urban contexts, and environmental parameters. Integrate research findings into design projects to inform decision-making and foster evidence-based design practices.
8. **Remote Internships and Experiential Learning:** Facilitate remote internships and experiential learning opportunities for students to gain real-world experience working with architectural firms, design studios, and construction companies. Use digital platforms and communication tools to connect students with industry professionals and mentors, enabling them to collaborate on projects and receive feedback remotely.
9. **Professional Development Workshops and Webinars:** Organize professional

development workshops, webinars, and guest lectures featuring leading practitioners, researchers, and educators in the field of architecture. Provide insights into emerging trends, technologies, and best practices to prepare students for the future of the profession.

10. **Continuous Evaluation and Feedback:** Implement digital assessment tools and feedback mechanisms to monitor student progress, evaluate learning outcomes, and provide constructive feedback. Use data analytics and learning analytics to identify areas for improvement and personalize learning experiences based on individual student needs and preferences.

By embracing digital technology and innovative pedagogical approaches, architectural education can evolve to meet the challenges and opportunities of the digital age, empowering students to become versatile, creative, and technologically proficient architects. International collaboration in architecture education is being transformed by digital technology in several ways, revolutionizing how students, educators, and institutions engage with one another across geographical boundaries. Here's how:

1. **Virtual Design Studios:** Digital technology enables the creation of virtual design studios where students from different countries collaborate on architectural projects in real-time. Through video conferencing, collaborative software, and cloud-based platforms, students can share ideas, critique designs, and work together on complex architectural challenges regardless of their physical location. This fosters cross-cultural exchange, diversity of perspectives, and interdisciplinary collaboration, enriching the learning experience.
2. **Online Courses and Workshops:** Institutions offer online courses, workshops, and seminars that attract students and faculty from around the world. Digital platforms facilitate the delivery of course materials, lectures, and interactive sessions, allowing participants to engage in learning activities remotely. International collaboration in these virtual spaces promotes knowledge sharing, networking, and professional development among students and educators across borders.
3. **Joint Research Projects:** Digital technology facilitates collaboration on joint research projects between institutions and scholars from different countries. Through online collaboration tools, shared databases, and virtual research environments, researchers can collaborate on architectural research topics, conduct data analysis, and publish findings collaboratively. International collaboration enhances the quality and impact of research outcomes, fostering innovation and advancing the field of architecture.
4. **Global Design Competitions:** Digital platforms host global design competitions and challenges that attract participants from diverse cultural backgrounds. These competitions provide opportunities for students to showcase their design skills, receive feedback from international jurors, and collaborate with peers from around the world. Digital submission platforms enable participants to submit design proposals remotely, making it accessible to a global audience and promoting cross-cultural exchange and dialogue.
5. **Cross-Institutional Partnerships:** Institutions form cross-institutional partnerships and academic exchange programs that facilitate student and faculty mobility. Digital technology streamlines the exchange process by

enabling online application procedures, virtual interviews, and digital document submission. Students can participate in international exchange programs, study abroad semesters, or joint degree programs without the need for physical travel, fostering cross-cultural understanding and academic collaboration.

6. **Virtual Study Abroad Programs**: Institutions offer virtual study abroad programs that allow students to experience architectural education in different cultural contexts without leaving their home country. Through virtual tours, online lectures, and collaborative projects with partner institutions abroad, students gain insights into global architectural practices, design traditions, and urban contexts. Virtual study abroad programs leverage digital technology to overcome barriers to physical mobility and promote internationalization.
7. **Multilingual Resources and Translation Tools:** Digital platforms provide multilingual resources, translation tools, and language support services to facilitate communication and collaboration among participants from different linguistic backgrounds. This enables international collaboration in architectural education without language barriers, allowing students and educators to engage in discussions, presentations, and collaborative projects in their preferred language.

Overall, international collaboration in architecture education is being transformed by digital technology, enabling students, educators, and institutions to overcome geographical barriers, foster cross-cultural exchange, and create a more interconnected and inclusive learning environment. Digital platforms and tools play a pivotal role in facilitating collaboration, communication, and knowledge sharing among participants from around the world, shaping the future of architectural education in the global digital age.

Conclusion

In conclusion, this paper highlights the transformative potential of international collaboration in shaping the future of architectural education in the digital age. By leveraging digital technologies and embracing a transnational approach to learning, students and educators can engage in cross-cultural exchanges, collaborative design projects, and innovative research initiatives that transcend geographical boundaries. Moving forward, it is imperative for stakeholders in architectural education to embrace international collaboration as a means of fostering creativity, diversity, and excellence in the profession.

References

Abdelmohsen, S., Massoud, P., and El-Khaldi, M. (2019). Digital collaboration in architecture education: a reflection on a remote and on-site international design workshop. *Journal of Technology and Design Education*, 29(2), 325–344.

Ansell, C., and Gash, A. (2008). Collaborative governance in theory and practice. *Journal of Public Administration Research and Theory*, 18(4), 543–571.

Carmona, A., and Garcia-Esparza, M. (2021). Digital platforms as mediators of transnational collaboration in architectural education. *Journal of Architectural Education*, 75(2), 273–288.

Duff, C. (2018). The impact of international collaboration on architectural education: a case study of the erasmus mundus joint master degree in architecture, engineering, and urbanism. *Journal of International Education Research*, 14(3), 143–156.

Groat, L., and Wang, D. (2013). Architectural Research Methods. John Wiley and Sons.

Jabi, W., Liebich, T., and Zeibak-Shini, N. (2018). Virtual reality in architectural education: defining possibilities at the onset

of an evolution. *International Journal of Architectural Computing*, 16(4), 331–344.
Kim, Y., Lee, H., and Chong, H. (2017). Overcoming cultural and language barriers in international collaboration: a case study of architectural education. *Journal of Studies in International Education*, 21(2), 142–160.
Kolarevic, B. (2020). Architecture in the Age of the Digital Factory. Routledge.
Kurbalija, J. (2019). Digital Diplomacy: Theory and Practice. DiploFoundation.
Salama, A. M. (2015). Spatial Design Education New Directions for Pedagogy in Architecture and Beyond. (1st Edn.), London: Routledge, pp. 58.
Watters, D., Ryan, M., and Gaffney, M. (2016). Transnational education crossing 'boundaries' in architectural education. *Journal of Education Policy*, 31(6), 729–744.

26 Three emerging challenges for architects

Mohammed Zakiul Islam[a]
Professor, Department of Architecture, Bangladesh University of Engineering and Technology, Dhaka, Bangladesh

Abstract

This paper discusses three emerging challenges that architects are currently facing in South-Asian countries such as Bangladesh.

There is a growing need for the involvement of architects in non-urban areas. Due to the threat of natural and artificial disasters, the demand for architects is now apparent in non-urban areas more than ever before.

Due to steady economic growth, our countries are observing mega projects in terms of both complexity and size that are not seen before. Internationally successful mega firms are commissioned to a majority of such projects and unfortunately, local architects are playing a minor role.

The urban population is rising at an unprecedented rate. Our cities are facing challenges that have not been seen before. Lack of public spaces, long commutes, dismal housing, polluting industries, and the negative impact of automobiles are just a few that can be mentioned. There is an absence of professionals to solve these urban challenges and Architects are one of the best to fill in this gap.

Our bachelor programs need to prepare future architects for the above challenges.

Keywords: South-Asia, challenges, architecture

Three Emerging Challenges for Architects

This paper has identified three emerging challenges for architects in South-Asian countries. This is not an exhaustive list of current challenges but prioritizes the three identified ones.

There is a growing need for architects to be involved in rural areas. Due to the threat of natural and artificial disasters, the demand for architects is now apparent in non-urban areas more than ever before.

Due to steady economic growth, our countries are observing mega projects in terms of complexity and size that have not been seen before. Internationally successful mega firms are commissioned to a majority of such projects, but unfortunately, local architects play a minor role.

The urban population is rising at an unprecedented rate. Our cities are facing challenges that have not been seen before. There is an absence of professionals to solve these urban challenges, and architects are among the best at filling this gap.

This paper proposes that the context of South Asian countries differs from that of other regions, and the architecture curriculum needs to prepare its graduates accordingly.

Urban Bias in the Bachelor of Architecture Curriculum

Most of the initial architecture schools in South Asia were formed either during foreign countries' rule or Western countries' influence. South Asia's rich history of architecture is older than that of many other regions of the world, but formal education in architecture was a new idea in this region. For example, the US Government

[a]zaki@arch.buet.ac.bd, zakiislam.mail@gmail.com

DOI: 10.1201/9781003571018-26

helped to establish the first school of architecture in Bangladesh in 1962. Sir JJ School of Art was founded in 1857. Later, an independent architecture department was established at the Sir JJ School in 1913. In 1952, the Department of Architecture became a department of the University of Mumbai, while the school became the Sir JJ College of Architecture (Clapcurve 2022). The National College of Arts (NCA) Lahore in Pakistan was established in 1875 under British rule. Architecture education in Sri Lanka started at the hands of Architect Justin Samarasekera and Lord Llewelyn-Davies (a professor at the Bartlett School of Architecture). The courses were aligned with the Royal Institute of British Architects (RIBA) standards (Source: University of Muratuwa's webpage).

One of the main aims of establishing these schools was to produce trained architects who would be involved in the then-current development works, which were mainly focused on the capital cities or other large cities of the respective country. It is observed that in colonial countries, the leading infrastructure development was highly concentrated in the capital city, whether it was colonized by the Dutch, for example, Jakarta, or the British, for example, Kolkata (Macionis, 2007). Therefore, non-urban areas were not the focus of the Bachelor of Architecture (B.Arch.) curriculum of the first schools.

Many years have passed, and all the countries of South Asia have gained their respective independence. However, the original curriculum's basic structure has mostly stayed the same. For example, the core five design studio courses of the original curriculum of the Department of Architecture of Bangladesh University of Engineering and Technology (BUET) are still the same, with ten design studio courses breaking each studio into two in a semester system. There was one design studio in one academic year, and currently, there are two design studio courses in two semesters in one academic year. The change is not significant.

However, it is high time that architects need to get involved in non-urban areas too. Architects' urgency to focus on the rural areas is apparent in Ar. Rem Koolhaas remarks, "We decided to focus on the 98% of the Earth's surface that is not occupied by cities" (Wainwright, 2020). "Countryside, The Future," titled exhibition at the Guggenheim Museum in 2020, is all about the non-urban areas of the world. The characteristics of the countryside in South Asian countries are different from those in other parts of the world. However, the "change" issue in the countryside is universal—most of the current B. Arch. Curriculums are not tuned to this change.

Architects' input is required in projects located in small towns outside of large cities. There is a growing need for architects to be involved in non-urban areas, too. In the past, most of the building projects were concentrated in cities in the south-Asian Countries. However, since our governments have prioritized the housing needs of different groups, the number of building projects in the non-urban areas has increased at a geometric rate. Following Sri Lanka's successful "One Million Houses Programme" (MHP) in 1984 (Abeyasekera and Gunasekeara, 2020), other countries of the region have also initiated similar projects. For example, The *Asrayon Prokolpo* by the Government of Bangladesh has already handed over 183,000 housing units to the ultra-poor of the society in the rural areas.

The needs and characteristics of these projects are different from those of their urban counterparts. They are different in terms of the user's lifestyle and the context. Unfortunately, most of the buildings that are being built in the rural areas are either carbon copies of their urban counterpart or very poorly designed. In many cases, architects are missing from the scene, and the buildings fail to take advantage of the rural context. Most architects in their schools are not trained to design such projects properly.

Influence of the Climate Change

Due to climate change, the world is witnessing more risk of natural disasters than ever before in history. The adverse effects of climate change are the most in areas with high populations, and South Asia is among the top on the list. According to the article by the World Wild Life (WWF) program, several factors are contributing to the increase in flood risk. These include changes in rainfall, storms, and temperatures driven by the climate crisis, as well as societal factors such as changes in land use and the development of floodplains (World Wild Life, 2024).

The influx of population makes people live in places prone to natural disasters. For example, people living in lowland areas are under threat of annual flood or river erosion. These are new challenges that people are facing. Architects have a role to play in this issue. Architects have a role to play because indigenous knowledge of the local users alone is not enough to tackle the climate change issues.

The Rural Area is Changing

The mindset of the rural people is changing due to infrastructure changes (roads, bridges, etc.) and the availability of digital resources (internet and use of cell phones). The remittance is going to the rural areas of Bangladesh, and that investment is changing the rural scenario of countries such as Bangladesh. In an article, Adnan Morshed draws our attention to three aspects of the changes in the rural side of Bangladesh: The rise of the middle class in rural areas, the next-generation remittance influence, and mobile technology in rural life (Morshed, 2022). According to an article by a finance company, there is a huge demand for housing finance in rural areas of Bangladesh, too. The Mortgage debt to GDP ratio is only 3.5% in Bangladesh, while Singapore has crossed 50%. The article claims that the housing loan market is highly constrained. Even within this constrained situation, the estimated annual housing finance demand among the low- and middle-income groups from 2017 to 2020 is 51.23 billion, which is half of the demand for urban areas in Bangladesh (Ahmed, 2021). Are the Bangladeshi architects ready to play their role in this 51 billion market?

Rise of Megaprojects

Due to steady economic growth, South Asian countries are observing mega projects in terms of both complexity and size that are new in this area. Internationally, successful mega firms are commissioned to a majority of such projects, but unfortunately, local architects play a minor role.

Our cities are seeing new kinds of buildings that were not seen before. These are large in terms of size and highly complex. South Asian lifestyles have changed. Nucleus families are on the rise, and urban lifestyles demand new types of projects. European and Northern American architects faced similar challenges in the late modern period in the 1970s (Jencks, 1993). South Asian architects must rise to this challenge, too.

For example, our cities are observing super general stores instead of *Katcha* bazaar or corner shops in the neighborhood. Large shopping malls are constructed that consume more electricity than a neighborhood. These projects' programs are new to our architects. Foreign architects have already designed such programs and are ahead in terms of experience.

Foreign architects are not a new phenomenon in South Asia. South Asian countries have experienced the architecture of foreign architects. Chandigarh city by Le Corbusier in India and *Sangshad Vaban* by Louise I. Kahn in Dhaka, Bangladesh, are examples of such foreign architects in this area. There are positive issues related to foreign architects working in our countries. Local architects gain experience by working with foreign architects. However, there are also not-so-positive examples too. According to

Rahul Mehrotra, there are some examples of projects designed by foreign architects that are context-less (2017).

The private developers also looking for famous foreign architects to use their branding value to sell the projects. Our local firms are yet to gain international branding. Another valid reason for not going for local architects is the lack of experience in handling such complex projects. Multi-disciplinary teams work on this type of project, and our local firms have yet to become experts in handling multi-disciplinary issues and teams.

As mentioned before, this project's complexity is also new for our architects to handle. It requires more than design skills. Different technical groups are involved. A recent report titled *"Architectural Education in Bangladesh: Evaluation of the Existing Curricula and Future Direction,"* commissioned by the Institute of Architects Bangladesh (IAB), reflects the lack of courses in general education—this lack of courses in the B. Arch. curricula may be the reason for our local firms' inefficiency in handling multi-disciplinary complex projects. The report identified four different types of general education: numeracy and scientific literacy (course examples, Calculus, Physics, Computer programming, and Geography), values, beliefs, and identity (History, Psychology, and Philosophy), community and society (Sociology, Finance, Economics, and Anthropology), communication (Technical writing, Film, and history, Language, Professional speaking). B. Arch curricula in South Asia need to include general education courses that will allow graduates to handle complex projects and increase their managerial skills.

Urban Issues

According to the United Nations Department of Economic and Social Affairs, 68% of the world's population is projected to live in urban areas by 2050. According to the World Bank data, the annual urban population growth rate of Bangladesh, India, Nepal, Pakistan, and Sri Lanka is 3%, 2%, 3.8%, 2.7%, and 1%, respectively. Apart from Sri Lanka's figures, all are high.

The usual trend in these countries is that a large proportion of the urban population is concentrated in a few metropolises. Due to colonialism, these countries only have a few developed cities. The colonial power preferred to concentrate power in one capital city instead of decentralizing. This is a trend that is usually observed in countries that were ruled by colonialism in the past.

Five of the top 20 cities with the highest population are in South Asia (Delhi, Dhaka, Mumbai, Kolkata, and Karachi). As a result, our cities are facing challenges that have not been seen before. Lack of public spaces, long commutes, dismal housing, polluting industries, and the negative impact of automobiles are just a few among many. More professionals are needed to solve these urban challenges.

Lack of Professionals

The number of planners in our countries is less than what is required. The first undergraduate program related to urban planning was established in 1991 in Bangladesh. It is a relatively new profession. Currently, there are only seven undergraduate programs related to planning in Bangladesh compared to over 35 bachelors in architecture programs. Even with the small number of experienced planners that we have, many of them need to be more direct experts in physical planning. Many of their educational backgrounds are in economics or social science.

It is difficult for a Foreigner to Design a City

The need for more professionals in the planning profession has made foreign planners commissioned to design our cities in South Asia. Planning a city is much more challenging if it is difficult for a foreign architect to design a building. One of the main reasons is that city planning needs to understand the local context, which is difficult for a consultant who comes for a short period. City planning requires a deep understanding and knowledge of the local context.

Research Gap

Research on our cities is limited compared to what is required. For example, density is a crucial issue for our cities. It is simply impractical to design our cities with Western standards of density. We need to formulate our standards and to do that; we need our research based on our cities.

Architects Can Fill in the Gap

Architects are the only professionals who can visualize what needs to be built. In the absence of proper planners and urban designers, architects of our countries may need to take on the responsibilities of planners for the time being. There are opportunities to include post-graduate programs or courses focused on urban issues for architects.

The Task in Hand

How can we prepare our architects to face the three identified challenges? The first reaction would be to include new courses in the B. Arch. curriculums. However, including new courses will be challenging because B. Arch. is already a five-year program. Adding new courses will increase the completion time.

Moreover, there will be new challenges in the future, too. We cannot keep adding courses because we face new challenges. It will be futile—instead, the B. Arch. The curriculum may focus on the process of learning. An architect will be prepared to learn herself when faced with a challenge.

References

Abeyasekera, A. L., and Gunasekeara, V. (2022). Square-Rule Paper Engineers: People, Power, and Participation in Sri Lanka's Million Houses Programme. Accountability Research Center.

Ahmed, A. (2021). Housing market: rising construction material prices and its impact *IDLC Monthly Business Review*, 17(12).

Clapcurve (2022). Architecture Colleges: Oldest Architecture Colleges in India. Retrieved from: https://www.clapcurve.com/oldest-architecture-colleges-in-india/.

Jencks, C. (1993). Architecture Today. London: Academy Editions.

Macionis, J. J. (2007). Cities and Urban Life. 4th Edn. New Jersey. Pearson Education, Inc.

Morshed, A. (2022). Researching Rural Transformation 2.0 in Bangladesh, Daily Stat News. Retrieved May 10, 2024, from https://www.thedailystar.net/opinion/views/the-grudging-urbanist/news/researching-rural-transformation-20-bangladesh-3165601.

SpaceSync (2017). How are foreign firms impacting Indian Architectural Practices? Retrieved from: https://studiospacesync.wordpress.com/2017/07/15/architectural-practice-india/.

University of Moratuwa (n.d.). History. University of Moratuwa, Architecture. Retrieved from: https://uom.lk/archi/history.

Wainwright, O. (2020). The Countryside is Where the Radical Changes are': Rem Koolhaas Goes Rural. Retrieved from: https://www.theguardian.com/artanddesign/2020/feb/11/rem-koolhaas-rural-countryside-the-future-guggenheim.

World Wild Life (2024). Is Climate Change Increasing the Risk of Disasters? Facts about Floods, Fires, Hurricanes, and Tornadoes. Retrieved from: https://www.worldwildlife.org/stories/is-climate-change-increasing-the-risk-of-disasters#:~:text=Climate%20change%20is%20increasing%20ocean,storm%20surge%2C%20and%20rainfall%20rates.

27 The national education policy 2020-transforming higher education through holistic, integrated, flexible multidisciplinary education

Shakila Shamsu[a]

Former OSD(NEP), Department of Higher Education, Ministry of Education, India

Abstract

The latest National Education Policy, rolled out in 2020, covers the education sector comprehensively across the various verticals of school and higher education and aims at a paradigm shift in the structures, processes and methods of educational transactions. While the reforms touch upon academic thrusts, curricular and pedagogical changes including assessment; it also looks at governance, quality assurance, management and regulation; cross cutting areas of language, technology and research in education, emerging demands of vocational and open learning systems and many more. In particular, within higher education the need for moving away from highly discipline specific silo-based undergraduate programmes to a more integrated approach is a major significant shift in all areas of education. In this article, we are outlining the major changes in higher education and the nature of undergraduate offerings.

Keywords: Multidisciplinary undergraduate education, integrative and flexible approaches, restructuring higher education, outcome based education, architectural education

Introduction

Independent India has seen three National Education Policies (NPE) 1968. NPE 1986 modified in 1992, and the NEP in 2020. While the previous two policies were enacted in the previous century, the latest NEP 2020 is the first education policy of this 21st century; and is expected to set in motion a wide range of reforms to be implemented in entirety by 2040.

In general, education policies are comprehensive in nature covering the entire sector, ranging from school to higher education. However, one must recognise that over the years, several committees/commissions have made significant recommendations on various aspects of education. Every new policy in any sector is a response to the contemporary social, industrial, economic, technological changes, both at the national and global levels. Policies must not be viewed, either as gap-filling of the previous policy or fault it of inadequacies; it is only that new policies seek to address the emerging future demands and imperatives. Policies are vision documents characterized by its aspiration and is forward looking. The NEP 2020 aims to transform India into a knowledge society, with educated skilled manpower of high standards, equipped to meet the 21st century challenges. As a signatory to the UN Development Goals, the NEP reaffirms its global commitment and particularly SDG-4 which seeks to "ensure inclusive and equitable quality education

[a]shaquil13@gmail.com

DOI: 10.1201/9781003571018-27

and promote lifelong learning opportunities for all" by 2030.

The 21st century is witnessing several drivers of change, that calls for a paradigm shift in entire education sector. We are living in a highly connected world, a global village – triggered by the immensely growing dynamical impact of ICT in all aspects of our lives. The emergence of disruptive technologies, such as, robotics, data science, machine learning, artificial intelligence, and generative AI, augmented and virtual reality (AR&VR) which are challenging existing job profiles and creating a whole new set of professions. Consequently, it is crucial that our students combine multiple competencies, capacities and the 21st century skills.

As a nation, we are currently fortunate to have the largest youth population, and we can be the main worldwide provider of skilled workforce and human resources. This human capital can become a soft power, only if the youth are equipped with the right knowledge, skills, competencies that make them productive individuals contributing to their individual growth, national development and become global citizens.

The Consultation Process

It is pertinent to know the consultation process of NEP 2020, as it differed from past efforts. The consultations were based on a collaborative, multi-stakeholder approach. Ministry of Human Resource Development (MHRD), now renamed as Ministry of Education (MoE), Government of India undertook an unprecedented collaborative, multi-stakeholder and multi-pronged consultation process which included online, grassroots and national level thematic deliberations.

It is possibly the first time in the history of India that the combined strength of technology and digital connectivity was leveraged inviting wider public participation, feedback, and deliberations from different stakeholders. Online competitions, online talks and platforms, grassroots consultations across nearly 2.5 lakh *Gram Panchayats*, 6600 blocks, 6000 urban local bodies, 676 districts and 36 states/union territories, video conferencing at sub-national administrative levels, field practitioner engagement through the multilateral discussion platforms, online survey, youth survey, thematic expert consultations by ministries and autonomous bodies, and focus group discussions were all part of the humungous exercise. The blending of digital and conventional discussions is a unique feature of this mammoth policy engagement.

It included presentations and dialogues with Hon'ble MPs and education ministers, GOI ministries, state governments, autonomous bodies, varied stakeholder cross-sections and representatives, the Central Advisory Board on Education and Parliamentary Committee, etc. The policy formulation exercise spanned through three Union HRD Ministers and three reports. Initially, the Ministry had constituted a 5-member Committee for Evolution of the NEP in October 2015, under the Chairmanship of late Shri T.S.R. Subramanian, former Cabinet Secretary to the Government of India, with the National University of Education Planning and Administration (NUEPA) now National Institute of Educational Planning and Administration (NIEPA) as its secretariat. The Committee submitted its report in May 2016, which was then followed by the Ministry Report titled 'Some Inputs for the Draft NEP, 2016'. The nine-member 'Committee for the Draft NEP' under the chairmanship of eminent scientist, Dr. K. Kasturirangan was constituted in June 2017, which conducted detailed and wide-ranging discussions, and examined an enormous body of inputs. After an in-depth rigorous exercise, the Committee submitted its Draft NEP 2019 Report in May 2019. Among the regulatory bodies which gave suggestions, the Council of Architecture is also one among them. The Draft NEP 2019 Report, a voluminous treatise, analyses the challenges and the ground realities across the different education verticals and proposed structural and academic reforms to mitigate and correct the existing deficiencies and imbalances. The recommendations seek to achieve the overarching goals of access, equity, quality in a mutually reinforcing manner; while ensuring affordability, and accountability. Stakeholder views, suggestions and comments including those of the states and UTs Governments and Government of India Ministries were invited on the Draft Report 2019. The NEP 2020 has been finalized on the basis of the Draft NEP 2019 Report and after curating a large number of over 2 lakh suggestions received. (https://www.education.gov.in/nep/consultation-process)

NEP 2020 - The Imperatives of Flexible Pathways in Tertiary Education

The NEP has covered the spectrum of education from early childhood education (ECE) to secondary stage, going into higher education encompassing technical education, all professional education of agriculture, health, law, open and distance education, teacher education and vocational education; cross-cutting themes of language, technology, and research in education. The recommendations are founded on 22 cardinal principles thereby ensuring consistency and universality.

Many of transformations that NEP 2020 have recommended, such as, rationalisation of the higher education institutions (HEIs); flexible holistic multidisciplinary 3–4-year undergraduate education with flexible designs of master's programmes promoting internationalization, open and distance learning, online and digital learning at all levels; integration of vocational education, emphasis on multilingualism, a robust teacher education, are enabled by flexible pathways. We need to understand why it is imperative to have an education system that is adaptable, emphasises on individualised and customised approaches to suit a variety of academic choices. In addition to the above-mentioned external drivers, it is the character of knowledge and learning the two core elements of education, which have completely metamorphosed in this century, making a compelling case of flexibility, both at individual and institutional levels. In the current knowledge driven society, knowledge domains are no longer insular and exclusionary, but are interconnected and interacting with each other. This has resulted in the blurring of disciplinary silos and the emergence of the notions of interdisciplinary, multidisciplinary, cross-disciplinary, transdisciplinary knowledge. Similarly, learning in the 21st century, has four aspects of learning to know, including, learning to do; learning to live together and with others, and learning to be, (Delors, Jacques, 1996). This amalgamation of various aspects of learning underscores the need for flexibility, individualisation, and customisation. Both knowledge and learning are not unidimensional but intrinsically multidimensional, making it imperative to have an enabling educational ecosystem, that nurtures and optimises individual potentials, provides multiple pathways and learning trajectories for achieving and attaining holistic development.

The flexibilities recommended in NEP 2020 are currently in different stages of implementation. These flexible, imaginative, and innovative approaches in curricula, teaching-learning, and assessment stems from the recognition of the transformed character of knowledge and learning situated within a technologically driven and connected world. The new curricular and pedagogical restructuring of 5 + 3 + 3 + 4 in school education, incorporates these flexibilities at the middle and secondary stages. In the middle stage from class/grade 6, there are no hard separations between curricular, co-curricular and extra-curricular subjects, between arts and sciences, vocational and academic streams. The National Curriculum Framework for School Education reflects this objective to give equal emphasis on all subjects, by outlining eight curricular areas that students must cover in the secondary stage (class 912)-language education, science education, social science education, mathematics education, art education, interdisciplinary areas, physical education and vocational education. These elasticities flow in a logical continuum to higher education.

Restructuring Undergraduate and Post Graduate Programmes

In the extant higher education scenario, we limit the choices of subjects opted by students in a single discipline, be it sciences, social sciences, applied sciences, humanities, engineering, architecture, design, management, medicine, law, agriculture, pharmacy, nursing etc. These silo-based disciplinary qualifications will be counter-productive in the future economy, as students are required to have both depth in a core discipline, alongside with a breadth of exposure to multiple knowledge domains and skills. This reflects the idea of holistic development through an integrated multidisciplinary approach. In an era where we commonly hear adages, like the jobs of tomorrow are unknown or job profiles will change every seven years, it becomes important to develop graduates who have both depth and breadth, now called as *'T'* shaped graduates, as against the current *'I'* shaped ones. Students will have the choice to choose varied set of courses as per his/her own dispositions, likings and interests, across a range of disciplines for their undergraduate studies. At the UG level, the credit combinations will cover six categories of courses. These are common courses of collaborative skills, communication skills, problem-solving and analytical skills, understanding environment and sustainability, digital skills, research skills, gender sensitisation, Indian knowledge systems, universal human values and so on. The second and third categories are the core disciplinary and multi/interdisciplinary/open elective courses which, *inter alia,* includes astronomy and astrophysics, biology, biochemistry, chemistry, computer science, data science, architecture and design, earth and environmental sciences, mathematics, physics, statistics, architecture, design, anthropology, communication and media, economics, history, linguistics, political science, psychology, social work, sociology, archaeology, comparative literature, arts and creative expressions, creative writing and literature, languages, philosophy, etc; multidisciplinary minor areas, such as, natural and physical sciences, mathematics, statistics, and computer applications, library, information, and media sciences, commerce and management, humanities and social sciences, humanities and social sciences. The fourth category is ability enhancement courses focused on language and communication skills; the fifth category is skill enhancement courses imparting practical skills, hands-on training, soft skills, etc., to enhance the employability of students and finally the sixth category is value added courses can be community projects, media and journalism, understanding India, environmental science, digital & technological solutions, health and wellness and many more innovative courses. All undergraduate students will mandatorily undergo internships/apprenticeships aimed at bridging the current disconnect between the two worlds of education and work. Internships/apprenticeships will help UG students to get an induction into actual work situations. These could be in an industry, or training in labs with faculty and researchers in their own or other HEIs/research institutions, local industry, business organizations, architectural firms, design studios, health and allied areas, local governments, media organizations, artists, crafts persons, and a wide variety of organizations etc. so that students will develop their practical and experiential learning, thereby improving their employability quotient. Even students who wish to exit at the end of the first year, will undergo a 4-credit internship to get a UG certificate.

The undergraduate degree in non-technical domains can be either of 3 or 4-years duration, with multiple entry, exit and re-entry options. At the end of each year the student can get a certification appropriate to the level completed-a UG certificate after completing 1 year (2 semesters) in a discipline or field including vocational and professional areas and having secured 40 credits; an UG diploma after 2 years (4 semesters) of study and having secured 80 credits; an Undergraduate Bachelor's degree with a specialisation after a 3-year (6 semesters) programme and having secured 120 credits; a bachelor's with honours with major in a given discipline and also minors after 4 years (8 semesters) of study with 160 credits. If the student completes a rigorous research project in their major area(s) of study in the 4th year of the degree, then he/she is awarded a bachelor's (honours with research). It is to be noted that the distribution of courses across semesters would be different for architecture education and the Council of Architecture (CoA) guidelines/regulations would have to factored. This new curricular structure of the undergraduate studies will have a number of flexibilities: flexibility to move from one discipline of study to another; flexibility for learners to choose the courses of their interest across various disciplines; flexibility in in the duration and pace of study through multiple entry and exit options depending upon the number of credits secured; flexibility for learners to move from one institution to another to enable them to have multi and/or interdisciplinary learning; as well as flexibility to switch to alternative modes of learning (offline, ODL, and online learning, and hybrid modes of learning).

The flexibility in the modes of learning allows a student to take online MOOCs course through SWAYAM, or through distance education, to the extent up to 40% of the total programme credits. So also, students who opt to exit during their study are allowed to re-enter the degree programme within the stipulated maximum period of seven years, through the facility of multiple entry and multiple exit (MEME) in keeping with the idea of lifelong learning. In view of the changes in the structure of the UG programme, the PG level also has three options for pursuing master's programme: two years masters for those completing 3 years degree; one-year masters for those completing 4 years degree; and five years integrated master's program. The eligibility for PhD shall be either a master's degree or a 4-year UG degree with research. This restructured flexible but integrated UG, PG and doctoral programmes will see students pursuing their studies in a diversified manner reflecting different trajectories.

The Enabling Instrumentalities

The challenge is to ensure that the right mechanisms are created to support the flexibilities envisaged. Three such instrumentalities are the Academic Bank of Credit (ABC), National Higher Education Qualifications Framework (NHEQF), and National Credit Framework (NCrF). To facilitate flexibility in curricular choices and maintaining the record of credits acquired by students across disciplines, institutions, modes of learning and at varying pace over a stretched duration of seven years; the instrumentality of the ABC has been created for digital storage of academic credits. The NHEQF provides the equivalence across the courses covered by the students in different institutions and helps credit mobility. The NCrF would guide the seamless mobility across institutions, academic streams of skills, general education and open and distance education. NCrF would enable different models of vocational education, training and skilling, internships and apprenticeships, by school education institutions as well as higher education institutions. We must recognise that we can achieve the target of 50% gross enrolment ratio (GER) in higher education by 2035, only by integrating vocational education as also open & distance education. The NEP goal of providing holistic education, development of various individual capacities, be it in academics, skills, arts, sports etc. would require strengthening and reinvigorating the idea of liberal arts education, skills & vocational education and also open and distance learning.

Outcome-Based Education

The idea of holistic and integrated learning mandates that we shift from a teacher-driven pedagogical approach to a learner-centric/learner-directed one, with focus on outcome-based education (OBE). The UG programme and courses will be guided by learning outcomes, course outcomes and programme outcomes. The learners will be equipped with clearly defined graduate attributes that captures a basket of learning outcomes related to knowledge, understanding, skills, and competencies. These include core understanding and application of subjects, generic skills of creativity, cognitive and critical thinking, complex problem-solving, communication skills, collaborative skills, digital and technological skills, research skills, analytical reasoning, multicultural competence and inclusivity, ethical and constitutional values, leadership readiness, learning how to learn skills, environmental awareness and gender sensitivity, autonomy, responsibility, and accountability, community engagement and service. The traditional classroom lecture, with passive or almost zero student engagement, and rote-testing is outmoded and redundant. Innovative ways of engaging students in active learning environments that allow for interactive lectures, group discussions, flip learning, projects and practical, seminars, field and studio activities, quizzes, community service, technology assisted and enabled learning and many more, will enable every student to flourish as per each one's capability and optimise one's own levels of performance and achievement. This flexible learner directed pedagogies will compel faculty to change their conventional styles of teaching and adopt the newer roles of a facilitator, an enabler, a mentor, and co-learner, who helps his/her students in learning how to learn and thus achieve a range of graduate outcomes. Since the learning is aimed at acquisition of a combination of deep disciplinary knowledge, core competencies and skills and application of knowledge to real world situations, the assimilation of these will require a different kind of assessment. A variety of assessment methods that are appropriate to a given disciplinary/subject area and a programme of study will be used to assess progress towards the course/programme learning outcomes. There will be a shift from summative assessment to regular and formative assessment, which is more competency-based, promotes learning and development, and tests higher-order skills of analysis, critical thinking, and conceptual clarity. Evaluation will be based on continuous assessment, in which the various in-session activities and the terminal examination will contribute to the final grade. In session or what we commonly refer as internal assessment will consist of class tests, mid-semester examinations, home- work assignments etc., as determined by the respective course faculty. The progress in achieving learning outcomes will be assessed in multiple methods, such as, examinations; closed- book and open-book tests; problem-based assignments; practical assignments, laboratory reports; observation of practical skills; individual and group project reports; oral presentations, including seminars; viva voce interviews; computerised adaptive assessment, examination on demand, modular certifications etc.

Rationalising the Higher Education Landscape

NEP 2020 recognises that we are home to one of the largest higher education systems in the world. The recent statistics, as per the latest All India Survey of Higher Education 2021-22, show that we have 1168 Universities, 45,473 Colleges and 12,002 Standalone Institutions. At the same, NEP factors the complex character of HEIs, which are largely differentiated by its jurisdictional limits (Central or State), nature of academic programmes offered (general, technical, professional etc.) funding patterns (government, government-aided), ownership (public or private), nature of governance (deemed, private or public), mode of education (regular conventional, open and distance mode, virtual), autonomy (unitary, affiliated, or autonomous). These characterisations are not necessarily mutually exclusive, posing challenges that impact quality and efficiency. To address these anomalies, the policy recommends mapping of HEIs into a rationalised institutional architecture, largely by the functions of teaching and research -multidisciplinary research-intensive universities; multidisciplinary teaching-intensive universities; and autonomous degree-granting colleges. Multidisciplinary Education and Research Universities (MERUs), which are model public universities for holistic and multidisciplinary education, aimed to attain the highest global standards in quality education are also to be set up. This does not necessarily mean new greenfield institutions but converting existing ones. A major shift from affiliation to autonomy for the large number of affiliated colleges is envisaged, over a period of 15 years. Through a suitable system of graded accreditation and graded autonomy, and in a phased manner, the system of 'affiliated colleges' will be phased out. Over a period, it is envisaged that every college would develop into either an autonomous degree-granting college, or a constituent college of a university. While such a rationalisation is recommended, care to ensure access and ensure equity in higher education by catering to the rural, unserved, and underserved areas and aspirational districts will be ensured. This new variable architecture of HEIs would help them carve their growth trajectories to suit the new thrusts aligned to the functions of collegiate and university education. Each HEI can embark on a strategic plan of action visualising its growth by formulating an institutional development plan. Such an exercise would help the HEIs to undertake a SWOT/SWOC analysis and decide on its own growth strategies and organisational goals.

Transforming Single Discipline to Multidisciplinary Institutions

To ensure multidisciplinary offerings, HEIs will need to enter into academic arrangements with other HEIs. Such arrangements would pave the way for innovative academic collaboration between institutions, through HEI clusters, leading to multidisciplinary education and research in different modes; merger of single-stream institutions with other multidisciplinary institutions under the same management or different managements; strengthening of institutions by adding departments in subjects, such as: languages, literature, music, philosophy, Indology, art, dance, theatre, education, mathematics, statistics, pure and applied sciences, sociology, economics, sports, translation and interpretation and other subjects as needed for a multidisciplinary institution. Such collaboration between two institutions can pave the way for the award of dual degree where students enrolled in an HEI can take up the first degree at the host institution and the second degree at the partnering institution. These differentiated structures can be adopted by technical and professional institutions, including agriculture, architecture, medicine, pharmacy, law, teacher education etc. as they also need to relook at their curricula for multidisciplinary offerings. This further extends as part of internationalisation too, where flexible options of joint degree, dual degrees and twinning programmes reflect the broader range of academic programmes.

Aligning Architecture Education to NEP 2020 Transformative Reforms

It would be relevant to look at the global trends in architecture education, before examining the reforms in our architecture education so as to be aligned to NEP. Architecture education must transform to keep up with the future emerging trends, https://www.architecturecourses.org/learn/future-trends-architectural-education

- Technology integration of AR, VR and building information modelling (BIM);
- Focus on sustainability and environment, with training in eco-friendly designs, renewable materials, energy-efficient systems, and resilient infrastructure;
- Interdisciplinary/multidisciplinary learning, in collaboration with engineering, environmental science, sociology, and psychology for a holistic understanding;
- Global perspectives and cultural diversity exposing students to global design practices and encouraging appreciation for diverse cultures, heritage preservation, and the integration of varied architectural styles;
- Experiential learning and design thinking, with site-based training skills, instilling problem-solving and innovative skills, fostering creativity and adaptability through hands on work in design studios, workshops, and real-world projects enabling students to apply design thinking;
- Social impact and community engagement, to instil a sense of social responsibility by encouraging students to design spaces for societal needs, promote inclusivity, and engage with communities;
- Adaptive curriculum and lifelong learning, with dynamic architectural education models that emphasize adaptability, encouraging lifelong learning, updated with technological advancements & to evolving industry trends.

These trends augur a significant shift in the nature of the pedagogy in architecture education and prepare future architects to acquire knowledge, skills, and practical experience.

In the light of NEP 2020, the CoA released the regulations for Minimum Standards in Architecture (MSAR) that has incorporated the curricular, academic, pedagogy and assessment related, skill oriented approach, embedding flexibility and integrating technology driven and industry relevant education https://dsel.education.gov.in/sites/default/files/update/PR_COA.pdf. The CoA also brought out an interim report on the way forward- reforms in architecture education,https://www.coa.gov.in/app/myauth/notification/COA%20Circular%20on%20Education%20Reforms%20-%2009.01.23.pdf. The ongoing transformations in architecture education on the lines on NEP recommendations, largely reflect the global emergent trends and are not just contemporary but futuristic. Accordingly, credit-based offerings along with introduction of new courses that include open electives from multidisciplinary areas, relevant technology and design thinking-based skill enhancement courses, industry-led and employment-oriented value-added courses are now provided. Equally aligned is the recognition of holistic and multidisciplinary education fostering institutional collaborations, multi-disciplinary approach with modularity and flexibility, which allows multiple modes of online and hybrid /blended models of learning, with lateral entry and exit options. NEP promotes multilingualism as it gives impetus to cognitive abilities and given our nation's linguistic plurality recommended that higher education can be offered in regional languages. Architecture education seems to have taken this on board and actively considered including regional languages within their UG curricula. Innovative pedagogies that are application oriented with greater focus on hands on practical, live projects, discussion, dialogue, problem solving, group work, study tours, apprenticeships, including industry professionals for teaching, are some other changes. Combining history, arts, culture with new emergences of disaster management, earthquake resistant, green buildings factoring a more holistic understanding is also a major academic reform. Further, as part of internationalisation, recommended in NEP 2020, the possibilities of dual degrees and joint degrees in architecture stream are also worth exploring. Micro-credentialing and large-scale digitisation are other futuristic trends that architecture must work on. Thus, it would be fair to state that architecture education is in transition, while adopting and embracing reforms that espoused in NEP 2020.

Conclusion

Education is a subject included in the concurrent list of the Constitution and state governments must judiciously work out state specific implementation strategies. Given the flexibilities in curricula, regional needs and relevance and local contexts must be factored suitably. Also, states are at different levels of preparedness and multiple models may be necessary; but these must be harmonised, to avoid hurdles in student mobility. At the same time, higher education institutions can broadly align to the national and state frameworks provided, while at the same time as autonomous entities must independently work out their own detailed action plans, with the active and collective involvement and participation of faculty, students and parents. We are living and looking at a future that is a disruptive VUCA world. National Education Policy 2020 has created an opportunity for diverse options needed to meet these disruptions and challenges, in which knowledge and learning calls for dynamism and accelerated adaptability.

References

Draft National Education Policy 2019, Report of the Committee to Draft NEP. https://mhrd.gov.in/sites/upload_files/mhrd/files/Draft_NEP_2019_EN_Revised.pdf

International Commission on Education for the Twenty-first Century, Delors, Jacques. https://unesdoc.unesco.org/ark:/48223/pf00001147663.

https://www.education.gov.in/nep/consultation-process

National Education Policy, 2020, MHRD, Government of India. https://www.mhrd.gov.in/sites/upload_files/mhrd/files/NEP_Final_English_0.pdf

Shamsu,, S. (2023). Catalysing and transforming the higher education landscape in India-The National Education Policy, 2020. Ed. Akhtar, Najma, Revolutionizing Education: Navigating the NEP Era, Publisher: V L Media Solutions.

Shamsu, S. (2021). Realistic implementation of the National Education Policy–2020 a roadmap, Ed. Book, Mittal, Pankaj and Pani, Rama Devi, 'National Education Policy– 2020: Transforming Higher Education in India', Association of Indian Universities, New Delhi (India), 2021.

https://www.architecturecourses.org/learn/future-trends-architectural-education

https://www.coa.gov.in/app/myauth/notification/COA%20Circular%20on%20Education%20Reforms%20-%2009.01.23.pdf

For Product Safety Concerns and Information please contact our EU
representative GPSR@taylorandfrancis.com
Taylor & Francis Verlag GmbH, Kaufingerstraße 24, 80331 München, Germany

www.ingramcontent.com/pod-product-compliance
Ingram Content Group UK Ltd.
Pitfield, Milton Keynes, MK11 3LW, UK
UKHW050058200926
475702UK00019B/2897

* 9 7 8 1 0 3 2 9 4 4 9 4 4 *